Topics in Random Matrix Theory

GRADUATE STUDIES
IN MATHEMATICS **132**

Topics in Random Matrix Theory

Terence Tao

AMERICAN
MATHEMATICAL
SOCIETY
Providence, Rhode Island

2010 *Mathematics Subject Classification.* Primary 60B20, 15B52, 15B35.

For additional information and updates on this book, visit
www.ams.org/bookpages/gsm-132

Library of Congress Cataloging-in-Publication Data

Tao, Terence, 1975–
 Topics in random matrix theory / Terence Tao.
   p. cm. – (Graduate studies in mathematics ; v. 132)
   Includes bibliographical references and index.
 ISBN 978-0-8218-7430-1 (alk. paper)| Softcover ISBN 978-1-4704-7459-1
 1. Random matrices. I. Title.

QA196.5.T36 2012
512.9′434–dc23
                                                              2011045194

To Garth Gaudry, who set me on the road;
To my family, for their constant support;
And to the readers of my blog, for their feedback and contributions.

Contents

Preface

In the winter of 2010, I taught a topics graduate course on random matrix theory, the lecture notes of which then formed the basis for this text. This course was inspired by recent developments in the subject, particularly with regard to the rigorous demonstration of universal laws for eigenvalue spacing distributions of Wigner matrices (see the recent survey [**Gu2009b**]). This course does not directly discuss these laws, but instead focuses on more foundational topics in random matrix theory upon which the most recent work has been based. For instance, the first part of the course is devoted to basic probabilistic tools such as concentration of measure and the central limit theorem, which are then used to establish basic results in random matrix theory, such as the Wigner semicircle law on the bulk distribution of eigenvalues of a Wigner random matrix, or the circular law on the distribution of eigenvalues of an iid matrix. Other fundamental methods, such as free probability, the theory of determinantal processes, and the method of resolvents, are also covered in the course.

This text begins in Chapter 1 with a review of the aspects of probability theory and linear algebra needed for the topics of discussion, but assumes some existing familiarity with both topics, as well as a first-year graduate-level understanding of measure theory (as covered for instance in my books [**Ta2011, Ta2010**]). If this text is used to give a graduate course, then Chapter 1 can largely be assigned as reading material (or reviewed as necessary), with the lectures then beginning with Section 2.1.

The core of the book is Chapter 2. While the focus of this chapter is ostensibly on random matrices, the first two sections of this chapter focus more on random *scalar* variables, in particular, discussing extensively the concentration of measure phenomenon and the central limit theorem in this setting. These facts will be used repeatedly when we then turn our attention to random matrices, and also many of the proof techniques used in the scalar setting (such as the moment method) can be adapted to the matrix context. Several of the key results in this chapter are developed through the exercises, and the book is designed for a student who is willing to work through these exercises as an integral part of understanding the topics covered here.

The material in Chapter 3 is related to the main topics of this text, but is optional reading (although the material on Dyson Brownian motion from Section 3.1 is referenced several times in the main text).

This text is *not* intended as a comprehensive introduction to random matrix theory, which is by now a vast subject. For instance, only a small amount of attention is given to the important topic of invariant matrix ensembles, and we do not discuss connections between random matrix theory and number theory, or to physics. For these topics we refer the reader to other texts such as [**AnGuZi2010**], [**DeGi2007**], [**De1999**], [**Fo2010**], [**Me2004**]. We hope, however, that this text can serve as a foundation for the reader to then tackle these more advanced texts.

Acknowledgments

I am greatly indebted to my students of the course on which this text was based, as well as many further commenters on my blog, including Ahmet Arivan, Joshua Batson, Florent Benaych-Georges, Sivaraman Balakrishnan, Alex Bloemendal, Kaihua Cai, Andres Caicedo, Emmanuel Candés, Jérôme Chauvet, Brian Davies, Ben Golub, Stephen Heilman, John Jiang, Li Jing, Rowan Killip, Sungjin Kim, Allen Knutson, Greg Kuperberg, Choongbum Lee, George Lowther, Rafe Mazzeo, Mark Meckes, William Meyerson, Samuel Monnier, Andreas Naive, Srivatsan Narayanan, Giovanni Peccati, Leonid Petrov, Anand Rajagopalan, Brian Simanek, James Smith, Mads Sørensen, David Speyer, Ambuj Tewari, Luca Trevisan, Qiaochu Yuan, and several anonymous contributors, for comments and corrections. These comments, as well as the original lecture notes for this course, can be viewed online at:

`terrytao.wordpress.com/category/teaching/254a-random-matrices`

The author is supported by a grant from the MacArthur Foundation, by NSF grant DMS-0649473, and by the NSF Waterman award.

Last, but not least, I am indebted to my co-authors Emmanuel Candés and Van Vu, for introducing me to the fascinating world of random matrix theory, and to the anonymous referees of this text for valuable feedback and suggestions.

Chapter 1

Preparatory material

1.1. A review of probability theory

Random matrix theory is the study of matrices whose entries are random variables (or equivalently, the study of random variables which take values in spaces of matrices). As such, probability theory is an obvious prerequisite for this subject. As such, we will begin by quickly reviewing some basic aspects of probability theory that we will need in the sequel.

We will certainly not attempt to cover all aspects of probability theory in this review. Aside from the basic foundations, we will be focusing primarily on those probabilistic concepts and operations that are useful for *bounding* the distribution of random variables, and on ensuring *convergence* of such variables as one sends a parameter n off to infinity.

We will assume familiarity with the foundations of measure theory, which can be found in any text book (including my own text [**Ta2011**]). This is also *not* intended to be a first introduction to probability theory, but is instead a revisiting of these topics from a graduate-level perspective (and in particular, after one has understood the foundations of measure theory). Indeed, it will be almost impossible to follow this text without already having a firm grasp of undergraduate probability theory.

1.1.1. Foundations. At a purely formal level, one could call probability theory the study of *measure spaces* with total measure one, but that would be like calling number theory the study of strings of digits which terminate. At a practical level, the opposite is true: just as number theorists study concepts (e.g., primality) that have the same meaning in every numeral system that models the natural numbers, we shall see that probability theorists study concepts (e.g., independence) that have the same meaning in every measure space that models a family of *events* or *random variables*. And indeed, just as the natural numbers can be defined abstractly without reference to any numeral system (e.g., by the *Peano axioms*), core concepts of probability theory, such as random variables, can also be defined abstractly, without explicit mention of a measure space; we will return to this point when we discuss *free probability* in Section 2.5.

For now, though, we shall stick to the standard measure-theoretic approach to probability theory. In this approach, we assume the presence of an ambient *sample space* Ω, which intuitively is supposed to describe all the possible outcomes of all the sources of randomness that one is studying. Mathematically, this sample space is a *probability space* $\Omega = (\Omega, \mathcal{B}, \mathbf{P})$—a set Ω, together with a σ-*algebra* $\mathcal{B}$ of subsets of Ω (the elements of which we will identify with the probabilistic concept of an *event*), and a *probability measure* $\mathbf{P}$ on the space of events, i.e., an assignment $E \mapsto \mathbf{P}(E)$ of a real number in $[0, 1]$ to every event E (known as the *probability* of that event), such that the whole space Ω has probability 1, and such that $\mathbf{P}$ is countably additive.

Elements of the sample space Ω will be denoted ω. However, for reasons that will be explained shortly, we will try to avoid actually referring to such elements unless absolutely required to.

If we were studying just a single random process, e.g., rolling a single die, then one could choose a very simple sample space; in this case, one could choose the finite set $\{1, \ldots, 6\}$, with the discrete σ-algebra $2^{\{1,\ldots,6\}} := \{A : A \subset \{1, \ldots, 6\}\}$ and the uniform probability measure. But if one later wanted to also study additional random processes (e.g., supposing one later wanted to roll a second die, and then add the two resulting rolls), one would have to change the sample space (e.g., to change it now to the product space $\{1, \ldots, 6\} \times \{1, \ldots, 6\}$). If one was particularly well organised, one could in principle work out in advance all of the random variables one would ever want or need, and then specify the sample space accordingly, before doing any actual probability theory. In practice, though, it is far more convenient to add new sources of randomness on the fly, if and when they are needed, and *extend* the sample space as necessary. This point is often glossed over in introductory probability texts, so let us spend a little time on it. We say that one probability space $(\Omega', \mathcal{B}', \mathcal{P}')$ *extends*[1] another $(\Omega, \mathcal{B}, \mathcal{P})$ if there is a surjective map $\pi : \Omega' \to \Omega$ which is *measurable* (i.e. $\pi^{-1}(E) \in \mathcal{B}'$ for every $E \in \mathcal{B}$) and *probability preserving* (i.e. $\mathbf{P}'(\pi^{-1}(E)) = \mathbf{P}(E)$ for every $E \in \mathcal{B}$). By definition, every event E in the original probability space is canonically identified with an event $\pi^{-1}(E)$ of the same probability in the extension.

Example 1.1.1. As mentioned earlier, the sample space $\{1, \ldots, 6\}$, that models the roll of a single die, can be extended to the sample space $\{1, \ldots, 6\} \times \{1, \ldots, 6\}$ that models the roll of the original die together with a new die, with the projection map $\pi : \{1, \ldots, 6\} \times \{1, \ldots, 6\} \to \{1, \ldots, 6\}$ being given by $\pi(x, y) := x$.

Another example of an extension map is that of a permutation; for instance, replacing the sample space $\{1, \ldots, 6\}$ by the isomorphic space $\{a, \ldots, f\}$ by mapping a to 1, etc. This extension is not actually adding any new sources of randomness, but is merely reorganising the existing randomness present in the sample space.

---

[1]Strictly speaking, it is the pair $((\Omega', \mathcal{B}', \mathcal{P}'), \pi)$ which is the extension of $(\Omega, \mathcal{B}, \mathcal{P})$, not just the space $(\Omega', \mathcal{B}', \mathcal{P}')$, but let us abuse notation slightly here.

In order to have the freedom to perform extensions every time we need to introduce a new source of randomness, we will try to adhere to the following important dogma[2]: **probability theory is only "allowed" to study concepts and perform operations which are preserved with respect to extension of the underlying sample space**. As long as one is adhering strictly to this dogma, one can insert as many new sources of randomness (or reorganise existing sources of randomness) as one pleases; but if one deviates from this dogma and uses specific properties of a single sample space, then one has left the category of probability theory and must now take care when doing any subsequent operation that could alter that sample space. This dogma is an important aspect of the *probabilistic way of thinking*, much as the insistence on studying concepts and performing operations that are invariant with respect to coordinate changes or other symmetries is an important aspect of the modern geometric way of thinking. With this probabilistic viewpoint, we shall soon see the sample space essentially disappear from view altogether, after a few foundational issues are dispensed with.

Let us now give some simple examples of what is and what is not a probabilistic concept or operation. The probability $\mathbf{P}(E)$ of an event is a probabilistic concept; it is preserved under extensions. Similarly, Boolean operations on events such as union, intersection, and complement are also preserved under extensions and are thus also probabilistic operations. The emptiness or non-emptiness of an event E is also probabilistic, as is the equality or non-equality[3] of two events E, F. On the other hand, the *cardinality* of an event is not a probabilistic concept; for instance, the event that the roll of a given die gives 4 has cardinality one in the sample space $\{1, \ldots, 6\}$, but has cardinality six in the sample space $\{1, \ldots, 6\} \times \{1, \ldots, 6\}$ when the values of an additional die are used to extend the sample space. Thus, in the probabilistic way of thinking, one should avoid thinking about events as having cardinality, except to the extent that they are either empty or non-empty. For a related reason, the notion of the underlying probability space being *complete* (i.e. every subset of a null set is again a null set) is not preserved by extensions, and is thus technically not a probabilistic notion. As such, we will downplay the role of completeness in our probability spaces.

Indeed, once one is no longer working at the foundational level, it is best to try to suppress the fact that events are being modeled as sets altogether.

---

[2]This is analogous to how differential geometry is only "allowed" to study concepts and perform operations that are preserved with respect to coordinate change, or how graph theory is only "allowed" to study concepts and perform operations that are preserved with respect to relabeling of the vertices, etc..

[3]Note how it was important here that we demanded the map π to be surjective in the definition of an extension.

To assist in this, we will choose notation that avoids explicit use of set theoretic notation. For instance, the union of two events E, F will be denoted $E \vee F$ rather than $E \cup F$, and will often be referred to by a phrase such as "the event that at least one of E or F holds". Similarly, the intersection $E \cap F$ will instead be denoted $E \wedge F$, or "the event that E and F both hold", and the complement $\Omega \backslash E$ will instead be denoted $\overline{E}$, or "the event that E does not hold" or "the event that E fails". In particular, the sure event Ω can now be referred to without any explicit mention of the sample space as $\overline{\emptyset}$. We will continue to use the subset notation $E \subset F$ (since the notation $E \leq F$ may cause confusion), but refer to this statement as "E is contained in F" or "E implies F" or "E holds only if F holds" rather than "E is a subset of F", again to downplay the role of set theory in modeling these events.

We record the trivial but fundamental *union bound*

$$(1.1) \qquad \mathbf{P}(\bigvee_i E_i) \leq \sum_i \mathbf{P}(E_i)$$

for any finite or countably infinite collection of events E_i. Taking complements, we see that if each event E_i fails with probability at most ε_i, then the joint event $\bigwedge_i E_i$ fails with probability at most $\sum_i \varepsilon_i$. Thus, if one wants to ensure that all the events E_i hold at once with a reasonable probability, one can try to do this by showing that the failure rate of the individual E_i is small compared to the number of events one is controlling. This is a reasonably efficient strategy so long as one expects the events E_i to be genuinely "different" from each other; if there are plenty of repetitions, then the union bound is poor (consider for instance the extreme case when E_i does not even depend on i).

We will sometimes refer to use of the union bound to bound probabilities as the *zeroth moment method*, to contrast it with the first moment method, second moment method, exponential moment method, and Fourier moment methods for bounding probabilities that we will encounter later in this course; see (1.22) below for an explanation of the terminology "zeroth moment method".

Let us formalise some specific cases of the union bound that we will use frequently in the course. In most of this course, there will be an integer parameter n, which will often be going off to infinity, and upon which most other quantities will depend; for instance, we will often be considering the spectral properties of $n \times n$ random matrices.

Definition 1.1.2 (Asymptotic notation). We use $X = O(Y)$, $Y = \Omega(X)$, $X \ll Y$, or $Y \gg X$ to denote the estimate $|X| \leq CY$ for some C independent of n and all $n \geq C$. If we need C to depend on a parameter, e.g., $C = C_k$, we will indicate this by subscripts, e.g., $X = O_k(Y)$. We write

$X = o(Y)$ if $|X| \leq c(n)Y$ for some c that goes to zero as $n \to \infty$. We write $X \sim Y$ or $X = \Theta(Y)$ if $X \ll Y \ll X$.

Given an event $E = E_n$ depending on such a parameter n, we have five notions (in decreasing order of confidence) that an event is likely to hold:

(i) An event E holds *surely* (or is *true*) if it is equal to the sure event $\overline{\emptyset}$.

(ii) An event E holds *almost surely* (or *with full probability*) if it occurs with probability 1: $\mathbf{P}(E) = 1$.

(iii) An event E holds *with overwhelming probability* if, for every fixed $A > 0$, it holds with probability $1 - O_A(n^{-A})$ (i.e., one has $\mathbf{P}(E) \geq 1 - C_A n^{-A}$ for some C_A independent of n).

(iv) An event E holds *with high probability* if it holds with probability $1 - O(n^{-c})$ for some $c > 0$ independent of n (i.e., one has $\mathbf{P}(E) \geq 1 - C n^{-c}$ for some C independent of n).

(v) An event E holds *asymptotically almost surely* if it holds with probability $1 - o(1)$, thus the probability of success goes to 1 in the limit $n \to \infty$.

Of course, all of these notions are probabilistic notions.

Given a family of events E_α depending on some parameter α, we say that each event in the family holds with overwhelming probability *uniformly in* α if the constant C_A in the definition of overwhelming probability is independent of α; one can similarly define uniformity in the concepts of holding with high probability or asymptotic almost sure probability.

From the union bound (1.1) we immediately have

Lemma 1.1.3 (Union bound).

(i) *If E_α is an arbitrary family of events that each hold surely, then $\bigwedge_\alpha E_\alpha$ holds surely.*

(ii) *If E_α is an at most countable family of events that each hold almost surely, then $\bigwedge_\alpha E_\alpha$ holds almost surely.*

(iii) *If E_α is a family of events of polynomial cardinality (i.e., cardinality $O(n^{O(1)})$) which hold with uniformly overwhelming probability, the $\bigwedge_\alpha E_\alpha$ holds with overwhelming probability.*

(iv) *If E_α is a family of events of sub-polynomial cardinality (i.e., cardinality $O(n^{o(1)})$) which hold with uniformly high probability, the $\bigwedge_\alpha E_\alpha$ holds with high probability. (In particular, the cardinality can be polylogarithmic in size, $O(\log^{O(1)} n)$.)*

(v) *If E_α is a family of events of uniformly bounded cardinality (i.e., cardinality $O(1)$) which each hold asymptotically almost surely, then $\bigwedge_\alpha E_\alpha$ holds asymptotically almost surely. (Note that uniformity of asymptotic almost sureness is automatic when the cardinality is bounded.)*

Note how as the certainty of an event gets stronger, the number of times one can apply the union bound increases. In particular, holding with overwhelming probability is practically as good as holding surely or almost surely in many of our applications (except when one has to deal with the metric entropy of an n-dimensional system, which can be exponentially large, and will thus require a certain amount of caution).

1.1.2. Random variables. An event E can be in just one of two states: the event can hold or fail, with some probability assigned to each. But we will usually need to consider the more general class of *random variables* which can be in multiple states.

Definition 1.1.4 (Random variable). Let $R = (R, \mathcal{R})$ be a *measurable space* (i.e., a set R, equipped with a σ-algebra of subsets of $\mathcal{R}$). A *random variable* taking values in R (or an *R-valued random variable*) is a measurable map X from the sample space to R, i.e., a function $X : \Omega \to R$ such that $X^{-1}(S)$ is an event for every $S \in \mathcal{R}$.

As the notion of a random variable involves the sample space, one has to pause to check that it invariant under extensions before one can assert that it is a probabilistic concept. But this is clear: if $X : \Omega \to R$ is a random variable, and $\pi : \Omega' \to \Omega$ is an extension of Ω, then $X' := X \circ \pi$ is also a random variable, which generates the same events in the sense that $(X')^{-1}(S) = \pi^{-1}(X^{-1}(S))$ for every $S \in \mathcal{R}$.

At this point let us make the convenient convention (which we have, in fact, been implicitly using already) that an event is identified with the *predicate* which is true on the event set and false outside of the event set. Thus, for instance, the event $X^{-1}(S)$ could be identified with the predicate "$X \in S$"; this is preferable to the set-theoretic notation $\{\omega \in \Omega : X(\omega) \in S\}$, as it does not require explicit reference to the sample space and is thus more obviously a probabilistic notion. We will often omit the quotes when it is safe to do so, for instance, $\mathbf{P}(X \in S)$ is shorthand for $\mathbf{P}(\text{``}X \in S\text{''})$.

Remark 1.1.5. On occasion, we will have to deal with *almost surely defined* random variables, which are only defined on a subset Ω' of Ω of full probability. However, much as measure theory and integration theory is largely unaffected by modification on sets of measure zero, many probabilistic concepts, in particular, probability, distribution, and expectation, are similarly unaffected by modification on events of probability zero. Thus, a lack of definedness on an event of probability zero will usually not cause difficulty, so long as there are at most countably many such events in which one of the probabilistic objects being studied is undefined. In such cases, one can usually resolve such issues by setting a random variable to some arbitrary value (e.g., 0) whenever it would otherwise be undefined.

We observe a few key subclasses and examples of random variables:

(i) *Discrete random variables*, in which $\mathcal{R} = 2^R$ is the discrete σ-algebra, and R is at most countable. Typical examples of R include a countable subset of the reals or complexes, such as the natural numbers or integers. If $R = \{0, 1\}$, we say that the random variable is *Boolean*, while if R is just a singleton set $\{c\}$ we say that the random variable is *deterministic*, and (by abuse of notation) we identify this random variable with c itself. Note that a Boolean random variable is nothing more than an *indicator function* $\mathbf{I}(E)$ of an event E, where E is the event that the Boolean function equals 1.

(ii) *Real-valued random variables*, in which R is the real line $\mathbf{R}$ and $\mathcal{R}$ is the *Borel σ-algebra*, generated by the open sets of $\mathbf{R}$. Thus for any real-valued random variable X and any interval I, we have the events "$X \in I$". In particular, we have the *upper tail event* "$X \geq \lambda$" and *lower tail event* "$X \leq \lambda$" for any threshold λ. (We also consider the events "$X > \lambda$" and "$X < \lambda$" to be tail events; in practice, there is very little distinction between the two types of tail events.)

(iii) *Complex random variables*, whose range is the complex plane $\mathbf{C}$ with the Borel σ-algebra. A typical event associated to a complex random variable X is the *small ball event* "$|X - z| < r$" for some complex number z and some (small) radius $r > 0$. We refer to real and complex random variables collectively as *scalar random variables*.

(iv) Given an R-valued random variable X, and a measurable map $f : R \to R'$, the R'-valued random variable $f(X)$ is indeed a random variable, and the operation of converting X to $f(X)$ is preserved under extension of the sample space and is thus probabilistic. This variable $f(X)$ can also be defined without reference to the sample space as the unique random variable for which the identity

$$\text{``} f(X) \in S \text{''} = \text{``} X \in f^{-1}(S) \text{''}$$

holds for all R'-measurable sets S.

(v) Given two random variables X_1 and X_2 taking values in R_1, R_2, respectively, one can form the *joint random variable* (X_1, X_2) with range $R_1 \times R_2$ with the product σ-algebra, by setting $(X_1, X_2)(\omega) := (X_1(\omega), X_2(\omega))$ for every $\omega \in \Omega$. One easily verifies that this is indeed a random variable, and that the operation of taking a joint random variable is a probabilistic operation. This variable can also be defined without reference to the sample space as the unique random variable for which one has $\pi_1(X_1, X_2) = X_1$ and $\pi_2(X_1, X_2) = X_2$, where $\pi_1 : (x_1, x_2) \mapsto x_1$ and $\pi_2 : (x_1, x_2) \mapsto x_2$ are the usual projection maps from $R_1 \times R_2$ to R_1, R_2, respectively. One can similarly define the joint random variable $(X_\alpha)_{\alpha \in A}$ for any family of random variables X_α in various ranges R_α. Note here that the set A of labels can be infinite or even uncountable, though of course one needs to endow infinite product spaces $\prod_{\alpha \in A} R_\alpha$ with the product σ-algebra to retain measurability.

(vi) Combining the previous two constructions, given any measurable binary operation $f : R_1 \times R_2 \to R'$ and random variables X_1, X_2 taking values in R_1, R_2, respectively, one can form the R'-valued random variable $f(X_1, X_2) := f((X_1, X_2))$, and this is a probabilistic operation. Thus, for instance, one can add or multiply together scalar random variables, and similarly for the matrix-valued random variables that we will consider shortly. Similarly for ternary and higher order operations. A technical issue: if one wants to perform an operation (such as division of two scalar random variables) which is not defined everywhere (e.g., division when the denominator is zero). In such cases, one has to adjoin an additional "undefined" symbol $\perp$ to the output range R'. In practice, this will not be a problem as long as all random variables concerned are defined (i.e., avoid $\perp$) almost surely.

(vii) *Vector-valued random variables*, which take values in a finite-dimensional vector space such as $\mathbf{R}^n$ or $\mathbf{C}^n$ with the Borel σ-algebra. One can view a vector-valued random variable $X = (X_1, \ldots, X_n)$ as the joint random variable of its scalar component random variables $X_1, \ldots, X_n$. (Here we are using the basic fact from measure theory that the Borel σ-algebra on $\mathbf{R}^n$ is the product σ-algebra of the individual Borel σ-algebras on $\mathbf{R}$.)

(viii) *Matrix-valued random variables* or *random matrices*, which take values in a space $M_{n \times p}(\mathbf{R})$ or $M_{n \times p}(\mathbf{C})$ of $n \times p$ real or complex-valued matrices, again with the Borel σ-algebra, where $n, p \geq 1$ are integers (usually we will focus on the square case $n = p$). Note here that the shape $n \times p$ of the matrix is deterministic; we will not consider in this course matrices whose shapes are themselves random variables. One can view a matrix-valued random variable $X = (X_{ij})_{1 \leq i \leq n; 1 \leq j \leq p}$ as the joint random variable of its scalar components X_{ij}. One can apply all the usual matrix operations (e.g., sum, product, determinant, trace, inverse, etc.) on random matrices to get a random variable with the appropriate range, though in some cases (e.g., with inverse) one has to adjoin the undefined symbol $\perp$ as mentioned earlier.

(ix) *Point processes*, which take values in the space $\mathfrak{N}(S)$ of subsets A of a space S (or more precisely, on the space of *multisets* of S, or even more precisely still as integer-valued locally finite measures on S), with the σ-algebra being generated by the counting functions $|A \cap B|$ for all precompact measurable sets B. Thus, if X is a point process in S, and B is a precompact measurable set, then the counting function $|X \cap B|$ is a discrete random variable in $\{0, 1, 2, \ldots\} \cup \{+\infty\}$. For us, the key example of a point process comes from taking the *spectrum* $\{\lambda_1, \ldots, \lambda_n\}$ of eigenvalues (counting multiplicity) of a random $n \times n$ matrix M_n. Point processes are discussed further in [**Ta2010b**, §2.6]. We will return to point processes (and define them more formally) later in this text.

Remark 1.1.6. A pedantic point: strictly speaking, one has to include the range $R = (R, \mathcal{R})$ of a random variable X as part of that variable (thus one should really be referring to the pair (X, R) rather than X). This leads to the annoying conclusion that, technically, Boolean random variables are not integer-valued, integer-valued random variables are not real-valued, and real-valued random variables are not complex-valued. To avoid this issue we shall abuse notation very slightly and identify any random variable $X = (X, R)$ to any *coextension* (X, R') of that random variable to a larger range space $R' \supset R$ (assuming of course that the σ-algebras are compatible). Thus, for instance, a real-valued random variable which happens to only take a countable number of values will now be considered a discrete random variable also.

Given a random variable X taking values in some range R, we define the *distribution* μ_X of X to be the probability measure on the measurable space $R = (R, \mathcal{R})$ defined by the formula

(1.2) $\mu_X(S) := \mathbf{P}(X \in S)$,

thus μ_X is the *pushforward* $X_* \mathbf{P}$ of the sample space probability measure $\mathbf{P}$ by X. This is easily seen to be a probability measure, and is also a probabilistic concept. The probability measure μ_X is also known as the *law* for X.

We write $X \equiv Y$ for $\mu_X = \mu_Y$; we also abuse notation slightly by writing $X \equiv \mu_X$.

We have seen that every random variable generates a probability distribution μ_X. The converse is also true:

Lemma 1.1.7 (Creating a random variable with a specified distribution). *Let μ be a probability measure on a measurable space $R = (R, \mathcal{R})$. Then (after extending the sample space Ω if necessary) there exists an R-valued random variable X with distribution μ.*

Proof. Extend Ω to $\Omega \times R$ by using the obvious projection map $(\omega, r) \mapsto \omega$ from $\Omega \times R$ back to Ω, and extending the probability measure $\mathbf{P}$ on Ω to the product measure $\mathbf{P} \times \mu$ on $\Omega \times R$. The random variable $X(\omega, r) := r$ then has distribution μ. $\qquad\square$

If X is a discrete random variable, μ_X is the discrete probability measure

$$(1.3) \qquad \mu_X(S) = \sum_{x \in S} p_x$$

where $p_x := \mathbf{P}(X = x)$ are non-negative real numbers that add up to 1. To put it another way, the distribution of a discrete random variable can be expressed as the sum of Dirac masses (defined below):

$$(1.4) \qquad \mu_X = \sum_{x \in R} p_x \delta_x.$$

We list some important examples of discrete distributions:

(i) *Dirac distributions* δ_{x_0}, in which $p_x = 1$ for $x = x_0$ and $p_x = 0$ otherwise;

(ii) *discrete uniform distributions*, in which R is finite and $p_x = 1/|R|$ for all $x \in R$;

(iii) (unsigned) *Bernoulli distributions*, in which $R = \{0, 1\}$, $p_1 = p$, and $p_0 = 1 - p$ for some parameter $0 \leq p \leq 1$;

(iv) the *signed Bernoulli distribution*, in which $R = \{-1, +1\}$ and $p_{+1} = p_{-1} = 1/2$;

(v) *lazy signed Bernoulli distributions*, in which $R = \{-1, 0, +1\}$, $p_{+1} = p_{-1} = \mu/2$, and $p_0 = 1 - \mu$ for some parameter $0 \leq \mu \leq 1$;

(vi) *geometric distributions*, in which $R = \{0, 1, 2, \ldots\}$ and $p_k = (1 - p)^k p$ for all natural numbers k and some parameter $0 \leq p \leq 1$; and

(vii) *Poisson distributions*, in which $R = \{0, 1, 2, \ldots\}$ and $p_k = \frac{\lambda^k e^{-\lambda}}{k!}$ for all natural numbers k and some parameter λ.

Now we turn to non-discrete random variables X taking values in some range R. We say that a random variable is *continuous* if $\mathbf{P}(X = x) = 0$ for all $x \in R$ (here we assume that all points are measurable). If R is already equipped with some reference measure dm (e.g., Lebesgue measure in the case of scalar, vector, or matrix-valued random variables), we say that the random variable is *absolutely continuous* if $\mathbf{P}(X \in S) = 0$ for all null sets S in R. By the *Radon-Nikodym theorem* (see e.g., [**Ta2010**, §1.10]), we can thus find a non-negative, absolutely integrable function $f \in L^1(R, dm)$ with $\int_R f \, dm = 1$ such that

$$(1.5) \qquad \mu_X(S) = \int_S f \, dm$$

for all measurable sets $S \subset R$. More succinctly, one has

$$(1.6) \qquad d\mu_X = f \, dm.$$

We call f the *probability density function* of the probability distribution μ_X (and thus, of the random variable X). As usual in measure theory, this function is only defined up to almost everywhere equivalence, but this will not cause any difficulties.

In the case of real-valued random variables X, the distribution μ_X can also be described in terms of the *cumulative distribution function*

$$(1.7) \qquad F_X(x) := \mathbf{P}(X \le x) = \mu_X((-\infty, x]).$$

Indeed, μ_X is the *Lebesgue-Stieltjes measure* of F_X, and (in the absolutely continuous case) the derivative of F_X exists and is equal to the probability density function almost everywhere. We will not use the cumulative distribution function much in this text, although we will be very interested in bounding tail events such as $\mathbf{P}(X > \lambda)$ or $\mathbf{P}(X < \lambda)$.

We give some basic examples of absolutely continuous scalar distributions:

(i) *uniform distributions*, in which $f := \frac{1}{m(I)} 1_I$ for some subset I of the reals or complexes of finite non-zero measure, e.g., an interval $[a, b]$ in the real line, or a disk in the complex plane.

(ii) The real *normal distribution* $N(\mu, \sigma^2) = N(\mu, \sigma^2)_{\mathbf{R}}$ of mean $\mu \in \mathbf{R}$ and variance $\sigma^2 > 0$, given by the density function $f(x) := \frac{1}{\sqrt{2\pi\sigma^2}} \exp(-(x - \mu)^2/2\sigma^2)$ for $x \in \mathbf{R}$. We isolate, in particular, the *standard (real) normal distribution* $N(0, 1)$. Random variables with normal distributions are known as *Gaussian random variables*.

(iii) The complex *normal distribution* $N(\mu, \sigma^2)_{\mathbf{C}}$ of mean $\mu \in \mathbf{C}$ and variance $\sigma^2 > 0$, given by the density function $f(z) := \frac{1}{\pi\sigma^2} \exp(-|z - \mu|^2/\sigma^2)$. Again, we isolate the standard complex normal distribution $N(0, 1)_{\mathbf{C}}$.

Later on, we will encounter several more scalar distributions of relevance to random matrix theory, such as the semicircular law or Marcenko-Pastur law. We will also of course encounter many matrix distributions (also known as *matrix ensembles*) as well as point processes.

Given an unsigned random variable X (i.e., a random variable taking values in $[0, +\infty]$), one can define the *expectation* or *mean* $\mathbf{E}X$ as the unsigned integral

$$(1.8) \qquad \mathbf{E}X := \int_0^\infty x \, d\mu_X(x),$$

which by the *Fubini-Tonelli theorem* (see e.g. [**Ta2011**, §1.7]) can also be rewritten as

$$(1.9) \qquad \mathbf{E}X = \int_0^\infty \mathbf{P}(X \geq \lambda) \, d\lambda.$$

The expectation of an unsigned variable lies in also $[0, +\infty]$. If X is a scalar random variable (which is allowed to take the value ∞) for which $\mathbf{E}|X| < \infty$, we say that X is *absolutely integrable*, in which case we can define its expectation as

$$(1.10) \qquad \mathbf{E}X := \int_\mathbf{R} x \, d\mu_X(x)$$

in the real case, or

$$(1.11) \qquad \mathbf{E}X := \int_\mathbf{C} z \, d\mu_X(z)$$

in the complex case. Similarly, for vector-valued random variables (note that in finite dimensions, all norms are equivalent, so the precise choice of norm used to define $|X|$ is not relevant here). If $X = (X_1, \ldots, X_n)$ is a vector-valued random variable, then X is absolutely integrable if and only if the components X_i are all absolutely integrable, in which case one has $\mathbf{E}X = (\mathbf{E}X_1, \ldots, \mathbf{E}X_n)$.

Examples 1.1.8. A deterministic scalar random variable c is its own mean. An indicator function $\mathbf{I}(E)$ has mean $\mathbf{P}(E)$. An unsigned Bernoulli variable (as defined previously) has mean p, while a signed or lazy signed Bernoulli variable has mean 0. A real or complex Gaussian variable with distribution $N(\mu, \sigma^2)$ has mean μ. A Poisson random variable has mean λ; a geometric random variable has mean p. A uniformly distributed variable on an interval $[a, b] \subset \mathbf{R}$ has mean $\frac{a+b}{2}$.

A fundamentally important property of expectation is that it is linear: if $X_1, \ldots, X_k$ are absolutely integrable scalar random variables and $c_1, \ldots, c_k$ are finite scalars, then $c_1 X_1 + \cdots + c_k X_k$ is also absolutely integrable and

$$(1.12) \qquad \mathbf{E}c_1 X_1 + \cdots + c_k X_k = c_1 \mathbf{E}X_1 + \cdots + c_k \mathbf{E}X_k.$$

By the Fubini-Tonelli theorem, the same result also applies to infinite sums $\sum_{i=1}^{\infty} c_i X_i$ provided that $\sum_{i=1}^{\infty} |c_i| \mathbf{E} |X_i|$ is finite.

We will use *linearity of expectation* so frequently in the sequel that we will often omit an explicit reference to it when it is being used. It is important to note that linearity of expectation requires no assumptions of independence or dependence[4] amongst the individual random variables X_i; this is what makes this property of expectation so powerful.

In the unsigned (or real absolutely integrable) case, expectation is also monotone: if $X \leq Y$ is true for some unsigned or real absolutely integrable X, Y, then $\mathbf{E} X \leq \mathbf{E} Y$. Again, we will usually use this basic property without explicit mentioning it in the sequel.

For an unsigned random variable, we have the obvious but very useful *Markov inequality*

$$(1.13) \qquad \mathbf{P}(X \geq \lambda) \leq \frac{1}{\lambda} \mathbf{E} X$$

for any $\lambda > 0$, as can be seen by taking expectations of the inequality $\lambda \mathbf{I}(X \geq \lambda) \leq X$. For signed random variables, Markov's inequality becomes

$$(1.14) \qquad \mathbf{P}(|X| \geq \lambda) \leq \frac{1}{\lambda} \mathbf{E} |X|.$$

Another fact related to Markov's inequality is that if X is an unsigned or real absolutely integrable random variable, then $X \geq \mathbf{E} X$ must hold with positive probability, and also $X \leq \mathbf{E} X$ must also hold with positive probability. Use of these facts or (1.13), (1.14), combined with monotonicity and linearity of expectation, is collectively referred to as the *first moment method*. This method tends to be particularly easy to use (as one does not need to understand dependence or independence), but by the same token often gives sub-optimal results (as one is not exploiting any independence in the system).

Exercise 1.1.1 (Borel-Cantelli lemma). Let $E_1, E_2, \ldots$ be a sequence of events such that $\sum_i \mathbf{P}(E_i) < \infty$. Show that almost surely, at most finitely many of the events E_i occur at once. State and prove a result to the effect that the condition $\sum_i \mathbf{P}(E_i) < \infty$ cannot be weakened.

If X is an absolutely integrable or unsigned scalar random variable, and F is a measurable function from the scalars to the unsigned extended reals $[0, +\infty]$, then one has the change of variables formula

$$(1.15) \qquad \mathbf{E} F(X) = \int_{\mathbf{R}} F(x) \, d\mu_X(x)$$

---

[4]We will define these terms in Section 1.1.3.

when X is real-valued and

$$(1.16) \qquad \mathbf{E}F(X) = \int_{\mathbf{C}} F(z) \, d\mu_X(z)$$

when X is complex-valued. The same formula applies to signed or complex F if it is known that $|F(X)|$ is absolutely integrable. Important examples of expressions such as $\mathbf{E}F(X)$ are *moments*

$$(1.17) \qquad \mathbf{E}|X|^k$$

for various $k \geq 1$ (particularly $k = 1, 2, 4$), *exponential moments*

$$(1.18) \qquad \mathbf{E}e^{tX}$$

for real t, X, and *Fourier moments* (or the *characteristic function*)

$$(1.19) \qquad \mathbf{E}e^{itX}$$

for real t, X, or

$$(1.20) \qquad \mathbf{E}e^{it \cdot X}$$

for complex or vector-valued t, X, where $\cdot$ denotes a real inner product. We shall also occasionally encounter the *resolvents*

$$(1.21) \qquad \mathbf{E}\frac{1}{X - z}$$

for complex z, though one has to be careful now with the absolute convergence of this random variable. Similarly, we shall also occasionally encounter *negative moments* $\mathbf{E}|X|^{-k}$ of X, particularly for $k = 2$. We also sometimes use the *zeroth moment* $\mathbf{E}|X|^0 = \mathbf{P}(X \neq 0)$, where we take the somewhat unusual convention that $x^0 := \lim_{k \to 0^+} x^k$ for non-negative x, thus $x^0 := 1$ for $x > 0$ and $0^0 := 0$. Thus, for instance, the union bound (1.1) can be rewritten (for finitely many i, at least) as

$$(1.22) \qquad \mathbf{E}|\sum_i c_i X_i|^0 \leq \sum_i |c_i|^0 \mathbf{E}|X_i|^0$$

for any scalar random variables X_i and scalars c_i (compare with (1.12)).

It will be important to know if a scalar random variable X is "usually bounded". We have several ways of quantifying this, in decreasing order of strength:

(i) X is *surely bounded* if there exists an $M > 0$ such that $|X| \leq M$ surely.

(ii) X is *almost surely bounded* if there exists an $M > 0$ such that $|X| \leq M$ almost surely.

(iii) X is *sub-Gaussian* if there exist $C, c > 0$ such that $\mathbf{P}(|X| \geq \lambda) \leq C \exp(-c\lambda^2)$ for all $\lambda > 0$.

(iv) X has *sub-exponential tail* if there exist $C, c, a > 0$ such that $\mathbf{P}(|X| \geq \lambda) \leq C \exp(-c\lambda^a)$ for all $\lambda > 0$.

(v) X has *finite k^{th} moment* for some $k \geq 1$ if there exists C such that $\mathbf{E}|X|^k \leq C$.

(vi) X is *absolutely integrable* if $\mathbf{E}|X| < \infty$.

(vii) X is *almost surely finite* if $|X| < \infty$ almost surely.

Exercise 1.1.2. Show that these properties genuinely are in decreasing order of strength, i.e., that each property on the list implies the next.

Exercise 1.1.3. Show that each of these properties are closed under vector space operations, thus, for instance, if X, Y have sub-exponential tail, show that $X + Y$ and cX also have sub-exponential tail for any scalar c.

Examples 1.1.9. The various species of Bernoulli random variable are surely bounded, and any random variable which is uniformly distributed in a bounded set is almost surely bounded. Gaussians and Poisson distributions are sub-Gaussian, while the geometric distribution merely has sub-exponential tail. *Cauchy distributions* (which have density functions of the form $f(x) = \frac{1}{\pi} \frac{\gamma}{(x-x_0)^2 + \gamma^2}$) are typical examples of *heavy-tailed* distributions which are almost surely finite, but do not have all moments finite (indeed, the Cauchy distribution does not even have finite first moment).

If we have a family of scalar random variables X_α depending on a parameter α, we say that the X_α are uniformly surely bounded (resp. uniformly almost surely bounded, uniformly sub-Gaussian, have uniform sub-exponential tails, or uniformly bounded k^{th} moment) if the relevant parameters M, C, c, a in the above definitions can be chosen to be independent of α.

Fix $k \geq 1$. If X has finite k^{th} moment, say $\mathbf{E}|X|^k \leq C$, then from Markov's inequality (1.14) one has

$$(1.23) \qquad\qquad \mathbf{P}(|X| \geq \lambda) \leq C\lambda^{-k},$$

thus we see that the higher the moments that we control, the faster the tail decay is. From the dominated convergence theorem we also have the variant

$$(1.24) \qquad\qquad \lim_{\lambda \to \infty} \lambda^k \mathbf{P}(|X| \geq \lambda) = 0.$$

However, this result is *qualitative* or *ineffective* rather than *quantitative* because it provides no *rate* of convergence of $\lambda^k \mathbf{P}(|X| \geq \lambda)$ to zero. Indeed, it is easy to construct a family X_α of random variables of uniformly bounded k^{th} moment, but for which the quantities $\lambda^k \mathbf{P}(|X_\alpha| \geq \lambda)$ do not converge uniformly to zero (e.g., take X_m to be m times the indicator of an event of probability m^{-k} for $m = 1, 2, \ldots$). Because of this issue, we will often have to strengthen the property of having a uniformly bounded moment, to that of obtaining a uniformly quantitative control on the decay in (1.24) for a family X_α of random variables; we will see examples of this in later lectures. However, this technicality does not arise in the important model case of *identically distributed* random variables, since in this case we trivially have uniformity in the decay rate of (1.24).

We observe some consequences of (1.23) and the preceding definitions:

Lemma 1.1.10. *Let $X = X_n$ be a scalar random variable depending on a parameter n.*

   (i) *If $|X_n|$ has uniformly bounded expectation, then for any $\varepsilon > 0$ independent of n, we have $|X_n| = O(n^\varepsilon)$ with high probability.*

   (ii) *If X_n has uniformly bounded k^{th} moment, then for any $A > 0$, we have $|X_n| = O(n^{A/k})$ with probability $1 - O(n^{-A})$.*

   (iii) *If X_n has uniform sub-exponential tails, then we have $|X_n| = O(\log^{O(1)} n)$ with overwhelming probability.*

Exercise 1.1.4. Show that a real-valued random variable X is sub-Gaussian if and only if there exists $C > 0$ such that $\mathbf{E}e^{tX} \leq C \exp(Ct^2)$ for all real t, and if and only if there exists $C > 0$ such that $\mathbf{E}|X|^k \leq (Ck)^{k/2}$ for all $k \geq 1$.

Exercise 1.1.5. Show that a real-valued random variable X has sub-exponential tails if and only if there exists $C > 0$ such that $\mathbf{E}|X|^k \leq \exp(Ck^C)$ for all positive integers k.

Once the second moment of a scalar random variable is finite, one can define the *variance*

$$(1.25) \qquad \mathbf{Var}(X) := \mathbf{E}|X - \mathbf{E}(X)|^2.$$

From Markov's inequality we thus have *Chebyshev's inequality*

$$(1.26) \qquad \mathbf{P}(|X - \mathbf{E}(X)| \geq \lambda) \leq \frac{\mathbf{Var}(X)}{\lambda^2}.$$

Upper bounds on $\mathbf{P}(|X - \mathbf{E}(X)| \geq \lambda)$ for λ large are known as *large deviation inequalities*. Chebyshev's inequality (1.26) gives a simple but still useful large deviation inequality, which becomes useful once λ exceeds the *standard deviation* $\mathbf{Var}(X)^{1/2}$ of the random variable. The use of Chebyshev's inequality, combined with a computation of variances, is known as the *second moment method*.

Exercise 1.1.6 (Scaling of mean and variance). If X is a scalar random variable of finite mean and variance, and a, b are scalars, show that $\mathbf{E}(a + bX) = a + b\mathbf{E}(X)$ and $\mathbf{Var}(a + bX) = |b|^2 \mathbf{Var}(X)$. In particular, if X has non-zero variance, then there exist scalars a, b such that $a + bX$ has mean zero and variance one.

Exercise 1.1.7. We say that a real number $\mathbf{M}(X)$ is a *median* of a real-valued random variable X if $\mathbf{P}(X > \mathbf{M}(X)), \mathbf{P}(X < \mathbf{M}(X)) \le 1/2$.

    (i) Show that a median always exists, and if X is absolutely continuous with strictly positive density function, then the median is unique.

    (ii) If X has finite second moment, show that

$$\mathbf{M}(X) = \mathbf{E}(X) + O(\mathbf{Var}(X)^{1/2})$$

for any median $\mathbf{M}(X)$.

Exercise 1.1.8 (Jensen's inequality). Let $F : \mathbf{R} \to \mathbf{R}$ be a convex function (thus $F((1-t)x + ty) \le (1-t)F(x) + tF(y)$ for all $x, y \in \mathbf{R}$ and $0 \le t \le 1$), and let X be a bounded real-valued random variable. Show that $\mathbf{E}F(X) \ge F(\mathbf{E}X)$. (*Hint:* Bound F from below using a tangent line at $\mathbf{E}X$.) Extend this inequality to the case when X takes values in $\mathbf{R}^n$ (and F has $\mathbf{R}^n$ as its domain.)

Exercise 1.1.9 (Paley-Zygmund inequality). Let X be a positive random variable with finite variance. Show that

$$\mathbf{P}(X \ge \lambda \mathbf{E}(X)) \ge (1 - \lambda)^2 \frac{(\mathbf{E}X)^2}{\mathbf{E}X^2}$$

for any $0 < \lambda < 1$.

If X is sub-Gaussian (or has sub-exponential tails with exponent $a > 1$), then from dominated convergence we have the Taylor expansion

$$(1.27) \qquad \mathbf{E}e^{tX} = 1 + \sum_{k=1}^{\infty} \frac{t^k}{k!} \mathbf{E}X^k$$

for any real or complex t, thus relating the exponential and Fourier moments with the kth moments.

1.1.3. Independence. When studying the behaviour of a single random variable X, the distribution μ_X captures all the probabilistic information one wants to know about X. The following exercise is one way of making this statement rigorous:

Exercise 1.1.10. Let X, X' be random variables (on sample spaces Ω, Ω', respectively) taking values in a range R, such that $X \equiv X'$. Show that after extending the spaces Ω, Ω', the two random variables X, X' are isomorphic, in the sense that there exists a probability space isomorphism $\pi : \Omega \to \Omega'$ (i.e., an invertible extension map whose inverse is also an extension map) such that $X = X' \circ \pi$.

However, once one studies *families* $(X_\alpha)_{\alpha \in A}$ of random variables X_α taking values in measurable spaces R_α (on a single sample space Ω), the distribution of the individual variables X_α are no longer sufficient to describe all the probabilistic statistics of interest; the *joint distribution* of the variables (i.e., the distribution of the tuple $(X_\alpha)_{\alpha \in A}$, which can be viewed as a single random variable taking values in the product measurable space $\prod_{\alpha \in A} R_\alpha$) also becomes relevant.

Example 1.1.11. Let (X_1, X_2) be drawn uniformly at random from the set $\{(-1, -1), (-1, +1), (+1, -1), (+1, +1)\}$. Then the random variables X_1, X_2, and $-X_1$ all individually have the same distribution, namely the signed Bernoulli distribution. However, the pairs (X_1, X_2), (X_1, X_1), and $(X_1, -X_1)$ all have different joint distributions: the first pair, by definition, is uniformly distributed in the set

$$\{(-1, -1), (-1, +1), (+1, -1), (+1, +1)\},$$

while the second pair is uniformly distributed in $\{(-1, -1), (+1, +1)\}$, and the third pair is uniformly distributed in $\{(-1, +1), (+1, -1)\}$. Thus, for instance, if one is told that X, Y are two random variables with the Bernoulli distribution, and asked to compute the probability that $X = Y$, there is insufficient information to solve the problem; if (X, Y) were distributed as (X_1, X_2), then the probability would be $1/2$, while if (X, Y) were distributed as (X_1, X_1), the probability would be 1, and if (X, Y) were distributed as $(X_1, -X_1)$, the probability would be 0. Thus one sees that one needs the joint distribution, and not just the individual distributions, to obtain a unique answer to the question.

There is, however, an important special class of families of random variables in which the joint distribution is determined by the individual distributions.

Definition 1.1.12 (Joint independence). A family $(X_\alpha)_{\alpha \in A}$ of random variables (which may be finite, countably infinite, or uncountably infinite) is said to be *jointly independent* if the distribution of $(X_\alpha)_{\alpha \in A}$ is the product measure of the distribution of the individual X_α.

A family $(X_\alpha)_{\alpha \in A}$ is said to be *pairwise independent* if the pairs (X_α, X_β) are jointly independent for all distinct $\alpha, \beta \in A$. More generally, $(X_\alpha)_{\alpha \in A}$ is said to be *k-wise independent* if $(X_{\alpha_1}, \ldots, X_{\alpha_{k'}})$ are jointly independent for all $1 \le k' \le k$ and all distinct $\alpha_1, \ldots, \alpha_{k'} \in A$.

We also say that X is *independent* of Y if (X, Y) are jointly independent.

A family of events $(E_\alpha)_{\alpha \in A}$ is said to be jointly independent if their indicators $(\mathbf{I}(E_\alpha))_{\alpha \in A}$ are jointly independent. Similarly for pairwise independence and k-wise independence.

From the theory of product measure, we have the following equivalent formulation of joint independence:

Exercise 1.1.11. Let $(X_\alpha)_{\alpha \in A}$ be a family of random variables, with each X_α taking values in a measurable space R_α.

(i) Show that the $(X_\alpha)_{\alpha \in A}$ are jointly independent if and only for every collection of distinct elements $\alpha_1, \ldots, \alpha_{k'}$ of A, and all measurable subsets $E_i \subset R_{\alpha_i}$ for $1 \le i \le k'$, one has

$$\mathbf{P}(X_{\alpha_i} \in E_i \text{ for all } 1 \le i \le k') = \prod_{i=1}^{k'} \mathbf{P}(X_{\alpha_i} \in E_i).$$

(ii) Show that the necessary and sufficient condition $(X_\alpha)_{\alpha \in A}$ being k-wise independent is the same, except that k' is constrained to be at most k.

In particular, a finite family $(X_1, \ldots, X_k)$ of random variables X_i, $1 \le i \le k$ taking values in measurable spaces R_i are jointly independent if and only if

$$\mathbf{P}(X_i \in E_i \text{ for all } 1 \le i \le k) = \prod_{i=1}^{k} \mathbf{P}(X_i \in E_i)$$

for all measurable $E_i \subset R_i$.

If the X_α are discrete random variables, one can take the E_i to be singleton sets in the above discussion.

From the above exercise we see that joint independence implies k-wise independence for any k, and that joint independence is preserved under permuting, relabeling, or eliminating some or all of the X_α. A single random variable is automatically jointly independent, and so 1-wise independence is vacuously true; pairwise independence is the first non-trivial notion of independence in this hierarchy.

Example 1.1.13. Let $\mathbf{F}_2$ be the field of two elements, let $V \subset \mathbf{F}_2^3$ be the subspace of triples $(x_1, x_2, x_3) \in \mathbf{F}_2^3$ with $x_1 + x_2 + x_3 = 0$, and let (X_1, X_2, X_3) be drawn uniformly at random from V. Then (X_1, X_2, X_3) are pairwise independent, but not jointly independent. In particular, X_3 is independent of each of X_1, X_2 separately, but is not independent of (X_1, X_2).

Exercise 1.1.12. This exercise generalises the above example. Let $\mathbf{F}$ be a finite field, and let V be a subspace of $\mathbf{F}^n$ for some finite n. Let $(X_1, \ldots, X_n)$ be drawn uniformly at random from V. Suppose that V is not contained in any coordinate hyperplane in $\mathbf{F}^n$.

(i) Show that each X_i, $1 \le i \le n$ is uniformly distributed in $\mathbf{F}$.

(ii) Show that for any $k \geq 2$, that $(X_1, \ldots, X_n)$ is k-wise independent if and only if V is not contained in any hyperplane which is definable using at most k of the coordinate variables.

(iii) Show that $(X_1, \ldots, X_n)$ is jointly independent if and only if $V = \mathbf{F}^n$.

Informally, we thus see that imposing constraints between k variables at a time can destroy k-wise independence, while leaving lower-order independence unaffected.

Exercise 1.1.13. Let $V \subset \mathbf{F}_2^3$ be the subspace of triples $(x_1, x_2, x_3) \in \mathbf{F}_2^3$ with $x_1 + x_2 = 0$, and let (X_1, X_2, X_3) be drawn uniformly at random from V. Then X_3 is independent of (X_1, X_2) (and in particular, is independent of x_1 and x_2 separately), but X_1, X_2 are not independent of each other.

Exercise 1.1.14. We say that one random variable Y (with values in R_Y) is *determined* by another random variable X (with values in R_X) if there exists a (deterministic) function $f : R_X \to R_Y$ such that $Y = f(X)$ is surely true (i.e., $Y(\omega) = f(X(\omega))$ for all $\omega \in \Omega$). Show that if $(X_\alpha)_{\alpha \in A}$ is a family of jointly independent random variables, and $(Y_\beta)_{\beta \in B}$ is a family such that each Y_β is determined by some subfamily $(X_\alpha)_{\alpha \in A_\beta}$ of the $(X_\alpha)_{\alpha \in A}$, with the A_β disjoint as β varies, then the $(Y_\beta)_{\beta \in B}$ are jointly independent also.

Exercise 1.1.15 (Determinism vs. independence). Let X, Y be random variables. Show that Y is deterministic if and only if it is simultaneously determined by X, and independent of X.

Exercise 1.1.16. Show that a complex random variable X is a complex Gaussian random variable (i.e., its distribution is a complex normal distribution) if and only if its real and imaginary parts $\text{Re}(X), \text{Im}(X)$ are independent real Gaussian random variables with the same variance. In particular, the variance of $\text{Re}(X)$ and $\text{Im}(X)$ will be half of the variance of X.

One key advantage of working with jointly independent random variables and events is that one can compute various probabilistic quantities quite easily. We give some key examples below.

Exercise 1.1.17. If $E_1, \ldots, E_k$ are jointly independent events, show that

$$(1.28) \qquad \mathbf{P}(\bigwedge_{i=1}^{k} E_i) = \prod_{i=1}^{k} \mathbf{P}(E_i)$$

and

$$(1.29) \qquad \mathbf{P}(\bigvee_{i=1}^{k} E_i) = 1 - \prod_{i=1}^{k} (1 - \mathbf{P}(E_i)).$$

Show that the converse statement (i.e., that (1.28) and (1.29) imply joint independence) is true for $k = 2$, but fails for higher k. Can one find a correct replacement for this converse for higher k?

Exercise 1.1.18.

(i) If $X_1, \ldots, X_k$ are jointly independent random variables taking values in $[0, +\infty]$, show that

$$\mathbf{E} \prod_{i=1}^{k} X_i = \prod_{i=1}^{k} \mathbf{E} X_i.$$

(ii) If $X_1, \ldots, X_k$ are jointly independent absolutely integrable scalar random variables taking values in $[0, +\infty]$, show that $\prod_{i=1}^{k} X_i$ is absolutely integrable, and

$$\mathbf{E} \prod_{i=1}^{k} X_i = \prod_{i=1}^{k} \mathbf{E} X_i.$$

Remark 1.1.14. The above exercise combines well with Exercise 1.1.14. For instance, if $X_1, \ldots, X_k$ are jointly independent sub-Gaussian variables, then from Exercises 1.1.14, 1.1.18 we see that

$$(1.30) \qquad \mathbf{E} \prod_{i=1}^{k} e^{t X_i} = \prod_{i=1}^{k} \mathbf{E} e^{t X_i}$$

for any complex t. This identity is a key component of the *exponential moment method*, which we will discuss in Section 2.1.

The following result is a key component of the second moment method.

Exercise 1.1.19 (Pairwise independence implies linearity of variance). If $X_1, \ldots, X_k$ are pairwise independent scalar random variables of finite mean and variance, show that

$$\mathbf{Var}(\sum_{i=1}^{k} X_i) = \sum_{i=1}^{k} \mathbf{Var}(X_i)$$

and more generally,

$$\mathbf{Var}(\sum_{i=1}^{k} c_i X_i) = \sum_{i=1}^{k} |c_i|^2 \mathbf{Var}(X_i)$$

for any scalars c_i (compare with (1.12), (1.22)).

The product measure construction allows us to extend Lemma 1.1.7:

Exercise 1.1.20 (Creation of new, independent random variables). Let $(X_\alpha)_{\alpha \in A}$ be a family of random variables (not necessarily independent or finite), and let $(\mu_\beta)_{\beta \in B}$ be a finite collection of probability measures μ_β on measurable spaces R_β. Then, after extending the sample space if necessary, one can find a family $(Y_\beta)_{\beta \in B}$ of independent random variables, such that each Y_β has distribution μ_β, and the two families $(X_\alpha)_{\alpha \in A}$ and $(Y_\beta)_{\beta \in B}$ are independent of each other.

Remark 1.1.15. It is possible to extend this exercise to the case when B is infinite using the *Kolmogorov extension theorem*, which can be found in any graduate probability text (see e.g. [**Ka2002**]). There is, however, the caveat that some (mild) topological hypotheses now need to be imposed on the range R_β of the variables Y_β. For instance, it is enough to assume that each R_β is a locally compact σ-compact metric space equipped with the Borel σ-algebra. These technicalities will, however, not be the focus of this course, and we shall gloss over them in the rest of the text.

We isolate the important case when $\mu_\beta = \mu$ is independent of β. We say that a family $(X_\alpha)_{\alpha \in A}$ of random variables is *independently and identically distributed*, or *iid* for short, if they are jointly independent and all the X_α have the same distribution.

Corollary 1.1.16. *Let $(X_\alpha)_{\alpha \in A}$ be a family of random variables (not necessarily independent or finite), let μ be a probability measure on a measurable space R, and let B be an arbitrary set. Then, after extending the sample space if necessary, one can find an iid family $(Y_\beta)_{\beta \in B}$ with distribution μ which is independent of $(X_\alpha)_{\alpha \in A}$.*

Thus, for instance, one can create arbitrarily large iid families of Bernoulli random variables, Gaussian random variables, etc., regardless of what other random variables are already in play. We thus see that the freedom to extend the underlying sample space allows us access to an unlimited source of randomness. This is in contrast to a situation studied in complexity theory and computer science, in which one does not assume that the sample space can be extended at will, and the amount of randomness one can use is therefore limited.

Remark 1.1.17. Given two probability measures μ_X, μ_Y on two measurable spaces R_X, R_Y, a *joining* or *coupling* of these measures is a random variable (X, Y) taking values in the product space $R_X \times R_Y$, whose individual components X, Y have distribution μ_X, μ_Y, respectively. Exercise 1.1.20 shows that one can always couple two distributions together in an independent manner; but one can certainly create non-independent couplings as well. The study of couplings (or joinings) is particularly important in ergodic theory, but this will not be the focus of this text.

1.1.4. Conditioning. Random variables are inherently non-deterministic in nature, and as such one has to be careful when applying deterministic laws of reasoning to such variables. For instance, consider the law of the excluded middle: a statement P is either true or false, but not both. If this statement is a random variable, rather than deterministic, then instead it is true with some probability p and false with some complementary probability $1 - p$. Also, applying set-theoretic constructions with random inputs can lead to sets, spaces, and other structures which are themselves random variables, which can be quite confusing and require a certain amount of technical care; consider, for instance, the task of rigorously defining a Euclidean space $\mathbf{R}^d$ when the dimension d is itself a random variable.

Now, one can always eliminate these difficulties by explicitly working with points ω in the underlying sample space Ω, and replacing every random variable X by its evaluation $X(\omega)$ at that point; this removes all the randomness from consideration, making everything deterministic (for fixed ω). This approach is rigorous, but goes against the "probabilistic way of thinking", as one now needs to take some care in extending the sample space.

However, if instead one only seeks to remove a *partial* amount of randomness from consideration, then one can do this in a manner consistent with the probabilistic way of thinking, by introducing the machinery of *conditioning*. By conditioning an event to be true or false, or conditioning a random variable to be fixed, one can turn that random event or variable into a deterministic one, while preserving the random nature of other events and variables (particularly those which are independent of the event or variable being conditioned upon).

We begin by considering the simpler situation of conditioning on an event.

Definition 1.1.18 (Conditioning on an event). Let E be an event (or statement) which holds with positive probability $\mathbf{P}(E)$. By *conditioning on the event E*, we mean the act of replacing the underlying sample space Ω with the subset of Ω where E holds, and replacing the underlying probability measure $\mathbf{P}$ by the *conditional probability measure* $\mathbf{P}(|E)$, defined by the formula

$$(1.31) \qquad \mathbf{P}(F|E) := \mathbf{P}(F \wedge E)/\mathbf{P}(E).$$

All events F on the original sample space can thus be viewed as events $(F|E)$ on the conditioned space, which we model set-theoretically as the set of all ω in E obeying F. Note that this notation is compatible with (1.31).

All random variables X on the original sample space can also be viewed as random variables X on the conditioned space, by restriction. We will refer to this conditioned random variable as $(X|E)$, and thus define conditional distribution $\mu_{(X|E)}$ and conditional expectation $\mathbf{E}(X|E)$ (if X is scalar) accordingly.

One can also condition on the complementary event $\overline{E}$, provided that this event holds with positive probility also.

By *undoing this conditioning*, we revert the underlying sample space and measure back to their original (or *unconditional*) values. Note that any random variable which has been defined both after conditioning on E, and conditioning on $\overline{E}$, can still be viewed as a combined random variable after undoing the conditioning.

Conditioning affects the underlying probability space in a manner which is different from extension, and so the act of conditioning is not guaranteed to preserve probabilistic concepts such as distribution, probability, or expectation. Nevertheless, the conditioned version of these concepts are closely related to their unconditional counterparts:

Exercise 1.1.21. If E and $\overline{E}$ both occur with positive probability, establish the identities

(1.32) $$\mathbf{P}(F) = \mathbf{P}(F|E)\mathbf{P}(E) + \mathbf{P}(F|\overline{E})\mathbf{P}(\overline{E})$$

for any (unconditional) event F and

(1.33) $$\mu_X = \mu_{(X|E)}\mathbf{P}(E) + \mu_{(X|\overline{E})}\mathbf{P}(\overline{E})$$

for any (unconditional) random variable X (in the original sample space). In a similar spirit, if X is a non-negative or absolutely integrable scalar (unconditional) random variable, show that $(X|E)$, $(X|\overline{E})$ are also non-negative and absolutely integrable on their respective conditioned spaces, and that

(1.34) $$\mathbf{E}X = \mathbf{E}(X|E)\mathbf{P}(E) + \mathbf{E}(X|\overline{E})\mathbf{P}(\overline{E}).$$

In the degenerate case when E occurs with full probability, conditioning to the complementary event $\overline{E}$ is not well defined, but show that in those cases we can still obtain the above formulae if we adopt the convention that any term involving the vanishing factor $\mathbf{P}(\overline{E})$ should be omitted. Similarly if E occurs with zero probability.

The above identities allow one to study probabilities, distributions, and expectations on the original sample space by conditioning to the two conditioned spaces.

From (1.32) we obtain the inequality

(1.35) $$\mathbf{P}(F|E) \leq \mathbf{P}(F)/\mathbf{P}(E),$$

thus conditioning can magnify probabilities by a factor of at most $1/\mathbf{P}(E)$. In particular:

(i) If F occurs unconditionally surely, it occurs surely conditioning on E also.

(ii) If F occurs unconditionally almost surely, it occurs almost surely conditioning on E also.

(iii) If F occurs unconditionally with overwhelming probability, it occurs with overwhelming probability conditioning on E also, provided that $\mathbf{P}(E) \geq cn^{-C}$ for some $c, C > 0$ independent of n.

(iv) If F occurs unconditionally with high probability, it occurs with high probability conditioning on E also, provided that $\mathbf{P}(E) \geq cn^{-a}$ for some $c > 0$ and some sufficiently small $a > 0$ independent of n.

(v) If F occurs unconditionally asymptotically almost surely, it occurs asymptotically almost surely conditioning on E also, provided that $\mathbf{P}(E) \geq c$ for some $c > 0$ independent of n.

Conditioning can distort the probability of events and the distribution of random variables. Most obviously, conditioning on E elevates the probability of E to 1, and sends the probability of the complementary event $\overline{E}$ to zero. In a similar spirit, if X is a random variable uniformly distributed on some finite set S, and S' is a non-empty subset of S, then conditioning to the event $X \in S'$ alters the distribution of X to now become the uniform distribution on S' rather than S (and conditioning to the complementary event produces the uniform distribution on $S\backslash S'$).

However, events and random variables that are *independent* of the event E being conditioned upon are essentially unaffected by conditioning. Indeed, if F is an event independent of E, then $(F|E)$ occurs with the same probability as F; and if X is a random variable independent of E (or equivalently, independently of the indicator $\mathbf{I}(E)$), then $(X|E)$ has the same distribution as X.

Remark 1.1.19. One can view conditioning to an event E and its complement $\overline{E}$ as the probabilistic analogue of the law of the excluded middle. In deterministic logic, given a statement P, one can divide into two separate cases, depending on whether P is true or false; and any other statement Q is unconditionally true if and only if it is conditionally true in both of these two cases. Similarly, in probability theory, given an event E, one can condition into two separate sample spaces, depending on whether E is conditioned to be true or false; and the unconditional statistics of any random variable or event are then a weighted average of the conditional statistics on the two sample spaces, where the weights are given by the probability of E and its complement.

Now we consider conditioning with respect to a discrete random variable Y, taking values in some range R. One can condition on any event $Y = y$, $y \in R$ which occurs with positive probability. It is then not difficult to establish the analogous identities to those in Exercise 1.1.21:

Exercise 1.1.22. Let Y be a discrete random variable with range R. Then we have

$$(1.36) \qquad \mathbf{P}(F) = \sum_{y \in R} \mathbf{P}(F|Y = y)\mathbf{P}(Y = y)$$

for any (unconditional) event F, and

$$(1.37) \qquad \mu_X = \sum_{y \in R} \mu_{(X|Y=y)}\mathbf{P}(Y = y)$$

for any (unconditional) random variable X (where the sum of non-negative measures is defined in the obvious manner), and for absolutely integrable or non-negative (unconditional) random variables X, one has

$$(1.38) \qquad \mathbf{E}X = \sum_{y \in R} \mathbf{E}(X|Y = y)\mathbf{P}(Y = y).$$

In all of these identities, we adopt the convention that any term involving $\mathbf{P}(Y = y)$ is ignored when $\mathbf{P}(Y = y) = 0$.

With the notation as in the above exercise, we define[5] the *conditional probability* $\mathbf{P}(F|Y)$ of an (unconditional) event F conditioning on Y to be the (unconditional) random variable that is defined to equal $\mathbf{P}(F|Y = y)$ whenever $Y = y$, and similarly, for any absolutely integrable or non-negative (unconditional) random variable X, we define the *conditional expectation* $\mathbf{E}(X|Y)$ to be the (unconditional) random variable that is defined to equal $\mathbf{E}(X|Y = y)$ whenever $Y = y$. Thus (1.36), (1.38) simplify to

$$(1.39) \qquad \mathbf{P}(F) = \mathbf{E}(\mathbf{P}(F|Y))$$

and

$$(1.40) \qquad \mathbf{E}(X) = \mathbf{E}(\mathbf{E}(X|Y)).$$

From (1.12) we have the *linearity of conditional expectation*

$$(1.41) \qquad \mathbf{E}(c_1 X_1 + \cdots + c_k X_k|Y) = c_1\mathbf{E}(X_1|Y) + \cdots + c_k\mathbf{E}(X_k|Y),$$

where the identity is understood to hold almost surely.

---

[5]Strictly speaking, since we are not defining conditional expectation when $\mathbf{P}(Y = y) = 0$, these random variables are only defined almost surely, rather than surely, but this will not cause difficulties in practice; see Remark 1.1.5.

Remark 1.1.20. One can interpret conditional expectation as a type of orthogonal projection; see, for instance, [**Ta2009**, §2.8]. But we will not use this perspective in this course. Just as conditioning on an event and its complement can be viewed as the probabilistic analogue of the law of the excluded middle, conditioning on a discrete random variable can be viewed as the probabilistic analogue of dividing into finitely or countably many cases. For instance, one could condition on the outcome $Y \in \{1, 2, 3, 4, 5, 6\}$ of a six-sided die, thus conditioning the underlying sample space into six separate subspaces. If the die is fair, then the unconditional statistics of a random variable or event would be an unweighted average of the conditional statistics of the six conditioned subspaces; if the die is weighted, one would take a weighted average instead.

Example 1.1.21. Let X_1, X_2 be iid signed Bernoulli random variables, and let $Y := X_1 + X_2$, thus Y is a discrete random variable taking values in $-2, 0, +2$ (with probability $1/4$, $1/2$, $1/4$, respectively). Then X_1 remains a signed Bernoulli random variable when conditioned to $Y = 0$, but becomes the deterministic variable $+1$ when conditioned to $Y = +2$, and similarly becomes the deterministic variable -1 when conditioned to $Y = -2$. As a consequence, the conditional expectation $\mathbf{E}(X_1|Y)$ is equal to 0 when $Y = 0$, $+1$ when $Y = +2$, and -1 when $Y = -2$; thus $\mathbf{E}(X_1|Y) = Y/2$. Similarly, $\mathbf{E}(X_2|Y) = Y/2$; summing and using the linearity of conditional expectation we obtain the obvious identity $\mathbf{E}(Y|Y) = Y$.

If X, Y are independent, then $(X|Y = y) \equiv X$ for all y (with the convention that those y for which $\mathbf{P}(Y = y) = 0$ are ignored), which implies, in particular (for absolutely integrable X), that

$$\mathbf{E}(X|Y) = \mathbf{E}(X)$$

(so in this case the conditional expectation is a deterministic quantity).

Example 1.1.22. Let X, Y be bounded scalar random variables (not necessarily independent), with Y discrete. Then we have

$$\mathbf{E}(XY) = \mathbf{E}(\mathbf{E}(XY|Y)) = \mathbf{E}(Y\mathbf{E}(X|Y))$$

where the latter equality holds since Y clearly becomes deterministic after conditioning on Y.

We will also need to condition with respect to continuous random variables (this is the probabilistic analogue of dividing into a potentially uncountable number of cases). To do this formally, we need to proceed a little differently from the discrete case, introducing the notion of a *disintegration* of the underlying sample space.

Definition 1.1.23 (Disintegration). Let Y be a random variable with range R. A *disintegration* $(R', (\mu_y)_{y \in R'})$ of the underlying sample space Ω with respect to Y is a subset R' of R of full measure in μ_Y (thus $Y \in R'$ almost surely), together with assignment of a probability measure $\mathbf{P}(|Y = y)$ on the subspace $\Omega_y := \{\omega \in \Omega : Y(\omega) = y\}$ of Ω for each $y \in R$, which is *measurable* in the sense that the map $y \mapsto \mathbf{P}(F|Y = y)$ is measurable for every event F, and such that

$$\mathbf{P}(F) = \mathbf{E}\mathbf{P}(F|Y)$$

for all such events, where $\mathbf{P}(F|Y)$ is the (almost surely defined) random variable defined to equal $\mathbf{P}(F|Y = y)$ whenever $Y = y$.

Given such a disintegration, we can then *condition* to the event $Y = y$ for any $y \in R'$ by replacing Ω with the subspace Ω_y (with the induced σ-algebra), but replacing the underlying probability measure $\mathbf{P}$ with $\mathbf{P}(|Y = y)$. We can thus condition (unconditional) events F and random variables X to this event to create conditioned events $(F|Y = y)$ and random variables $(X|Y = y)$ on the conditioned space, giving rise to conditional probabilities $\mathbf{P}(F|Y = y)$ (which is consistent with the existing notation for this expression) and conditional expectations $\mathbf{E}(X|Y = y)$ (assuming absolute integrability in this conditioned space). We then set $\mathbf{E}(X|Y)$ to be the (almost surely defined) random variable defined to equal $\mathbf{E}(X|Y = y)$ whenever $Y = y$.

A disintegration is also known as a *regular conditional probability* in the literature.

Example 1.1.24 (Discrete case). If Y is a discrete random variable, one can set R' to be the *essential range* of Y, which in the discrete case is the set of all $y \in R$ for which $\mathbf{P}(Y = y) > 0$. For each $y \in R'$, we define $\mathbf{P}(|Y = y)$ to be the conditional probability measure relative to the event $Y = y$, as defined in Definition 1.1.18. It is easy to verify that this is indeed a disintegration; thus the continuous notion of conditional probability generalises the discrete one.

Example 1.1.25 (Independent case). Starting with an initial sample space Ω, and a probability measure μ on a measurable space R, one can adjoin a random variable Y taking values in R with distribution μ that is independent of all previously existing random variables, by extending Ω to $\Omega \times R$ as in Lemma 1.1.7. One can then disintegrate Y by taking $R' := R$ and letting μ_y be the probability measure on $\Omega_y = \Omega \times \{y\}$ induced by the obvious isomorphism between $\Omega \times \{y\}$ and Ω; this is easily seen to be a disintegration. Note that if X is any random variable from the original space Ω, then $(X|Y = y)$ has the same distribution as X for any $y \in R$.

Example 1.1.26. Let $\Omega = [0, 1]^2$ with Lebesgue measure, and let (X_1, X_2) be the coordinate random variables of Ω, thus X_1, X_2 are iid with the uniform distribution on $[0, 1]$. Let Y be the random variable $Y := X_1 + X_2$ with range $R = \mathbf{R}$. Then one can disintegrate Y by taking $R' = [0, 2]$ and letting μ_y be normalised Lebesgue measure on the diagonal line segment $\{(x_1, x_2) \in [0, 1]^2 : x_1 + x_2 = y\}$.

Exercise 1.1.23 (Almost uniqueness of disintegrations). Let $(R', (\mu_y)_{y \in R'})$, $(\tilde{R}', (\tilde{\mu}_y)_{y \in \tilde{R}'})$ be two disintegrations of the same random variable Y. Show that for any event F, one has $\mathbf{P}(F|Y = y) = \tilde{\mathbf{P}}(F|Y = y)$ for μ_Y-almost every $y \in R$, where the conditional probabilities $\mathbf{P}(|Y = y)$ and $\tilde{\mathbf{P}}(|Y = y)$ are defined using the disintegrations $(R', (\mu_y)_{y \in R'})$, $(\tilde{R}', (\tilde{\mu}_y)_{y \in \tilde{R}'})$, respectively. (*Hint:* Argue by contradiction, and consider the set of y for which $\mathbf{P}(F|Y = y)$ exceeds $\tilde{\mathbf{P}}(F|Y = y)$ (or vice versa) by some fixed $\varepsilon > 0$.)

Similarly, for a scalar random variable X, show that for μ_Y-almost every $y \in R$, that $(X|Y = y)$ is absolutely integrable with respect to the first disintegration if and only if it is absolutely integrable with respect to the second disintegration, and one has $\mathbf{E}(X|Y = y) = \tilde{\mathbf{E}}(X|Y = y)$ in such cases.

Remark 1.1.27. Under some mild topological assumptions on the underlying sample space (and on the measurable space R), one can always find at least one disintegration for every random variable Y, by using tools such as the Radon-Nikodym theorem; see [**Ta2009**, Theorem 2.9.21]. In practice, we will not invoke these general results here (as it is not natural for us to place topological conditions on the sample space), and instead construct disintegrations by hand in specific cases, for instance, by using the construction in Example 1.1.25. See, e.g., [**Ka2002**] for a more comprehensive discussion of these topics; fortunately for us, these subtle issues will not have any significant impact on our discussion.

Remark 1.1.28. Strictly speaking, disintegration is not a probabilistic concept; there is no canonical way to extend a disintegration when extending the sample space. However, due to the (almost) uniqueness and existence results alluded to earlier, this will not be a difficulty in practice. Still, we will try to use conditioning on continuous variables sparingly, in particular, containing their Use inside the *proofs* of various lemmas, rather than in their *statements*, due to their slight incompatibility with the "probabilistic way of thinking".

Exercise 1.1.24 (Fubini-Tonelli theorem). Let $(R', (\mu_y)_{y \in R'})$ be a disintegration of a random variable Y taking values in a measurable space R, and let X be a non-negative (resp. absolutely integrable) scalar random variable. Show that for μ_Y-almost all $y \in R$, $(X|Y = y)$ is a non-negative (resp.

absolutely integrable) random variable, and one has the identity[6]

(1.42) $$\mathbf{E}(\mathbf{E}(X|Y)) = \mathbf{E}(X),$$

where $\mathbf{E}(X|Y)$ is the (almost surely defined) random variable that equals $\mathbf{E}(X|Y = y)$ whenever $y \in R'$. More generally, show that

(1.43) $$\mathbf{E}(\mathbf{E}(X|Y)f(Y)) = \mathbf{E}(Xf(Y)),$$

whenever $f : R \to \mathbf{R}$ is a non-negative (resp. bounded) measurable function. (One can essentially take (1.43), together with the fact that $\mathbf{E}(X|Y)$ is determined by Y, as a *definition* of the conditional expectation $\mathbf{E}(X|Y)$, but we will not adopt this approach here.)

A typical use of conditioning is to deduce a probabilistic statement from a deterministic one. For instance, suppose one has a random variable X, and a parameter y in some range R, and an event $E(X, y)$ that depends on both X and y. Suppose we know that $\mathbf{P}E(X, y) \leq \varepsilon$ for every $y \in R$. Then, we can conclude that whenever Y is a random variable in R independent of X, we also have $\mathbf{P}E(X, Y) \leq \varepsilon$, regardless of what the actual distribution of Y is. Indeed, if we condition Y to be a fixed value y (using the construction in Example 1.1.25, extending the underlying sample space if necessary), we see that $\mathbf{P}(E(X, Y)|Y = y) \leq \varepsilon$ for each y; and then one can integrate out the conditioning using (1.42) to obtain the claim.

The act of conditioning a random variable to be fixed is occasionally also called *freezing*.

1.1.5. Convergence. In a first course in undergraduate real analysis, we learn what it means for a sequence x_n of scalars to converge to a limit x; for every $\varepsilon > 0$, we have $|x_n - x| \leq \varepsilon$ for all sufficiently large n. Later on, this notion of convergence is generalised to metric space convergence, and generalised further to topological space convergence; in these generalisations, the sequence x_n can lie in some other space than the space of scalars (though one usually insists that this space is independent of n).

Now suppose that we have a sequence X_n of random variables, all taking values in some space R; we will primarily be interested in the scalar case when R is equal to $\mathbf{R}$ or $\mathbf{C}$, but will also need to consider fancier random variables, such as point processes or empirical spectral distributions. In what sense can we say that X_n "converges" to a random variable X, also taking values in R?

It turns out that there are several different notions of convergence which are of interest. For us, the four most important (in decreasing order of

---

[6]Note that one first needs to show that $\mathbf{E}(X|Y)$ is measurable before one can take the expectation.

strength) will be *almost sure convergence, convergence in probability, convergence in distribution,* and *tightness of distribution.*

Definition 1.1.29 (Modes of convergence). Let $R = (R, d)$ be a σ-compact[7] metric space (with the Borel σ-algebra), and let X_n be a sequence of random variables taking values in R. Let X be another random variable taking values in R.

   (i) X_n *converges almost surely* to X if, for almost every $\omega \in \Omega$, $X_n(\omega)$ converges to $X(\omega)$, or equivalently

$$\mathbf{P}(\limsup_{n \to \infty} d(X_n, X) \leq \varepsilon) = 1$$

   for every $\varepsilon > 0$.

   (ii) X_n *converges in probability* to X if, for every $\varepsilon > 0$, one has

$$\liminf_{n \to \infty} \mathbf{P}(d(X_n, X) \leq \varepsilon) = 1,$$

   or equivalently if $d(X_n, X) \leq \varepsilon$ holds asymptotically almost surely for every $\varepsilon > 0$.

   (iii) X_n *converges in distribution* to X if, for every bounded continuous function $F : R \to \mathbf{R}$, one has

$$\lim_{n \to \infty} \mathbf{E}F(X_n) = \mathbf{E}F(X).$$

   (iv) X_n has a *tight sequence of distributions* if, for every $\varepsilon > 0$, there exists a compact subset K of R such that $\mathbf{P}(X_n \in K) \geq 1 - \varepsilon$ for all sufficiently large n.

Remark 1.1.30. One can relax the requirement that R be a σ-compact metric space in the definitions, but then some of the nice equivalences and other properties of these modes of convergence begin to break down. In our applications, though, we will only need to consider the σ-compact metric space case. Note that all of these notions are probabilistic (i.e., they are preserved under extensions of the sample space).

Exercise 1.1.25 (Implications and equivalences). Let X_n, X be random variables taking values in a σ-compact metric space R.

   (i) Show that if X_n converges almost surely to X, then X_n converges in probability to X. (*Hint:* Use *Fatou's lemma.*)

   (ii) Show that if X_n converges in distribution to X, then X_n has a tight sequence of distributions.

   (iii) Show that if X_n converges in probability to X, then X_n converges in distribution to X. (*Hint:* First show tightness, then use the fact

---

[7] A metric space is σ-*compact* if it is the countable union of compact sets.

that on compact sets, continuous functions are uniformly continuous.)

(iv) Show that X_n converges in distribution to X if and only if μ_{X_n} converges to μ_X in the *vague topology* (i.e., $\int f \, d\mu_{X_n} \to \int f \, d\mu_X$ for all continuous functions $f : R \to \mathbf{R}$ of compact support).

(v) Conversely, if X_n has a tight sequence of distributions, and μ_{X_n} is convergent in the vague topology, show that X_n is convergent in distribution to another random variable (possibly after extending the sample space). What happens if the tightness hypothesis is dropped?

(vi) If X is deterministic, show that X_n converges in probability to X if and only if X_n converges in distribution to X.

(vii) If X_n has a tight sequence of distributions, show that there is a subsequence of the X_n which converges in distribution. (This is known as *Prokhorov's theorem*).

(viii) If X_n converges in probability to X, show that there is a subsequence of the X_n which converges almost surely to X.

(ix) X_n converges in distribution to X if and only if $\liminf_{n\to\infty} \mathbf{P}(X_n \in U) \geq \mathbf{P}(X \in U)$ for every open subset U of R, or equivalently if $\limsup_{n\to\infty} \mathbf{P}(X_n \in K) \leq \mathbf{P}(X \in K)$ for every closed subset K of R.

Exercise 1.1.26 (Skorokhod representation theorem). Let μ_n be a sequence of probability measures on $\mathbf{C}$ that converge in the vague topology to another probability measure μ. Show (after extending the probability space if necessary) that there exist random variables X_n with distribution μ_n that converge almost surely to a random variable X with distribution μ.

Remark 1.1.31. The relationship between almost sure convergence and convergence in probability may be clarified by the following observation. If E_n is a sequence of events, then the indicators $\mathbf{I}(E_n)$ converge in probability to zero iff $\mathbf{P}(E_n) \to 0$ as $n \to \infty$, but converge almost surely to zero iff $\mathbf{P}(\bigcup_{n \geq N} E_n) \to 0$ as $N \to \infty$.

Example 1.1.32. Let Y be a random variable drawn uniformly from $[0, 1]$. For each $n \geq 1$, let E_n be the event that the decimal expansion of Y begins with the decimal expansion of n, e.g., every real number in $[0.25, 0.26)$ lies in E_{25}. (Let us ignore the annoying $0.999\ldots = 1.000\ldots$ ambiguity in the decimal expansion here, as it will almost surely not be an issue.) Then the indicators $\mathbf{I}(E_n)$ converge in probability and in distribution to zero, but do not converge almost surely.

If y_n is the nth digit of Y, then the y_n converge in distribution (to the uniform distribution on $\{0, 1, \ldots, 9\}$), but do not converge in probability or almost surely. Thus we see that the latter two notions are sensitive not only to the distribution of the random variables, but how they are positioned in the sample space.

The limit of a sequence converging almost surely or in probability is clearly unique up to almost sure equivalence, whereas the limit of a sequence converging in distribution is only unique up to equivalence in distribution. Indeed, convergence in distribution is really a statement about the distributions μ_{X_n}, μ_X rather than of the random variables X_n, X themselves. In particular, for convergence in distribution one does not care about how correlated or dependent the X_n are with respect to each other, or with X; indeed, they could even live on different sample spaces Ω_n, Ω and we would still have a well-defined notion of convergence in distribution, even though the other two notions cease to make sense (except when X is deterministic, in which case we can recover convergence in probability by Exercise 1.1.25(vi)).

Exercise 1.1.27 (Borel-Cantelli lemma). Suppose that X_n, X are random variables such that $\sum_n \mathbf{P}(d(X_n, X) \geq \varepsilon) < \infty$ for every $\varepsilon > 0$. Show that X_n converges almost surely to X.

Exercise 1.1.28 (Convergence and moments). Let X_n be a sequence of scalar random variables, and let X be another scalar random variable. Let $k, \varepsilon > 0$.

(i) If $\sup_n \mathbf{E}|X_n|^k < \infty$, show that X_n has a tight sequence of distributions.

(ii) If $\sup_n \mathbf{E}|X_n|^k < \infty$ and X_n converges in distribution to X, show that $\mathbf{E}|X|^k \leq \liminf_{n \to \infty} \mathbf{E}|X_n|^k$.

(iii) If $\sup_n \mathbf{E}|X_n|^{k+\varepsilon} < \infty$ and X_n converges in distribution to X, show that $\mathbf{E}|X|^k = \lim_{n \to \infty} \mathbf{E}|X_n|^k$.

(iv) Give a counterexample to show that (iii) fails when $\varepsilon = 0$, even if we upgrade convergence in distribution to almost sure convergence.

(v) If the X_n are uniformly bounded and real-valued, and $\mathbf{E}X^k = \lim_{n \to \infty} \mathbf{E}X_n^k$ for every $k = 0, 1, 2, \ldots$, then X_n converges in distribution to X. (*Hint:* Use the *Weierstrass approximation theorem*. Alternatively, use the analytic nature of the moment generating function $\mathbf{E}e^{tX}$ and analytic continuation.)

(vi) If the X_n are uniformly bounded and complex-valued, and $\mathbf{E}X^k\overline{X}^l = \lim_{n \to \infty} \mathbf{E}X_n^k\overline{X_n}^l$ for every $k, l = 0, 1, 2, \ldots$, then X_n converges in distribution to X. Give a counterexample to show that the claim fails if one only considers the cases when $l = 0$.

There are other interesting modes of convergence on random variables and on distributions, such as convergence in *total variation norm*, in the *Lévy-Prokhorov metric*, or in *Wasserstein metric*, but we will not need these concepts in this text.

1.2. Stirling's formula

In this section we derive *Stirling's formula*, which is a useful approximation for $n!$ when n is large. This formula (and related formulae for binomial coefficients) $\binom{n}{m}$ will be useful for estimating a number of combinatorial quantities in this text, and also in allowing one to analyse discrete random walks accurately.

From Taylor expansion we have $x^n/n! \le e^x$ for any $x \ge 0$. Specialising this to $x = n$ we obtain a crude lower bound

(1.44)
$$n! \ge n^n e^{-n}.$$

In the other direction, we trivially have

(1.45)
$$n! \le n^n,$$

so we know already that $n!$ is within[8] an exponential factor of n^n.

One can do better by starting with the identity

$$\log n! = \sum_{m=1}^{n} \log m$$

and viewing the right-hand side as a Riemann integral approximation to $\int_1^n \log x \, dx$. Indeed, a simple area comparison (cf. the *integral test*) yields the inequalities

$$\int_1^n \log x \, dx \le \sum_{m=1}^{n} \log m \le \log n + \int_1^n \log x \, dx$$

which leads to the inequalities

(1.46)
$$en^n e^{-n} \le n! \le en \times n^n e^{-n},$$

so the lower bound in (1.44) was only off[9] by a factor of n or so.

One can improve these bounds further by using the *trapezoid rule* as follows. On any interval $[m, m+1]$, $\log x$ has a second derivative of $O(1/m^2)$,

---

[8]One can also obtain a cruder version of this fact that avoids Taylor expansion, by observing the trivial lower bound $n! \ge (n/2)^{\lfloor n/2 \rfloor}$ coming from considering the second half of the product $n! = 1 \cdots \cdot n$.

[9]This illustrates a general principle, namely that one can often get a non-terrible bound for a series (in this case, the Taylor series for e^n) by using the largest term in that series (which is $n^n/n!$).

which by Taylor expansion leads to the approximation

$$\int_m^{m+1} \log x \, dx = \frac{1}{2} \log m + \frac{1}{2} \log(m+1) + \epsilon_m$$

for some error $\epsilon_m = O(1/m^2)$.

The error is absolutely convergent; by the integral test, we have $\sum_{m=1}^n \epsilon_m = C + O(1/n)$ for some absolute constant $C := \sum_{m=1}^\infty \epsilon_m$. Performing this sum, we conclude that

$$\int_1^n \log x \, dx = \sum_{m=1}^{n-1} \log m + \frac{1}{2} \log n + C + O(1/n)$$

which after some rearranging leads to the asymptotic

(1.47) $$n! = (1 + O(1/n))e^{1-C} \sqrt{n} n^n e^{-n},$$

so we see that $n!$ actually lies roughly at the geometric mean of the two bounds in (1.46).

This argument does not easily reveal what the constant C actually is (though it can in principle be computed numerically to any specified level of accuracy by this method). To find this out, we take a different tack, interpreting the factorial via the *Gamma function* $\Gamma : \mathbf{R} \to \mathbf{R}$ as follows. Repeated integration by parts reveals the identity[10]

(1.48) $$n! = \int_0^\infty t^n e^{-t} \, dt.$$

So to estimate $n!$, it suffices to estimate the integral in (1.48). Elementary calculus reveals that the integrand $t^n e^{-t}$ achieves its maximum at $t = n$, so it is natural to make the substitution $t = n + s$, obtaining

$$n! = \int_{-n}^\infty (n+s)^n e^{-n-s} \, ds$$

which we can simplify a little bit as

$$n! = n^n e^{-n} \int_{-n}^\infty (1 + \frac{s}{n})^n e^{-s} \, ds,$$

pulling out the now-familiar factors of $n^n e^{-n}$. We combine the integrand into a single exponential,

$$n! = n^n e^{-n} \int_{-n}^\infty \exp(n \log(1 + \frac{s}{n}) - s) \, ds.$$

From Taylor expansion we see that

$$n \log(1 + \frac{s}{n}) = s - \frac{s^2}{2n} + \dots,$$

---

[10]The right-hand side of (1.48), by definition, is $\Gamma(n+1)$.

so we heuristically have

$$\exp(n\log(1+\frac{s}{n})-s)\approx\exp(-s^2/2n).$$

To achieve this approximation rigorously, we first scale s by $\sqrt{n}$ to remove the n in the denominator. Making the substitution $s=\sqrt{n}x$, we obtain

$$n! = \sqrt{n}n^n e^{-n}\int_{-\sqrt{n}}^{\infty}\exp(n\log(1+\frac{x}{\sqrt{n}})-\sqrt{n}x)\,dx,$$

thus extracting the factor of $\sqrt{n}$ that we know from (1.47) has to be there.

Now, Taylor expansion tells us that for fixed x, we have the pointwise convergence

(1.49)
$$\exp(n\log(1+\frac{x}{\sqrt{n}})-\sqrt{n}x)\to\exp(-x^2/2)$$

as $n\to\infty$. To be more precise, as the function $n\log(1+\frac{x}{\sqrt{n}})$ equals 0 with derivative $\sqrt{n}$ at the origin, and has second derivative $\frac{-1}{(1+x/\sqrt{n})^2}$, we see from two applications of the fundamental theorem of calculus that

$$n\log(1+\frac{x}{\sqrt{n}})-\sqrt{n}x = -\int_0^x\frac{(x-y)dy}{(1+y/\sqrt{n})^2}.$$

This gives a uniform lower bound

$$n\log(1+\frac{x}{\sqrt{n}})-\sqrt{n}x \le -cx^2$$

for some $c>0$ when $|x|\le\sqrt{n}$, and

$$n\log(1+\frac{x}{\sqrt{n}})-\sqrt{n}x \le -cx\sqrt{n}$$

for $|x|>\sqrt{n}$. This is enough to keep the integrands $\exp(n\log(1+\frac{x}{\sqrt{n}})-\sqrt{n}x)$ dominated by an absolutely integrable function. By (1.49) and the Lebesgue dominated convergence theorem, we thus have

$$\int_{-\sqrt{n}}^{\infty}\exp(n\log(1+\frac{x}{\sqrt{n}})-\sqrt{n}x)\,dx \to \int_{-\infty}^{\infty}\exp(-x^2/2)\,dx.$$

A classical computation (based, for instance, on computing $\int_{-\infty}^{\infty}\int_{-\infty}^{\infty}\exp(-(x^2+y^2)/2)\,dxdy$ in both Cartesian and polar coordinates) shows that

$$\int_{-\infty}^{\infty}\exp(-x^2/2)\,dx = \sqrt{2\pi}$$

and so we conclude *Stirling's formula*

(1.50)
$$n! = (1+o(1))\sqrt{2\pi n}n^n e^{-n}.$$

Remark 1.2.1. The dominated convergence theorem does not immediately give any effective rate on the decay $o(1)$ (though such a rate can eventually be extracted by a quantitative version of the above argument. But one can combine (1.50) with (1.47) to show that the error rate is of the form $O(1/n)$. By using fancier versions of the trapezoid rule (e.g., *Simpson's rule*) one can obtain an asymptotic expansion of the error term in $1/n$; see [**KeVa2007**].

Remark 1.2.2. The derivation of (1.50) demonstrates some general principles concerning the estimation of exponential integrals $\int e^{\phi(x)} \, dx$ when ϕ is large. First, the integral is dominated by the local maxima of ϕ. Then, near these maxima, $e^{\phi(x)}$ usually behaves like a rescaled Gaussian, as can be seen by Taylor expansion (though more complicated behaviour emerges if the second derivative of ϕ degenerates). So one can often understand the asymptotics of such integrals by a change of variables designed to reveal the Gaussian behaviour. This technique is known as *Laplace's method*. A similar set of principles also holds for oscillatory exponential integrals $\int e^{i\phi(x)} \, dx$; these principles are collectively referred to as the *method of stationary phase*.

One can use Stirling's formula to estimate binomial coefficients. Here is a crude bound:

Exercise 1.2.1 (Entropy formula). Let n be large, let $0 < \gamma < 1$ be fixed, and let $1 \le m \le n$ be an integer of the form $m = (\gamma + o(1))n$. Show that $\binom{n}{m} = \exp((h(\gamma) + o(1))n)$, where $h(\gamma)$ is the *entropy function*

$$h(\gamma) := \gamma \log \frac{1}{\gamma} + (1 - \gamma) \log \frac{1}{1 - \gamma}.$$

For m near $n/2$, one also has the following more precise bound:

Exercise 1.2.2 (Refined entropy formula). Let n be large, and let $1 \le m \le n$ be an integer of the form $m = n/2 + k$ for some $k = o(n^{2/3})$. Show that

$$(1.51) \qquad \binom{n}{m} = (\sqrt{\frac{2}{\pi}} + o(1)) \frac{2^n}{\sqrt{n}} \exp(-2k^2/n).$$

Note the Gaussian-type behaviour in k. This can be viewed as an illustration of the central limit theorem (see Section 2.2) when summing iid Bernoulli variables $X_1, \ldots, X_n \in \{0, 1\}$, where each X_i has a $1/2$ probability of being either 0 or 1. Indeed, from (1.51) we see that

$$\mathbf{P}(X_1 + \cdots + X_n = n/2 + k) = (\sqrt{\frac{2}{\pi}} + o(1)) \frac{1}{\sqrt{n}} \exp(-2k^2/n)$$

when $k = o(n^{2/3})$, which suggests that $X_1 + \cdots + X_n$ is distributed roughly like the Gaussian $N(n/2, n/4)$ with mean $n/2$ and variance $n/4$.

1.3. Eigenvalues and sums of Hermitian matrices

Let A be a Hermitian $n \times n$ matrix. By the *spectral theorem* for Hermitian matrices (which, for sake of completeness, we prove below), one can diagonalise A using a sequence[11]

$$\lambda_1(A) \geq \ldots \geq \lambda_n(A)$$

of n real eigenvalues, together with an orthonormal basis of eigenvectors $u_1(A), \ldots, u_n(A) \in \mathbf{C}^n$. The set $\{\lambda_1(A), \ldots, \lambda_n(A)\}$ is known as the *spectrum* of A.

A basic question in linear algebra asks the extent to which the eigenvalues $\lambda_1(A), \ldots, \lambda_n(A)$ and $\lambda_1(B), \ldots, \lambda_n(B)$ of two Hermitian matrices A, B constrain the eigenvalues $\lambda_1(A+B), \ldots, \lambda_n(A+B)$ of the sum. For instance, the linearity of trace

$$\operatorname{tr}(A + B) = \operatorname{tr}(A) + \operatorname{tr}(B),$$

when expressed in terms of eigenvalues, gives the trace constraint

$$(1.52) \qquad \lambda_1(A + B) + \cdots + \lambda_n(A + B) = \lambda_1(A) + \cdots + \lambda_n(A)$$
$$+ \lambda_1(B) + \cdots + \lambda_n(B);$$

the identity

$$(1.53) \qquad \lambda_1(A) = \sup_{|v|=1} v^* A v$$

(together with the counterparts for B and $A + B$) gives the inequality

$$(1.54) \qquad \lambda_1(A + B) \leq \lambda_1(A) + \lambda_1(B),$$

and so forth.

The complete answer to this problem is a fascinating one, requiring a strangely recursive description (once known as *Horn's conjecture*, which is now solved), and connected to a large number of other fields of mathematics, such as geometric invariant theory, intersection theory, and the combinatorics of a certain gadget known as a "honeycomb". See [**KnTa2001**] for a survey of this topic.

In typical applications to random matrices, one of the matrices (say, B) is "small" in some sense, so that $A+B$ is a perturbation of A. In this case, one does not need the full strength of the above theory, and instead relies on a simple aspect of it pointed out in [**HeRo1995**], [**To1994**], which generates several of the *eigenvalue inequalities* relating A, B, and $A + B$, of which

---

[11]The eigenvalues are uniquely determined by A, but the eigenvectors have a little ambiguity to them, particularly if there are repeated eigenvalues; for instance, one could multiply each eigenvector by a complex phase $e^{i\theta}$. In this text we are arranging eigenvalues in descending order; of course, one can also arrange eigenvalues in increasing order, which causes some slight notational changes in the results below.

(1.52) and (1.54) are examples[12]. These eigenvalue inequalities can mostly be deduced from a number of *minimax* characterisations of eigenvalues (of which (1.53) is a typical example), together with some basic facts about intersections of subspaces. Examples include the *Weyl inequalities*

$$(1.55) \qquad \lambda_{i+j-1}(A + B) \le \lambda_i(A) + \lambda_j(B),$$

valid whenever $i, j \ge 1$ and $i + j - 1 \le n$, and the *Ky Fan inequality*

$$(1.56) \qquad \begin{aligned} \lambda_1(A + B) + \cdots + \lambda_k(A + B) \le \\ \lambda_1(A) + \cdots + \lambda_k(A) + \lambda_1(B) + \cdots + \lambda_k(B). \end{aligned}$$

One consequence of these inequalities is that the spectrum of a Hermitian matrix is *stable* with respect to small perturbations.

We will also establish some closely related inequalities concerning the relationships between the eigenvalues of a matrix, and the eigenvalues of its minors.

Many of the inequalities here have analogues for the singular values of non-Hermitian matrices (by exploiting the augmented matrix (2.80)). However, the situation is markedly different when dealing with *eigenvalues* of *non-Hermitian* matrices; here, the spectrum can be far more unstable, if *pseudospectrum* is present. Because of this, the theory of the eigenvalues of a random non-Hermitian matrix requires an additional ingredient, namely upper bounds on the prevalence of pseudospectrum, which after recentering the matrix is basically equivalent to establishing lower bounds on least singular values. See Section 2.8.1 for further discussion of this point.

We will work primarily here with Hermitian matrices, which can be viewed as self-adjoint transformations on complex vector spaces such as $\mathbf{C}^n$. One can of course specialise the discussion to real symmetric matrices, in which case one can restrict these complex vector spaces to their real counterparts $\mathbf{R}^n$. The specialisation of the complex theory below to the real case is straightforward and is left to the interested reader.

1.3.1. Proof of spectral theorem. To prove the spectral theorem, it is convenient to work more abstractly, in the context of self-adjoint operators on finite-dimensional Hilbert spaces:

Theorem 1.3.1 (Spectral theorem). *Let V be a finite-dimensional complex Hilbert space of some dimension n, and let $T : V \to V$ be a self-adjoint operator. Then there exists an orthonormal basis $v_1, \ldots, v_n \in V$ of V and eigenvalues $\lambda_1, \ldots, \lambda_n \in \mathbf{R}$ such that $Tv_i = \lambda_i v_i$ for all $1 \le i \le n$.*

---

[12]Actually, this method eventually generates *all* of the eigenvalue inequalities, but this is a non-trivial fact to prove; see [**KnTaWo2004**]

The spectral theorem as stated in the introduction then follows by specialising to the case $V = \mathbf{C}^n$ and ordering the eigenvalues.

Proof. We induct on the dimension n. The claim is vacuous for $n = 0$, so suppose that $n \geq 1$ and that the claim has already been proven for $n = 1$.

Let v be a unit vector in V (thus $v^* v = 1$) that maximises the form $\text{Re}(v^* T v)$; this maximum exists by compactness. By the method of Lagrange multipliers, v is a critical point of $\text{Re}(v^* T v) - \lambda v^* v$ for some $\lambda \in \mathbf{R}$. Differentiating in an arbitrary direction $w \in V$, we conclude that

$$\text{Re}(v^* T w + w^* T v - \lambda v^* w - \lambda w^* v) = 0;$$

this simplifies using self-adjointness to

$$\text{Re}(w^* (T v - \lambda v)) = 0.$$

Since $w \in V$ was arbitrary, we conclude that $T v = \lambda v$, thus v is a unit eigenvector of T. By self-adjointness, this implies that the orthogonal complement $v^\perp := \{w \in V : v^* w = 0\}$ of v is preserved by T. Restricting T to this lower-dimensional subspace and applying the induction hypothesis, we can find an orthonormal basis of eigenvectors of T on $v^\perp$. Adjoining the new unit vector v to the orthonormal basis, we obtain the claim. $\qquad\square$

Suppose we have a self-adjoint transformation $A : \mathbf{C}^n \to \mathbf{C}^n$, which of course can be identified with a Hermitian matrix. Using the orthogonal eigenbasis provided by the spectral theorem, we can perform an orthonormal change of variables to set that eigenbasis to be the standard basis $e_1, \ldots, e_n$, so that the matrix of A becomes diagonal. This is very useful when dealing with just a single matrix A; for instance, it makes the task of computing functions of A, such as A^k or $\exp(tA)$, much easier. However, when one has *several* Hermitian matrices in play (e.g., $A, B, A+B$), then it is usually not possible to standardise all the eigenbases simultaneously (i.e., to simultaneously diagonalise all the matrices), except when the matrices all commute. Nevertheless, one can still normalise *one* of the eigenbases to be the standard basis, and this is still useful for several applications, as we shall soon see.

Exercise 1.3.1. Suppose that the eigenvalues $\lambda_1(A) > \cdots > \lambda_n(A)$ of an $n \times n$ Hermitian matrix are distinct. Show that the associated eigenbasis $u_1(A), \ldots, u_n(A)$ is unique up to rotating each individual eigenvector $u_j(A)$ by a complex phase $e^{i\theta_j}$. In particular, the *spectral projections* $P_j(A) := u_j(A)^* u_j(A)$ are unique. What happens when there is eigenvalue multiplicity?

1.3.2. Minimax formulae. The ith eigenvalue functional $A \mapsto \lambda_i(A)$ is not a linear functional (except in dimension one). It is not even a convex functional (except when $i = 1$) or a concave functional (except when $i = n$). However, it is the next best thing, namely it is a *minimax* expression of linear functionals[13]. More precisely, we have

Theorem 1.3.2 (Courant-Fischer minimax theorem). *Let A be an $n \times n$ Hermitian matrix. Then we have*

$$(1.57) \qquad \lambda_i(A) = \sup_{\dim(V)=i} \inf_{v \in V : |v|=1} v^* A v$$

and

$$(1.58) \qquad \lambda_i(A) = \inf_{\dim(V)=n-i+1} \sup_{v \in V : |v|=1} v^* A v$$

for all $1 \leq i \leq n$, where V ranges over all subspaces of $\mathbf{C}^n$ with the indicated dimension.

Proof. It suffices to prove (1.57), as (1.58) follows by replacing A by $-A$ (noting that $\lambda_i(-A) = -\lambda_{n-i+1}(A)$).

We first verify the $i = 1$ case, i.e., (1.53). By the spectral theorem, we can assume that A has the standard eigenbasis $e_1, \ldots, e_n$, in which case we have

$$(1.59) \qquad v^* A v = \sum_{i=1}^{n} \lambda_i |v_i|^2$$

whenever $v = (v_1, \ldots, v_n)$. The claim (1.53) is then easily verified.

To prove the general case, we may again assume A has the standard eigenbasis. By considering the space V spanned by $e_1, \ldots, e_i$, we easily see the inequality

$$\lambda_i(A) \leq \sup_{\dim(V)=i} \inf_{v \in V : |v|=1} v^* A v,$$

so we only need to prove the reverse inequality. In other words, for every i-dimensional subspace V of $\mathbf{C}^n$, we have to show that V contains a unit vector v such that

$$v^* A v \leq \lambda_i(A).$$

Let W be the space spanned by $e_i, \ldots, e_n$. This space has codimension $i-1$, so it must have non-trivial intersection with V. If we let v be a unit vector in $V \cap W$, the claim then follows from (1.59). $\square$

---

[13]Note that a convex functional is the same thing as a max of linear functionals, while a concave functional is the same thing as a min of linear functionals.

Remark 1.3.3. By homogeneity, one can replace the restriction $|v| = 1$ with $v \neq 0$ provided that one replaces the quadratic form v^*Av with the *Rayleigh quotient* v^*Av/v^*v.

A closely related formula is as follows. Given an $n \times n$ Hermitian matrix A and an m-dimensional subspace V of $\mathbf{C}^n$, we define the *partial trace* $\operatorname{tr}(A \mid_V)$ to be the expression

$$\operatorname{tr}(A \mid_V) := \sum_{i=1}^{m} v_i^* A v_i$$

where $v_1, \ldots, v_m$ is any orthonormal basis of V. It is easy to see that this expression is independent of the choice of orthonormal basis, and so the partial trace is well-defined.

Proposition 1.3.4 (Extremal partial trace). *Let A be an $n \times n$ Hermitian matrix. Then for any $1 \leq k \leq n$, one has*

$$\lambda_1(A) + \cdots + \lambda_k(A) = \sup_{\dim(V)=k} \operatorname{tr}(A \mid_V)$$

and

$$\lambda_{n-k+1}(A) + \cdots + \lambda_n(A) = \inf_{\dim(V)=k} \operatorname{tr}(A \mid_V).$$

As a corollary, we see that $A \mapsto \lambda_1(A) + \cdots + \lambda_k(A)$ is a convex function, and $A \mapsto \lambda_{n-k+1}(A) + \cdots + \lambda_n(A)$ is a concave function.

Proof. Again, by symmetry it suffices to prove the first formula. As before, we may assume, without loss of generality, that A has the standard eigenbasis $e_1, \ldots, e_n$ corresponding to $\lambda_1(A), \ldots, \lambda_n(A)$, respectively. By selecting V to be the span of $e_1, \ldots, e_k$ we have the inequality

$$\lambda_1(A) + \cdots + \lambda_k(A) \leq \sup_{\dim(V)=k} \operatorname{tr}(A \mid_V),$$

so it suffices to prove the reverse inequality. For this we induct on the dimension n. If V has dimension k, then it has a $k-1$-dimensional subspace V' that is contained in the span of $e_2, \ldots, e_n$. By the induction hypothesis applied to the restriction of A to this span (which has eigenvalues $\lambda_2(A), \ldots, \lambda_n(A)$), we have

$$\lambda_2(A) + \cdots + \lambda_k(A) \geq \operatorname{tr}(A \mid_{V'}).$$

On the other hand, if v is a unit vector in the orthogonal complement of V' in V, we see from (1.53) that

$$\lambda_1(A) \geq v^*Av.$$

Adding the two inequalities we obtain the claim. $\qquad\square$

Specialising Proposition 1.3.4 to the case when V is a coordinate subspace (i.e., the span of k of the basis vectors $e_1, \ldots, e_n$), we conclude the *Schur-Horn inequalities*

$$
\begin{aligned}
\lambda_{n-k+1}(A) + \cdots + \lambda_n(A) &\leq a_{i_1 i_1} + \cdots + a_{i_k i_k} \\
&\leq \lambda_1(A) + \cdots + \lambda_k(A)
\end{aligned}
$$
(1.60)

for any $1 \leq i_1 < \cdots < i_k \leq n$, where $a_{11}, a_{22}, \ldots, a_{nn}$ are the diagonal entries of A.

Exercise 1.3.2. Show that the inequalities (1.60) are equivalent to the assertion that the diagonal entries $\mathrm{diag}(A) = (a_{11}, a_{22}, \ldots, a_{nn})$ lies in the *permutahedron* of $\lambda_1(A), \ldots, \lambda_n(A)$, defined as the convex hull of the $n!$ permutations of $(\lambda_1(A), \ldots, \lambda_n(A))$ in $\mathbf{R}^n$.

Remark 1.3.5. It is a theorem of Schur and Horn [**Ho1954**] that these are the complete set of inequalities connecting the diagonal entries $\mathrm{diag}(A) = (a_{11}, a_{22}, \ldots, a_{nn})$ of a Hermitian matrix to its spectrum. To put it another way, the image of any *coadjoint orbit* $\mathcal{O}_A := \{UAU^* : U \in U(n)\}$ of a matrix A with a given spectrum $\lambda_1, \ldots, \lambda_n$ under the diagonal map $\mathrm{diag} : A \mapsto \mathrm{diag}(A)$ is the permutahedron of $\lambda_1, \ldots, \lambda_n$. Note that the vertices of this permutahedron can be attained by considering the diagonal matrices inside this coadjoint orbit, whose entries are then a permutation of the eigenvalues. One can interpret this diagonal map diag as the *moment map* associated with the conjugation action of the standard maximal torus of $U(n)$ (i.e., the diagonal unitary matrices) on the coadjoint orbit. When viewed in this fashion, the Schur-Horn theorem can be viewed as the special case of the more general *Atiyah convexity theorem* [**At1982**] (also proven independently by Guillemin and Sternberg [**GuSt1982**]) in symplectic geometry. Indeed, the topic of eigenvalues of Hermitian matrices turns out to be quite profitably viewed as a question in symplectic geometry (and also in algebraic geometry, particularly when viewed through the machinery of *geometric invariant theory*).

There is a simultaneous generalisation of Theorem 1.3.2 and Proposition 1.3.4:

Exercise 1.3.3 (Wielandt minimax formula). Let $1 \leq i_1 < \cdots < i_k \leq n$ be integers. Define a *partial flag* to be a nested collection $V_1 \subset \cdots \subset V_k$ of subspaces of $\mathbf{C}^n$ such that $\dim(V_j) = i_j$ for all $1 \leq j \leq k$. Define the associated *Schubert variety* $X(V_1, \ldots, V_k)$ to be the collection of all k-dimensional subspaces W such that $\dim(W \cap V_j) \geq j$. Show that for any $n \times n$ matrix A,

$$
\lambda_{i_1}(A) + \cdots + \lambda_{i_k}(A) = \sup_{V_1, \ldots, V_k} \inf_{W \in X(V_1, \ldots, V_k)} \mathrm{tr}(A \mid_W).
$$

1.3.3. Eigenvalue inequalities. Using the above minimax formulae, we can now quickly prove a variety of eigenvalue inequalities. The basic idea is to exploit the linearity relationship

$$(1.61) \qquad v^*(A + B)v = v^*Av + v^*Bv$$

for any unit vector v, and more generally,

$$(1.62) \qquad \operatorname{tr}((A + B) \mid_V) = \operatorname{tr}(A \mid_V) + \operatorname{tr}(B \mid_V)$$

for any subspace V.

For instance, as mentioned before, the inequality (1.54) follows immediately from (1.53) and (1.61). Similarly, for the Ky Fan inequality (1.56), one observes from (1.62) and Proposition 1.3.4 that

$$\operatorname{tr}((A + B) \mid_W) \le \operatorname{tr}(A \mid_W) + \lambda_1(B) + \cdots + \lambda_k(B)$$

for any k-dimensional subspace W. Substituting this into Proposition 1.3.4 gives the claim. If one uses Exercise 1.3.3 instead of Proposition 1.3.4, one obtains the more general *Lidskii inequality*

$$(1.63) \qquad \begin{aligned} &\lambda_{i_1}(A + B) + \cdots + \lambda_{i_k}(A + B) \\ &\le \lambda_{i_1}(A) + \cdots + \lambda_{i_k}(A) + \lambda_1(B) + \cdots + \lambda_k(B) \end{aligned}$$

for any $1 \le i_1 < \cdots < i_k \le n$.

In a similar spirit, using the inequality

$$|v^*Bv| \le \|B\|_{\mathrm{op}} = \max(|\lambda_1(B)|, |\lambda_n(B)|)$$

for unit vectors v, combined with (1.61) and (1.57), we obtain the eigenvalue stability inequality

$$(1.64) \qquad |\lambda_i(A + B) - \lambda_i(A)| \le \|B\|_{\mathrm{op}},$$

thus the spectrum of $A + B$ is close to that of A if B is small in operator norm. In particular, we see that the map $A \mapsto \lambda_i(A)$ is Lipschitz continuous on the space of Hermitian matrices, for fixed $1 \le i \le n$.

More generally, suppose one wants to establish the Weyl inequality (1.55). From (1.57) that it suffices to show that every $i + j - 1$-dimensional subspace V contains a unit vector v such that

$$v^*(A + B)v \le \lambda_i(A) + \lambda_j(B).$$

But from (1.57), one can find a subspace U of codimension $i - 1$ such that $v^*Av \le \lambda_i(A)$ for all unit vectors v in U, and a subspace W of codimension $j - 1$ such that $v^*Bv \le \lambda_j(B)$ for all unit vectors v in W. The intersection $U \cap W$ has codimension at most $i + j - 2$ and so has a non-trivial intersection with V; and the claim follows.

Remark 1.3.6. More generally, one can generate an eigenvalue inequality whenever the intersection numbers of three Schubert varieties of compatible dimensions is non-zero; see [**HeRo1995**]. In fact, this generates a complete set of inequalities; see [**Klyachko**]. One can in fact restrict attention to those varieties whose intersection number is exactly one; see [**KnTaWo2004**]. Finally, in those cases, the fact that the intersection is one can be proven by entirely elementary means (based on the standard inequalities relating the dimension of two subspaces V, W to their intersection $V \cap W$ and sum $V + W$); see [**BeCoDyLiTi2010**]. As a consequence, the methods in this section can, in principle, be used to derive all possible eigenvalue inequalities for sums of Hermitian matrices.

Exercise 1.3.4. Verify the inequalities (1.63) and (1.55) by hand in the case when A and B commute (and are thus simultaneously diagonalisable), without the use of minimax formulae.

Exercise 1.3.5. Establish the dual Lidskii inequality

$$\lambda_{i_1}(A + B) + \cdots + \lambda_{i_k}(A + B) \geq \lambda_{i_1}(A) + \cdots + \lambda_{i_k}(A)$$
$$+ \lambda_{n-k+1}(B) + \cdots + \lambda_n(B)$$

for any $1 \leq i_1 < \cdots < i_k \leq n$ and the dual Weyl inequality

$$\lambda_{i+j-n}(A + B) \geq \lambda_i(A) + \lambda_j(B)$$

whenever $1 \leq i, j, i + j - n \leq n$.

Exercise 1.3.6. Use the Lidskii inequality to establish the more general inequality

$$\sum_{i=1}^{n} c_i \lambda_i(A + B) \leq \sum_{i=1}^{n} c_i \lambda_i(A) + \sum_{i=1}^{n} c_i^* \lambda_i(B)$$

whenever $c_1, \ldots, c_n \geq 0$, and $c_1^* \geq \cdots \geq c_n^* \geq 0$ is the decreasing rearrangement of $c_1, \ldots, c_n$. (*Hint:* Express c_i as the integral of $\mathbf{I}(c_i \geq \lambda)$ as λ runs from 0 to infinity. For each fixed λ, apply (1.63).) Combine this with Hölder's inequality to conclude the *p-Weilandt-Hoffman inequality*

(1.65) $\|(\lambda_i(A + B) - \lambda_i(A))_{i=1}^{n}\|_{\ell_n^p} \leq \|B\|_{S^p}$

for any $1 \leq p \leq \infty$, where

$$\|(a_i)_{i=1}^{n}\|_{\ell_n^p} := \left(\sum_{i=1}^{n} |a_i|^p \right)^{1/p}$$

is the usual ℓ^p norm (with the usual convention that $\|(a_i)_{i=1}^{n}\|_{\ell_n^\infty} := \sup_{1 \leq i \leq p} |a_i|$), and

(1.66) $\|B\|_{S^p} := \|(\lambda_i(B))_{i=1}^{n}\|_{\ell_n^p}$

is the *p-Schatten norm* of B.

Exercise 1.3.7. Show that the p-Schatten norms are indeed norms on the space of Hermitian matrices for every $1 \le p \le \infty$.

Exercise 1.3.8. Show that for any $1 \le p \le \infty$ and any Hermitian matrix $A = (a_{ij})_{1 \le i,j \le n}$, one has

(1.67) $$\|(a_{ii})_{i=1}^n\|_{\ell_n^p} \le \|A\|_{S^p}.$$

Exercise 1.3.9. Establish the *non-commutative Hölder inequality*

$$|\operatorname{tr}(AB)| \le \|A\|_{S^p} \|B\|_{S^{p'}}$$

whenever $1 \le p, p' \le \infty$ with $1/p + 1/p' = 1$, and A, B are $n \times n$ Hermitian matrices. (*Hint:* Diagonalise one of the matrices and use the preceding exercise.)

The most important[14] p-Schatten norms are the ∞-*Schatten norm* $\|A\|_{S^\infty} = \|A\|_{\mathrm{op}}$, which is just the operator norm, and the 2-Schatten norm $\|A\|_{S^2} = (\sum_{i=1}^n \lambda_i(A)^2)^{1/2}$, which is also the *Frobenius norm* (or *Hilbert-Schmidt norm*)

$$\|A\|_{S^2} = \|A\|_F := \operatorname{tr}(AA^*)^{1/2} = \left(\sum_{i=1}^n \sum_{j=1}^n |a_{ij}|^2\right)^{1/2}$$

where a_{ij} are the coefficients of A. Thus we see that the $p = 2$ case of the Weilandt-Hoffman inequality can be written as

(1.68) $$\sum_{i=1}^n |\lambda_i(A+B) - \lambda_i(A)|^2 \le \|B\|_F^2.$$

We will give an alternate proof of this inequality, based on eigenvalue deformation, in the next section.

1.3.4. Eigenvalue deformation. From the Weyl inequality (1.64), we know that the eigenvalue maps $A \mapsto \lambda_i(A)$ are Lipschitz continuous on Hermitian matrices (and thus also on real symmetric matrices). It turns out that we can obtain better regularity, provided that we avoid repeated eigenvalues. Fortunately, repeated eigenvalues are rare:

Exercise 1.3.10 (Dimension count). Suppose that $n \ge 2$. Show that the space of Hermitian matrices with at least one repeated eigenvalue has codimension 3 in the space of all Hermitian matrices, and the space of real symmetric matrices with at least one repeated eigenvalue has codimension 2 in the space of all real symmetric matrices. (When $n = 1$, repeated eigenvalues of course do not occur.)

---

[14]The 1-Schatten norm S^1, also known as the *nuclear norm* or *trace class norm*, is important in a number of applications, such as matrix completion, but will not be used in this text.

Let us say that a Hermitian matrix has *simple spectrum* if it has no repeated eigenvalues. We thus see from the above exercise and (1.64) that the set of Hermitian matrices with simple spectrum forms an open dense set in the space of all Hermitian matrices, and similarly for real symmetric matrices; thus simple spectrum is the *generic* behaviour of such matrices. Indeed, the unexpectedly high codimension of the non-simple matrices (naively, one would expect a codimension 1 set for a collision between, say, $\lambda_i(A)$ and $\lambda_{i+1}(A)$) suggests a *repulsion* phenomenon: because it is unexpectedly rare for eigenvalues to be equal, there must be some "force" that "repels" eigenvalues of Hermitian (and to a lesser extent, real symmetric) matrices from getting too close to each other. We now develop some machinery to make this intuition more precise.

We first observe that when A has simple spectrum, the zeroes of the characteristic polynomial $\lambda \mapsto \det(A - \lambda I)$ are simple (i.e., the polynomial has nonzero derivative at those zeroes). From this and the inverse function theorem, we see that each of the eigenvalue maps $A \mapsto \lambda_i(A)$ are smooth on the region where A has simple spectrum. Because the eigenvectors $u_i(A)$ are determined (up to phase) by the equations $(A - \lambda_i(A)I)u_i(A) = 0$ and $u_i(A)^*u_i(A) = 1$, another application of the inverse function theorem tells us that we can (locally[15]) select the maps $A \mapsto u_i(A)$ to also be smooth.

Now suppose that $A = A(t)$ depends smoothly on a time variable t, so that (when A has simple spectrum) the eigenvalues $\lambda_i(t) = \lambda_i(A(t))$ and eigenvectors $u_i(t) = u_i(A(t))$ also depend smoothly on t. We can then differentiate the equations

$$(1.69) \qquad\qquad Au_i = \lambda_i u_i$$

and

$$(1.70) \qquad\qquad u_i^* u_i = 1$$

to obtain various equations of motion for λ_i and u_i in terms of the derivatives of A.

Let's see how this works. Taking first derivatives of (1.69), (1.70) using the product rule, we obtain

$$(1.71) \qquad\qquad \dot{A}u_i + A\dot{u}_i = \dot{\lambda}_i u_i + \lambda_i \dot{u}_i$$

and

$$(1.72) \qquad\qquad \dot{u}_i^* u_i + u_i^* \dot{u}_i = 0.$$

---

[15]There may be topological obstructions to smoothly selecting these vectors globally, but this will not concern us here as we will be performing a local analysis only. In some applications, it is more convenient not to work with the $u_i(A)$ at all due to their phase ambiguity, and work instead with the *spectral projections* $P_i(A) := u_i(A)u_i(A)^*$, which do not have this ambiguity.

The equation (1.72) simplifies to $\dot{u}_i^* u_i = 0$, thus $\dot{u}_i$ is orthogonal to u_i. Taking inner products of (1.71) with u_i, we conclude the *Hadamard first variation formula*

$$\dot{\lambda}_i = u_i^* \dot{A} u_i. \tag{1.73}$$

This can already be used to give alternate proofs of various eigenvalue identities. For instance, if we apply this to $A(t) := A + tB$, we see that

$$\frac{d}{dt}\lambda_i(A + tB) = u_i(A + tB)^* B u_i(A + tB)$$

whenever $A + tB$ has simple spectrum. The right-hand side can be bounded in magnitude by $\|B\|_{\mathrm{op}}$, and so we see that the map $t \mapsto \lambda_i(A + tB)$ is Lipschitz continuous, with Lipschitz constant $\|B\|_{\mathrm{op}}$ whenever $A + tB$ has simple spectrum, which happens for generic A, B (and all t) by Exercise 1.3.10. By the fundamental theorem of calculus, we thus conclude (1.64).

Exercise 1.3.11. Use a similar argument to the one above to establish (1.68) without using minimax formulae or Lidskii's inequality.

Exercise 1.3.12. Use a similar argument to the one above to deduce Lidskii's inequality (1.63) from Proposition 1.3.4 rather than Exercise 1.3.3.

One can also compute the second derivative of eigenvalues:

Exercise 1.3.13. Suppose that $A = A(t)$ depends smoothly on t. By differentiating (1.71) and (1.72), establish the *Hadamard second variation formula*[16]

$$\frac{d^2}{dt^2}\lambda_k = u_k^* \ddot{A} u_k + 2 \sum_{j \neq k} \frac{|u_j^* \dot{A} u_k|^2}{\lambda_k - \lambda_j} \tag{1.74}$$

whenever A has simple spectrum and $1 \leq k \leq n$.

Remark 1.3.7. In the proof of the *four moment theorem* [**TaVu2009b**] on the fine spacing of Wigner matrices, one also needs the variation formulae for the third, fourth, and fifth derivatives of the eigenvalues (the first four derivatives match up with the four moments mentioned in the theorem, and the fifth derivative is needed to control error terms). Fortunately, one does not need the precise formulae for these derivatives (which, as one can imagine, are quite complicated), but only their general form, and in particular, an upper bound for these derivatives in terms of more easily computable quantities.

---

[16]If one interprets the second derivative of the eigenvalues as being proportional to a "force" on those eigenvalues (in analogy with *Newton's second law*), (1.74) is asserting that each eigenvalue λ_j "repels" the other eigenvalues λ_k by exerting a force that is inversely proportional to their separation (and also proportional to the square of the matrix coefficient of $\dot{A}$ in the eigenbasis). See [**Ta2009b**, §1.5] for more discussion.

1.3.5. Minors. In the previous sections, we perturbed $n \times n$ Hermitian matrices $A = A_n$ by adding a (small) $n \times n$ Hermitian correction matrix B to them to form a new $n \times n$ Hermitian matrix $A + B$. Another important way to perturb a matrix is to pass to a *principal minor*, for instance to the top left $n - 1 \times n - 1$ minor A_{n-1} of A_n. There is an important relationship between the eigenvalues of the two matrices:

Exercise 1.3.14 (Cauchy interlacing law). For any $n \times n$ Hermitian matrix A_n with top left $n - 1 \times n - 1$ minor A_{n-1}, then

$$(1.75) \qquad \lambda_{i+1}(A_n) \leq \lambda_i(A_{n-1}) \leq \lambda_i(A_n)$$

for all $1 \leq i < n$. (*Hint:* Use the Courant-Fischer minimax theorem, Theorem 1.3.2.) Show furthermore that the space of A_n for which equality holds in one of the inequalities in (1.75) has codimension 2 (for Hermitian matrices) or 1 (for real symmetric matrices).

Remark 1.3.8. If one takes successive minors $A_{n-1}, A_{n-2}, \ldots, A_1$ of an $n \times n$ Hermitian matrix A_n, and computes their spectra, then (1.75) shows that this triangular array of numbers forms a pattern known as a *Gelfand-Tsetlin pattern*.

One can obtain a more precise formula for the eigenvalues of A_n in terms of those for A_{n-1}:

Exercise 1.3.15 (Eigenvalue equation). Let A_n be an $n \times n$ Hermitian matrix with top left $n - 1 \times n - 1$ minor A_{n-1}. Suppose that λ is an eigenvalue of A_n distinct from all the eigenvalues of A_{n-1} (and thus simple, by (1.75)). Show that

$$(1.76) \qquad \sum_{j=1}^{n-1} \frac{|u_j(A_{n-1})^* X|^2}{\lambda_j(A_{n-1}) - \lambda} = a_{nn} - \lambda$$

where a_{nn} is the bottom right entry of A, and $X = (a_{nj})_{j=1}^{n-1} \in \mathbf{C}^{n-1}$ is the right column of A (minus the bottom entry). (*Hint:* Expand out the eigenvalue equation $A_n u = \lambda u$ into the $\mathbf{C}^{n-1}$ and $\mathbf{C}$ components.) Note the similarities between (1.76) and (1.74).

Observe that the function $\lambda \to \sum_{j=1}^{n-1} \frac{|u_j(A_{n-1})^* X|^2}{\lambda_j(A_{n-1}) - \lambda}$ is a rational function of λ which is increasing away from the eigenvalues of A_{n-1}, where it has a pole (except in the rare case when the inner product $u_{j-1}(A_{n-1})^* X$ vanishes, in which case it can have a removable singularity). By graphing this function one can see that the interlacing formula (1.75) can also be interpreted as a manifestation of the intermediate value theorem.

The identity (1.76) suggests that under typical circumstances, an eigenvalue λ of A_n can only get close to an eigenvalue $\lambda_j(A_{n-1})$ if the associated inner product $u_j(A_{n-1})^* X$ is small. This type of observation is useful to achieve *eigenvalue repulsion*—to show that it is unlikely that the gap between two adjacent eigenvalues is small. We shall see examples of this in later sections.

1.3.6. Singular values. The theory of eigenvalues of $n \times n$ Hermitian matrices has an analogue in the theory of singular values of $p \times n$ non-Hermitian matrices. We first begin with the counterpart to the spectral theorem, namely the *singular value decomposition*.

Theorem 1.3.9 (Singular value decomposition). *Let $0 \leq p \leq n$, and let A be a linear transformation from an n-dimensional complex Hilbert space U to a p-dimensional complex Hilbert space V. (In particular, A could be an $p \times n$ matrix with complex entries, viewed as a linear transformation from $\mathbf{C}^n$ to $\mathbf{C}^p$.) Then there exist non-negative real numbers*

$$\sigma_1(A) \geq \cdots \geq \sigma_p(A) \geq 0$$

(known as the singular values *of A) and orthonormal sets $u_1(A), \ldots, u_p(A) \in U$ and $v_1(A), \ldots, v_p(A) \in V$ (known as* singular vectors *of A), such that*

$$Au_j = \sigma_j v_j; \quad A^* v_j = \sigma_j u_j$$

for all $1 \leq j \leq p$, where we abbreviate $u_j = u_j(A)$, etc.

Furthermore, $Au = 0$ whenever u is orthogonal to all $u_1(A), \ldots, u_p(A)$.

We adopt the convention that $\sigma_i(A) = 0$ for $i > p$. The above theorem only applies to matrices with at least as many rows as columns, but one can also extend the definition to matrices with more columns than rows by adopting the convention $\sigma_i(A^*) := \sigma_i(A)$ (it is easy to check that this extension is consistent on square matrices). All of the results below extend (with minor modifications) to the case when there are more columns than rows, but we have not displayed those extensions here in order to simplify the notation.

Proof. We induct on p. The claim is vacuous for $p = 0$, so suppose that $p \geq 1$ and that the claim has already been proven for $p - 1$.

We follow a similar strategy to the proof of Theorem 1.3.1. We may assume that A is not identically zero, as the claim is obvious otherwise. The function $u \mapsto \|Au\|^2$ is continuous on the unit sphere of U, so there exists a unit vector u_1 which maximises this quantity. If we set $\sigma_1 := \|Au_1\| > 0$, one easily verifies that u_1 is a critical point of the map $u \mapsto \|Au\|^2 - \sigma_1^2 \|u\|^2$, which then implies that $A^* Au_1 = \sigma_1^2 u_1$. Thus, if we set $v_1 := Au_1/\sigma_1$, then $Au_1 = \sigma_1 v_1$ and $A^* v_1 = \sigma_1 u_1$. This implies that A maps the orthogonal complement $u_1^\perp$ of u_1 in U to the orthogonal complement $v_1^\perp$ of v_1 in V. By induction hypothesis, the restriction of A to $u_1^\perp$ (and $v_1^\perp$) then admits a singular value decomposition with singular values $\sigma_2 \geq \cdots \geq \sigma_p \geq 0$ and singular vectors $u_2, \ldots, u_p \in u_1^\perp$, $v_2, \ldots, v_p \in v_1^\perp$ with the stated properties. By construction we see that $\sigma_2, \ldots, \sigma_p$ are less than or equal to σ_1. If we now adjoin σ_1, u_1, v_1 to the other singular values and vectors we obtain the claim. $\square$

Exercise 1.3.16. Show that the singular values $\sigma_1(A) \geq \cdots \geq \sigma_p(A) \geq 0$ of a $p \times n$ matrix A are unique. If we have $\sigma_1(A) > \cdots > \sigma_p(A) > 0$, show that the singular vectors are unique up to rotation by a complex phase.

By construction (and the above uniqueness claim) we see that $\sigma_i(UAV) = \sigma_i(A)$ whenever A is a $p \times n$ matrix, U is a unitary $p \times p$ matrix, and V is a unitary $n \times n$ matrix. Thus the singular spectrum of a matrix is invariant under left and right unitary transformations.

Exercise 1.3.17. If A is a $p \times n$ complex matrix for some $1 \leq p \leq n$, show that the augmented matrix

$$\tilde{A} := \begin{pmatrix} 0 & A \\ A^* & 0 \end{pmatrix}$$

is a $p+n \times p+n$ Hermitian matrix whose eigenvalues consist of $\pm\sigma_1(A), \ldots, \pm\sigma_p(A)$, together with $n - p$ copies of the eigenvalue zero. (This generalises Exercise 2.3.17.) What is the relationship between the singular vectors of A and the eigenvectors of $\tilde{A}$?

Exercise 1.3.18. If A is an $n \times n$ Hermitian matrix, show that the singular values $\sigma_1(A), \ldots, \sigma_n(A)$ of A are simply the absolute values $|\lambda_1(A)|, \ldots, |\lambda_n(A)|$ of A, arranged in descending order. Show that the same claim also holds when A is a *normal matrix* (that is, when A commutes with its adjoint). What is the relationship between the singular vectors and eigenvectors of A?

Remark 1.3.10. When A is not normal, the relationship between eigenvalues and singular values is more subtle. We will discuss this point in later sections.

Exercise 1.3.19. If A is a $p \times n$ complex matrix for some $1 \leq p \leq n$, show that AA^* has eigenvalues $\sigma_1(A)^2, \ldots, \sigma_p(A)^2$, and A^*A has eigenvalues $\sigma_1(A)^2, \ldots, \sigma_p(A)^2$ together with $n - p$ copies of the eigenvalue zero. Based on this observation, give an alternate proof of the singular value decomposition theorem using the spectral theorem for (positive semi-definite) Hermitian matrices.

Exercise 1.3.20. Show that the rank of a $p \times n$ matrix is equal to the number of non-zero singular values.

Exercise 1.3.21. Let A be a $p \times n$ complex matrix for some $1 \leq p \leq n$. Establish the Courant-Fischer minimax formula

(1.77) $$\sigma_i(A) = \sup_{\dim(V)=i} \inf_{v \in V; |v|=1} |Av|$$

for all $1 \leq i \leq p$, where the supremum ranges over all subspaces of $\mathbf{C}^n$ of dimension i.

One can use the above exercises to deduce many inequalities about singular values from analogous ones about eigenvalues. We give some examples below.

Exercise 1.3.22. Let A, B be $p \times n$ complex matrices for some $1 \leq p \leq n$.

(i) Establish the Weyl inequality $\sigma_{i+j-1}(A+B) \leq \sigma_i(A) + \sigma_j(B)$ whenever $1 \leq i, j, i+j-1 \leq p$.

(ii) Establish the Lidskii inequality
$$\sigma_{i_1}(A+B) + \cdots + \sigma_{i_k}(A+B) \leq \sigma_{i_1}(A) + \cdots + \sigma_{i_k}(A)$$
$$+ \sigma_1(B) + \cdots + \sigma_k(B)$$
whenever $1 \leq i_1 < \ldots < i_k \leq p$.

(iii) Show that for any $1 \leq k \leq p$, the map $A \mapsto \sigma_1(A) + \cdots + \sigma_k(A)$ defines a norm on the space $\mathbf{C}^{p \times n}$ of complex $p \times n$ matrices (this norm is known as the kth *Ky Fan norm*).

(iv) Establish the Weyl inequality $|\sigma_i(A+B) - \sigma_i(A)| \leq \|B\|_{\text{op}}$ for all $1 \leq i \leq p$.

(v) More generally, establish the q-Weilandt-Hoffman inequality $\|(\sigma_i(A+B) - \sigma_i(A))_{1 \leq i \leq p}\|_{\ell_p^q} \leq \|B\|_{S^q}$ for any $1 \leq q \leq \infty$, where $\|B\|_{S^q} := \|(\sigma_i(B))_{1 \leq i \leq p}\|_{\ell_p^q}$ is the q-Schatten norm of B. (Note that this is consistent with the previous definition of the Schatten norms.)

(vi) Show that the q-Schatten norm is indeed a norm on $\mathbf{C}^{p \times n}$ for any $1 \leq q \leq \infty$.

(vii) If A' is formed by removing one row from A, show that $\lambda_{i+1}(A) \leq \lambda_i(A') \leq \lambda_i(A)$ for all $1 \leq i < p$.

(viii) If $p < n$ and A' is formed by removing one column from A, show that $\lambda_{i+1}(A) \leq \lambda_i(A') \leq \lambda_i(A)$ for all $1 \leq i < p$ and $\lambda_p(A') \leq \lambda_p(A)$. What changes when $p = n$?

Exercise 1.3.23. Let A be a $p \times n$ complex matrix for some $1 \leq p \leq n$. Observe that the linear transformation $A : \mathbf{C}^n \to \mathbf{C}^p$ naturally induces a linear transformation $A^{\wedge k} : \bigwedge^k \mathbf{C}^n \to \bigwedge^k \mathbf{C}^p$ from k-forms on $\mathbf{C}^n$ to k-forms on $\mathbf{C}^p$. We give $\bigwedge^k \mathbf{C}^n$ the structure of a Hilbert space by declaring the basic forms $e_{i_1} \wedge \ldots \wedge e_{i_k}$ for $1 \leq i_1 < \cdots < i_k \leq n$ to be orthonormal. For any $1 \leq k \leq p$, show that the operator norm of $A^{\wedge k}$ is equal to $\sigma_1(A) \ldots \sigma_k(A)$.

Exercise 1.3.24. Let A be a $p \times n$ matrix for some $1 \leq p \leq n$, let B be a $r \times p$ matrix, and let C be a $n \times s$ matrix for some $r, s \geq 1$. Show that $\sigma_i(BA) \leq \|B\|_{op}\sigma_i(A)$ and $\sigma_i(AC) \leq \sigma_i(A)\|C\|_{op}$ for any $1 \leq i \leq p$.

Exercise 1.3.25. Let $A = (a_{ij})_{1 \leq i \leq p; 1 \leq j \leq n}$ be a $p \times n$ matrix for some $1 \leq p \leq n$, let $i_1, \ldots, i_k \in \{1, \ldots, p\}$ be distinct, and let $j_1, \ldots, j_k \in \{1, \ldots, n\}$ be distinct. Show that

$$a_{i_1 j_1} + \cdots + a_{i_k j_k} \leq \sigma_1(A) + \cdots + \sigma_k(A).$$

Using this, show that if $j_1, \ldots, j_p \in \{1, \ldots, n\}$ are distinct, then

$$\|(a_{ij_i})_{i=1}^{p}\|_{\ell_p^q} \leq \|A\|_{S^q}$$

for every $1 \leq q \leq \infty$.

Exercise 1.3.26. Establish the Hölder inequality

$$|\operatorname{tr}(AB^*)| \leq \|A\|_{S^q}\|B\|_{S^{q'}}$$

whenever A, B are $p \times n$ complex matrices and $1 \leq q, q' \leq \infty$ are such that $1/q + 1/q' = 1$.

Chapter 2

Random matrices

2.1. Concentration of measure

Suppose we have a large number of scalar random variables $X_1, \ldots, X_n$, which each have bounded size on average (e.g., their mean and variance could be $O(1)$). What can one then say about their sum $S_n := X_1 + \cdots + X_n$? If each individual summand X_i varies in an interval of size $O(1)$, then their sum of course varies in an interval of size $O(n)$. However, a remarkable phenomenon, known as *concentration of measure*, asserts that assuming a sufficient amount of independence between the component variables $X_1, \ldots, X_n$, this sum sharply concentrates in a much narrower range, typically in an interval of size $O(\sqrt{n})$. This phenomenon is quantified by a variety of *large deviation inequalities* that give upper bounds (often exponential in nature) on the probability that such a combined random variable deviates significantly from its mean. The same phenomenon applies not only to linear expressions such as $S_n = X_1 + \cdots + X_n$, but more generally to non-linear combinations $F(X_1, \ldots, X_n)$ of such variables, provided that the non-linear function F is sufficiently regular (in particular, if it is Lipschitz, either separately in each variable, or jointly in all variables).

The basic intuition here is that it is difficult for a large number of independent variables $X_1, \ldots, X_n$ to "work together" to simultaneously pull a sum $X_1 + \cdots + X_n$ or a more general combination $F(X_1, \ldots, X_n)$ too far away from its mean. Independence here is the key; concentration of measure results typically fail if the X_i are too highly correlated with each other.

There are many applications of the concentration of measure phenomenon, but we will focus on a specific application which is useful in the random matrix theory topics we will be studying, namely on controlling the behaviour of random n-dimensional vectors with independent components, and in particular, on the distance between such random vectors and a given subspace.

Once one has a sufficient amount of independence, the concentration of measure tends to be sub-Gaussian in nature; thus the probability that one is at least λ standard deviations from the mean tends to drop off like $C \exp(-c\lambda^2)$ for some $C, c > 0$. In particular, one is $O(\log^{1/2} n)$ standard deviations from the mean with high probability, and $O(\log^{1/2+\varepsilon} n)$ standard deviations from the mean with overwhelming probability. Indeed, concentration of measure is our primary tool for ensuring that various events hold with overwhelming probability (other moment methods can give high probability, but have difficulty ensuring overwhelming probability).

This is only a brief introduction to the concentration of measure phenomenon. A systematic study of this topic can be found in [**Le2001**].

2.1.1. Linear combinations, and the moment method. We begin with the simple setting of studying a sum $S_n := X_1 + \cdots + X_n$ of random variables. As we shall see, these linear sums are particularly amenable to the moment method, though to use the more powerful moments, we will require more powerful independence assumptions (and, naturally, we will need more moments to be finite or bounded). As such, we will take the opportunity to use this topic (large deviation inequalities for sums of random variables) to give a tour of the *moment method*, which we will return to when we consider the analogous questions for the bulk spectral distribution of random matrices.

In this section we shall concern ourselves primarily with bounded random variables; in the next section we describe the basic *truncation method* that can allow us to extend from the bounded case to the unbounded case (assuming suitable decay hypotheses).

The zeroth moment method (1.22) gives a crude upper bound when S is non-zero,

$$(2.1) \qquad \mathbf{P}(S_n \neq 0) \leq \sum_{i=1}^{n} \mathbf{P}(X_i \neq 0)$$

but in most cases this bound is worse than the trivial bound $\mathbf{P}(S_n \neq 0) \leq 1$. This bound, however, will be useful when performing the *truncation trick*, which we will discuss below.

The first moment method is somewhat better, giving the bound

$$\mathbf{E}|S_n| \leq \sum_{i=1}^{n} \mathbf{E}|X_i|$$

which, when combined with Markov's inequality (1.14), gives the rather weak large deviation inequality

$$(2.2) \qquad \mathbf{P}(|S_n| \geq \lambda) \leq \frac{1}{\lambda} \sum_{i=1}^{n} \mathbf{E}|X_i|.$$

As weak as this bound is, this bound is sometimes sharp. For instance, if the X_i are all equal to a single signed Bernoulli variable $X \in \{-1, +1\}$, then $S_n = nX$. In particular, in this case we have $|S_n| = n$, and so (2.2) is sharp when $\lambda = n$. The problem here is a complete lack of independence; the X_i are all simultaneously positive or simultaneously negative, causing huge fluctuations in the value of S_n.

Informally, one can view (2.2) as the assertion that S_n typically has size $S_n = O(\sum_{i=1}^{n} |X_i|)$.

The first moment method also shows that

$$\mathbf{E}S_n = \sum_{i=1}^{n} \mathbf{E}X_i$$

and so we can normalise out the means using the identity

$$S_n - \mathbf{E}S_n = \sum_{i=1}^{n} X_i - \mathbf{E}X_i.$$

Replacing the X_i by $X_i - \mathbf{E}X_i$ (and S_n by $S_n - \mathbf{E}S_n$) we may thus assume for simplicity that all the X_i have mean zero.

Now we consider what the second moment method gives us. We square S_n and take expectations to obtain

$$\mathbf{E}|S_n|^2 = \sum_{i=1}^{n}\sum_{j=1}^{n} \mathbf{E}X_i\overline{X_j}.$$

If we assume that the X_i are pairwise independent (in addition to having mean zero), then $\mathbf{E}X_i\overline{X_j}$ vanishes unless $i = j$, in which case this expectation is equal to $\mathbf{Var}(X_i)$. We thus have

$$(2.3) \qquad \mathbf{Var}(S_n) = \sum_{i=1}^{n} \mathbf{Var}(X_i)$$

which, when combined with Chebyshev's inequality (1.26) (and the mean zero normalisation), yields the large deviation inequality

$$(2.4) \qquad \mathbf{P}(|S_n| \geq \lambda) \leq \frac{1}{\lambda^2}\sum_{i=1}^{n} \mathbf{Var}(X_i).$$

Without the normalisation that the X_i have mean zero, we obtain

$$(2.5) \qquad \mathbf{P}(|S_n - \mathbf{E}S_n| \geq \lambda) \leq \frac{1}{\lambda^2}\sum_{i=1}^{n} \mathbf{Var}(X_i).$$

Informally, this is the assertion that S_n typically has size $S_n = \mathbf{E}S_n + O((\sum_{i=1}^{n} \mathbf{Var}(X_i))^{1/2})$, if we have pairwise independence. Note also that we do not need the full strength of the pairwise independence assumption; the slightly weaker hypothesis of being pairwise uncorrelated[1] would have sufficed.

The inequality (2.5) is sharp in two ways. First, we cannot expect any significant concentration in any range narrower than the standard deviation $O((\sum_{i=1}^{n} \mathbf{Var}(X_i))^{1/2})$, as this would likely contradict (2.3). Second, the quadratic-type decay in λ in (2.5) is sharp given the pairwise independence

---

[1]In other words, we only need to assume that the covariances $\mathbf{Cov}(X_i, X_j) := \mathbf{E}(X_i - \mathbf{E}X_i)\overline{(X_j - \mathbf{E}X_j)}$ vanish for all distinct i, j.

hypothesis. For instance, suppose that $n = 2^m - 1$, and that $X_j := (-1)^{a_j \cdot Y}$, where Y is drawn uniformly at random from the cube $\{0, 1\}^m$, and $a_1, \ldots, a_n$ are an enumeration of the non-zero elements of $\{0, 1\}^m$. Then a little Fourier analysis shows that each X_j for $1 \leq j \leq n$ has mean zero, variance 1, and are pairwise independent in j; but S is equal to $(n+1)\mathbf{I}(Y = 0) - 1$, which is equal to n with probability $1/(n+1)$; this is despite the standard deviation of S being just $\sqrt{n}$. This shows that (2.5) is essentially (i.e., up to constants) sharp here when $\lambda = n$.

Now we turn to higher moments. Let us assume that the X_i are normalised to have mean zero and variance at most 1, and are also almost surely bounded in magnitude by some[2] K: $|X_i| \leq K$. To simplify the exposition very slightly we will assume that the X_i are real-valued; the complex-valued case is very analogous (and can also be deduced from the real-valued case) and is left to the reader.

Let us also assume that the $X_1, \ldots, X_n$ are k-wise independent for some even positive integer k. With this assumption, we can now estimate the kth moment

$$\mathbf{E}|S_n|^k = \sum_{1 \leq i_1, \ldots, i_k \leq n} \mathbf{E}X_{i_1} \ldots X_{i_k}.$$

To compute the expectation of the product, we can use the k-wise independence, but we need to divide into cases (analogous to the $i \neq j$ and $i = j$ cases in the second moment calculation above) depending on how various indices are repeated. If one of the X_{i_j} only appear once, then the entire expectation is zero (since X_{i_j} has mean zero), so we may assume that each of the X_{i_j} appear at least twice. In particular, there are at most $k/2$ distinct X_j which appear. If exactly $k/2$ such terms appear, then from the unit variance assumption we see that the expectation has magnitude at most 1; more generally, if $k/2 - r$ terms appear, then from the unit variance assumption and the upper bound by K we see that the expectation has magnitude at most K^{2r}. This leads to the upper bound

$$\mathbf{E}|S_n|^k \leq \sum_{r=0}^{k/2} K^{2r} N_r$$

where N_r is the number of ways one can select integers $i_1, \ldots, i_k$ in $\{1, \ldots, n\}$ such that each i_j appears at least twice, and such that exactly $k/2 - r$ integers appear.

We are now faced with the purely combinatorial problem of estimating N_r. We will use a somewhat crude bound. There are $\binom{n}{k/2-r} \leq n^{k/2-r}/(k/2-r)!$ ways to choose $k/2 - r$ integers from $\{1, \ldots, n\}$. Each of the integers i_j

---

[2]Note that we must have $K \geq 1$ to be consistent with the unit variance hypothesis.

has to come from one of these $k/2 - r$ integers, leading to the crude bound

$$N_r \leq \frac{n^{k/2-r}}{(k/2-r)!}(k/2-r)^k$$

which after using a crude form $n! \geq n^n e^{-n}$ of Stirling's formula (see Section 1.2) gives

$$N_r \leq (en)^{k/2-r}(k/2)^{k/2+r},$$

and so

$$\mathbf{E}|S_n|^k \leq (enk/2)^{k/2} \sum_{r=0}^{k/2} (\frac{K^2 k}{en})^r.$$

If we make the mild assumption

(2.6) $K^2 \leq n/k,$

then from the geometric series formula we conclude that

$$\mathbf{E}|S_n|^k \leq 2(enk/2)^{k/2}$$

(say), which leads to the large deviation inequality

(2.7) $\mathbf{P}(|S_n| \geq \lambda\sqrt{n}) \leq 2 \left(\frac{\sqrt{ek/2}}{\lambda} \right)^k.$

This should be compared with (2.2), (2.5). As k increases, the rate of decay in the λ parameter improves, but to compensate for this, the range that S_n concentrates in grows slowly, to $O(\sqrt{nk})$ rather than $O(\sqrt{n})$.

Remark 2.1.1. Note how it was important here that k was even. Odd moments, such as $\mathbf{E}S_n^3$, can be estimated, but due to the lack of the absolute value sign, these moments do not give much usable control on the distribution of the S_n. One could be more careful in the combinatorial counting than was done here, but the net effect of such care is only to improve the explicit constants such as $\sqrt{e/2}$ appearing in the above bounds.

Now suppose that the $X_1, \ldots, X_n$ are not just k-wise independent for any fixed k, but are in fact jointly independent. Then we can apply (2.7) for any k obeying (2.6). We can optimise in k by setting $\sqrt{nk}$ to be a small multiple of λ, and conclude the Gaussian-type bound[3]

(2.8) $\mathbf{P}(|S_n| \geq \lambda\sqrt{n}) \leq C \exp(-c\lambda^2)$

for some absolute constants $C, c > 0$, provided that $|\lambda| \leq c\sqrt{n}/\sqrt{K}$ for some small c. Thus we see that while control of each individual moment $\mathbf{E}|S_n|^k$ only gives polynomial decay in λ, by using all the moments simultaneously one can obtain square-exponential decay (i.e., sub-Gaussian type decay).

---

[3]Note that the bound (2.8) is trivial for $|\lambda| \gg \sqrt{n}$, so we may assume that λ is small compared to this quantity.

By using Stirling's formula (see Exercise 1.2.2) one can show that the quadratic decay in (2.8) cannot be improved; see Exercise 2.1.2 below.

It was a little complicated to manage such large moments $\mathbf{E}|S_n|^k$. A slicker way to proceed (but one which exploits the joint independence and commutativity more strongly) is to work instead with the *exponential moments* $\mathbf{E}\exp(tS_n)$, which can be viewed as a sort of generating function for the power moments. A useful lemma in this regard is

Lemma 2.1.2 (Hoeffding's lemma). *Let X be a scalar variable taking values in an interval $[a, b]$. Then for any $t > 0$,*

(2.9) $$\mathbf{E}e^{tX} \leq e^{t\mathbf{E}X}\left(1 + O(t^2\mathbf{Var}(X)\exp(O(t(b-a))))\right).$$

In particular,

(2.10) $$\mathbf{E}e^{tX} \leq e^{t\mathbf{E}X}\exp\left(O(t^2(b-a)^2)\right).$$

Proof. It suffices to prove the first inequality, as the second then follows using the bound $\mathbf{Var}(X) \leq (b-a)^2$ and from various elementary estimates.

By subtracting the mean from X, a, b we may normalise $\mathbf{E}(X) = 0$. By dividing X, a, b (and multiplying t to balance) we may assume that $b-a = 1$, which implies that $X = O(1)$. We then have the Taylor expansion

$$e^{tX} = 1 + tX + O(t^2X^2\exp(O(t)))$$

which, on taking expectations, gives

$$\mathbf{E}e^{tX} = 1 + O(t^2\mathbf{Var}(X)\exp(O(t))$$

and the claim follows. $\square$

Exercise 2.1.1. Show that the $O(t^2(b-a)^2)$ factor in (2.10) can be replaced with $t^2(b-a)^2/8$, and that this is sharp. (*Hint:* Use Jensen's inequality, Exercise 1.1.8.)

We now have the fundamental *Chernoff bound*:

Theorem 2.1.3 (Chernoff inequality). *Let $X_1, \ldots, X_n$ be independent scalar random variables with $|X_i| \leq K$ almost surely, with mean μ_i and variance σ_i^2. Then for any $\lambda > 0$, one has*

(2.11) $$\mathbf{P}(|S_n - \mu| \geq \lambda\sigma) \leq C\max(\exp(-c\lambda^2), \exp(-c\lambda\sigma/K))$$

for some absolute constants $C, c > 0$, where $\mu := \sum_{i=1}^n \mu_i$ and $\sigma^2 := \sum_{i=1}^n \sigma_i^2$.

Proof. By taking real and imaginary parts we may assume that the X_i are real. By subtracting off the mean (and adjusting K appropriately) we may assume that $\mu_i = 0$ (and so $\mu = 0$); dividing the X_i (and σ_i) through by K

we may assume that $K = 1$. By symmetry it then suffices to establish the upper tail estimate

$$\mathbf{P}(S_n \geq \lambda\sigma) \leq C \max(\exp(-c\lambda^2), \exp(-c\lambda\sigma))$$

(with slightly different constants C, c).

To do this, we shall first compute the exponential moments

$$\mathbf{E} \exp(tS_n)$$

where $0 \leq t \leq 1$ is a real parameter to be optimised later. Expanding out the exponential and using the independence hypothesis, we conclude that

$$\mathbf{E} \exp(tS_n) = \prod_{i=1}^{n} \mathbf{E} \exp(tX_i).$$

To compute $\mathbf{E} \exp(tX)$, we use the hypothesis that $|X| \leq 1$ and (2.9) to obtain

$$\mathbf{E} \exp(tX) \leq \exp(O(t^2\sigma_i^2)).$$

Thus we have

$$\mathbf{E} \exp(tS_n) = \exp(O(t^2\sigma^2)),$$

and thus by Markov's inequality (1.13)

$$\mathbf{P}(S_n \geq \lambda\sigma) \leq \exp(O(t^2\sigma^2) - t\lambda\sigma).$$

If we optimise this in t, subject to the constraint $0 \leq t \leq 1$, we obtain the claim.   □

Informally, the Chernoff inequality asserts that S_n is sharply concentrated in the range $\mu + O(\sigma)$. The bounds here are fairly sharp, at least when λ is not too large:

Exercise 2.1.2. Let $0 < p < 1/2$ be fixed independently of n, and let $X_1, \ldots, X_n$ be iid copies of a Bernoulli random variable that equals 1 with probability p, thus $\mu_i = p$ and $\sigma_i^2 = p(1 - p)$, and so $\mu = np$ and $\sigma^2 = np(1 - p)$. Using Stirling's formula (Section 1.2), show that

$$\mathbf{P}(|S_n - \mu| \geq \lambda\sigma) \geq c \exp(-C\lambda^2)$$

for some absolute constants $C, c > 0$ and all $\lambda \leq c\sigma$. What happens when λ is much larger than σ?

Exercise 2.1.3. Show that the term $\exp(-c\lambda\sigma/K)$ in (2.11) can be replaced with $(\lambda K/\sigma)^{-c\lambda\sigma/K}$ (which is superior when $\lambda K \gg \sigma$). (*Hint:* Allow t to exceed 1.) Compare this with the results of Exercise 2.1.2.

Exercise 2.1.4 (Hoeffding's inequality). Let $X_1, \ldots, X_n$ be independent real variables, with X_i taking values in an interval $[a_i, b_i]$, and let $S_n := X_1 + \cdots + X_n$. Show that one has

$$\mathbf{P}(|S_n - \mathbf{E}S_n| \geq \lambda\sigma) \leq C\exp(-c\lambda^2)$$

for some absolute constants $C, c > 0$, where $\sigma^2 := \sum_{i=1}^n |b_i - a_i|^2$.

Remark 2.1.4. As we can see, the exponential moment method is very slick compared to the power moment method. Unfortunately, due to its reliance on the identity $e^{X+Y} = e^X e^Y$, this method relies very strongly on commutativity of the underlying variables, and as such will not be as useful when dealing with non-commutative random variables, and in particular, with random matrices[4]. Nevertheless, we will still be able to apply the Chernoff bound to good effect to various components of random matrices, such as rows or columns of such matrices.

The full assumption of joint independence is not completely necessary for Chernoff-type bounds to be present. It suffices to have a *martingale difference sequence*, in which each X_i can depend on the preceding variables $X_1, \ldots, X_{i-1}$, but which always has mean zero even when the preceding variables are conditioned out. More precisely, we have *Azuma's inequality*:

Theorem 2.1.5 (Azuma's inequality). *Let $X_1, \ldots, X_n$ be a sequence of scalar random variables with $|X_i| \leq 1$ almost surely. Assume also that we have[5] the martingale difference property*

$$(2.12) \qquad \mathbf{E}(X_i | X_1, \ldots, X_{i-1}) = 0$$

almost surely for all $i = 1, \ldots, n$. Then for any $\lambda > 0$, the sum $S_n := X_1 + \cdots + X_n$ obeys the large deviation inequality

$$(2.13) \qquad \mathbf{P}(|S_n| \geq \lambda\sqrt{n}) \leq C\exp(-c\lambda^2)$$

for some absolute constants $C, c > 0$.

A typical example of S_n here is a dependent random walk, in which the magnitude and probabilities of the ith step are allowed to depend on the outcome of the preceding $i - 1$ steps, but where the mean of each step is always fixed to be zero.

Proof. Again, we can reduce to the case when the X_i are real, and it suffices to establish the upper tail estimate

$$\mathbf{P}(S_n \geq \lambda\sqrt{n}) \leq C\exp(-c\lambda^2).$$

---

[4]See, however, Section 3.2 for a partial resolution of this issue.

[5]Here we assume the existence of a suitable disintegration in order to define the conditional expectation, though in fact it is possible to state and prove Azuma's inequality without this disintegration.

Note that $|S_n| \leq n$ almost surely, so we may assume, without loss of generality, that $\lambda \leq \sqrt{n}$.

Once again, we consider the exponential moment $\mathbf{E} \exp(tS_n)$ for some parameter $t > 0$. We write $S_n = S_{n-1} + X_n$, so that

$$\mathbf{E} \exp(tS_n) = \mathbf{E} \exp(tS_{n-1}) \exp(tX_n).$$

We do not have independence between S_{n-1} and X_n, so cannot split the expectation as in the proof of Chernoff's inequality. Nevertheless, we can use conditional expectation as a substitute. We can rewrite the above expression as

$$\mathbf{E}\mathbf{E}(\exp(tS_{n-1}) \exp(tX_n)|X_1, \ldots, X_{n-1}).$$

The quantity S_{n-1} is deterministic once we condition on $X_1, \ldots, X_{n-1}$, and so we can pull it out of the conditional expectation:

$$\mathbf{E} \exp(tS_{n-1})\mathbf{E}(\exp(tX_n)|X_1, \ldots, X_{n-1}).$$

Applying (2.10) to the conditional expectation, we have

$$\mathbf{E}(\exp(tX_n)|X_1, \ldots, X_{n-1}) \leq \exp(O(t^2))$$

and

$$\mathbf{E} \exp(tS_n) \leq \exp(O(t^2))\mathbf{E} \exp(tS_{n-1}).$$

Iterating this argument gives

$$\mathbf{E} \exp(tS_n) \leq \exp(O(nt^2))$$

and thus by Markov's inequality (1.13),

$$\mathbf{P}(S_n \geq \lambda\sqrt{n}) \leq \exp(O(nt^2) - t\lambda\sqrt{n}).$$

Optimising in t gives the claim. $\square$

Exercise 2.1.5. Suppose we replace the hypothesis $|X_i| \leq 1$ in Azuma's inequality with the more general hypothesis $|X_i| \leq c_i$ for some scalars $c_i > 0$. Show that we still have (2.13), but with $\sqrt{n}$ replaced by $(\sum_{i=1}^n c_i^2)^{1/2}$.

Remark 2.1.6. The exponential moment method is also used frequently in harmonic analysis to deal with lacunary exponential sums, or sums involving Radamacher functions (which are the analogue of lacunary exponential sums for characteristic 2). Examples here include *Khintchine's inequality* (and the closely related *Kahane's inequality*); see e.g. [**Wo2003**], [**Ka1985**]. The exponential moment method also combines very well with log-Sobolev inequalities, as we shall see below (basically because the logarithm inverts the exponential), as well as with the closely related *hypercontractivity* inequalities.

2.1.2. The truncation method. To summarise the discussion so far, we have identified a number of large deviation inequalities to control a sum $S_n = X_1 + \cdots + X_n$:

(i) The zeroth moment method bound (2.1), which requires no moment assumptions on the X_i but is only useful when X_i is usually zero, and has no decay in λ.

(ii) The first moment method bound (2.2), which only requires absolute integrability on the X_i, but has only a linear decay in λ.

(iii) The second moment method bound (2.5), which requires second moment and pairwise independence bounds on X_i, and gives a quadratic decay in λ.

(iv) Higher moment bounds (2.7), which require boundedness and k-wise independence, and give a kth power decay in λ (or quadratic-exponential decay, after optimising in k).

(v) Exponential moment bounds such as (2.11) or (2.13), which require boundedness and joint independence (or martingale behaviour), and give quadratic-exponential decay in λ.

We thus see that the bounds with the strongest decay in λ require strong boundedness and independence hypotheses. However, one can often partially extend these strong results from the case of bounded random variables to that of unbounded random variables (provided one still has sufficient control on the decay of these variables) by a simple but fundamental trick, known as the *truncation method*. The basic idea here is to take each random variable X_i and split it as $X_i = X_{i,\leq N} + X_{i,>N}$, where N is a truncation parameter to be optimised later (possibly in a manner depending on n),

$$X_{i,\leq N} := X_i \mathbf{I}(|X_i| \leq N)$$

is the restriction of X_i to the event that $|X_i| \leq N$ (thus $X_{i,\leq N}$ vanishes when X_i is too large), and

$$X_{i,>N} := X_i \mathbf{I}(|X_i| > N)$$

is the complementary event. One can similarly split $S_n = S_{n,\leq N} + S_{n,>N}$ where

$$S_{n,\leq N} = X_{1,\leq N} + \cdots + X_{n,\leq N}$$

and

$$S_{n,>N} = X_{1,>N} + \cdots + X_{n,>N}.$$

The idea is then to estimate the tail of $S_{n,\leq N}$ and $S_{n,>N}$ by two different means. With $S_{n,\leq N}$, the point is that the variables $X_{i,\leq N}$ have been made bounded by fiat, and so the more powerful large deviation inequalities can now be put into play. With $S_{n,>N}$, in contrast, the underlying variables

$X_{i,>N}$ are certainly not bounded, but they tend to have small zeroth and first moments, and so the bounds based on those moment methods tend to be powerful here[6].

Let us begin with a simple application of this method.

Theorem 2.1.7 (Weak law of large numbers). *Let* $X_1, X_2, \ldots$ *be iid scalar random variables with* $X_i \equiv X$ *for all* i, *where* X *is absolutely integrable. Then* S_n/n *converges in probability to* $\mathbf{E}X$.

Proof. By subtracting $\mathbf{E}X$ from X we may assume, without loss of generality, that X has mean zero. Our task is then to show that $\mathbf{P}(|S_n| \geq \varepsilon n) = o(1)$ for all fixed $\varepsilon > 0$.

If X has finite variance, then the claim follows from (2.5). If X has infinite variance, we cannot apply (2.5) directly, but we may perform the truncation method as follows. Let N be a large parameter to be chosen later, and split $X_i = X_{i,\leq N} + X_{i,>N}$, $S_n = S_{n,\leq N} + S_{n,>N}$ (and $X = X_{\leq N} + X_{>N}$) as discussed above. The variable $X_{\leq N}$ is bounded and thus has bounded variance; also, from the dominated convergence theorem we see that $|\mathbf{E}X_{\leq N}| \leq \varepsilon/4$ (say) if N is large enough. From (2.5) we conclude that

$$\mathbf{P}(|S_{n,\leq N}| \geq \varepsilon n/2) = o(1)$$

(where the rate of decay here depends on N and ε). Meanwhile, to deal with the tail $X_{>N}$ we use (2.2) to conclude that

$$\mathbf{P}(|S_{n,>N}| \geq \varepsilon n/2) \leq \frac{2}{\varepsilon}\mathbf{E}|X_{>N}|.$$

But by the dominated convergence theorem (or monotone convergence theorem), we may make $\mathbf{E}|X_{>N}|$ as small as we please (say, smaller than $\delta > 0$) by taking N large enough. Summing, we conclude that

$$\mathbf{P}(|S_n| \geq \varepsilon n) = \frac{2}{\varepsilon}\delta + o(1);$$

since δ is arbitrary, we obtain the claim. $\square$

A more sophisticated variant of this argument[7] gives

Theorem 2.1.8 (Strong law of large numbers). *Let* $X_1, X_2, \ldots$ *be iid scalar random variables with* $X_i \equiv X$ *for all* i, *where* X *is absolutely integrable. Then* S_n/n *converges almost surely to* $\mathbf{E}X$.

---

[6]Readers who are familiar with harmonic analysis may recognise this type of "divide and conquer argument" as an *interpolation argument*; see [**Ta2010**, §1.11].

[7]See [**Ta2009**, §1.4] for a more detailed discussion of this argument.

Proof. We may assume, without loss of generality, that X is real, since the complex case then follows by splitting into real and imaginary parts. By splitting X into positive and negative parts, we may furthermore assume that X is non-negative[8]. In particular, S_n is now non-decreasing in n.

Next, we apply a sparsification trick. Let $0 < \varepsilon < 1$. Suppose that we knew that, almost surely, S_{n_m}/n_m converged to $\mathbf{E}X$ for $n = n_m$ of the form $n_m := \lfloor (1+\varepsilon)^m \rfloor$ for some integer m. Then, for all other values of n, we see that asymptotically, S_n/n can only fluctuate by a multiplicative factor of $1 + O(\varepsilon)$, thanks to the monotone nature of S_n. Because of this and countable additivity, we see that it suffices to show that S_{n_m}/n_m converges to $\mathbf{E}X$. Actually, it will be enough to show that almost surely, one has $|S_{n_m}/n_m - \mathbf{E}X| \le \varepsilon$ for all but finitely many m.

Fix ε. As before, we split $X = X_{>N_m} + X_{\le N_m}$ and $S_{n_m} = S_{n_m,>N_m} + S_{n_m,\le N_m}$, but with the twist that we now allow $N = N_m$ to depend on m. Then for N_m large enough we have $|\mathbf{E}X_{\le N_m} - \mathbf{E}X| \le \varepsilon/2$ (say), by dominated convergence. Applying (2.5) as before, we see that

$$\mathbf{P}(|S_{n_m,\le N_m}/n_m - \mathbf{E}X| > \varepsilon) \le \frac{C_\varepsilon}{n_m}\mathbf{E}|X_{\le N_m}|^2$$

for some C_ε depending only on ε (the exact value is not important here). To handle the tail, we will not use the first moment bound (2.2) as done previously, but now turn to the zeroth-moment bound (2.1) to obtain

$$\mathbf{P}(S_{n_m,>N_m} \ne 0) \le n_m\mathbf{P}(|X| > N_m);$$

summing, we conclude

$$\mathbf{P}(|S_{n_m}/n_m - \mathbf{E}X| > \varepsilon) \le \frac{C_\varepsilon}{n_m}\mathbf{E}|X_{\le N_m}|^2 + n_m\mathbf{P}(|X| > N_m).$$

Applying the Borel-Cantelli lemma (Exercise 1.1.1), we see that we will be done as long as we can choose N_m such that

$$\sum_{m=1}^\infty \frac{1}{n_m}\mathbf{E}|X_{\le N_m}|^2$$

and

$$\sum_{m=1}^\infty n_m\mathbf{P}(|X| > N_m)$$

are both finite. But this can be accomplished by setting $N_m := n_m$ and interchanging the sum and expectations (writing $\mathbf{P}(|X| > N_m)$ as $\mathbf{EI}(|X| > N_m)$) and using the lacunary nature of the n_m (which, in particular, shows that $\sum_{m:n_m\le X} n_m = O(X)$ and $\sum_{m:n_m\ge X} n_m^{-1} = O(X^{-1})$ for any $X > 0$). $\qquad\square$

---

[8]Of course, by doing so, we can no longer normalise X to have mean zero, but for us the non-negativity will be more convenient than the zero mean property.

To give another illustration of the truncation method, we extend a version of the Chernoff bound to the sub-Gaussian case.

Proposition 2.1.9. Let $X_1, \ldots, X_n \equiv X$ be iid copies of a sub-Gaussian random variable X, thus X obeys a bound of the form

$$(2.14) \qquad\qquad \mathbf{P}(|X| \geq t) \leq C \exp(-ct^2)$$

for all $t > 0$ and some $C, c > 0$. Let $S_n := X_1 + \cdots + X_n$. Then for any sufficiently large A (independent of n) we have

$$\mathbf{P}(|S_n - n\mathbf{E}X| \geq An) \leq C_A \exp(-c_A n)$$

for some constants C_A, c_A depending on A, C, c. Furthermore, c_A grows linearly in A as $A \to \infty$.

Proof. By subtracting the mean from X we may normalise $\mathbf{E}X = 0$. We perform a dyadic decomposition

$$X_i = X_{i,0} + \sum_{m=1}^{\infty} X_{i,m}$$

where $X_{i,0} := X_i \mathbf{I}(X_i \leq 1)$ and $X_{i,m} := X_i \mathbf{I}(2^{m-1} < X_i \leq 2^m)$. We similarly split

$$S_n = S_{n,0} + \sum_{m=1}^{\infty} S_{n,m}$$

where $S_{n,m} = \sum_{i=1}^{n} X_{i,m}$. Then by the union bound and the pigeonhole principle we have

$$\mathbf{P}(|S_n| \geq An) \leq \sum_{m=0}^{\infty} \mathbf{P}\left(|S_{n,m}| \geq \frac{A}{100(m+1)^2}n\right)$$

(say). Each $X_{i,m}$ is clearly bounded in magnitude by 2^m; from the sub-Gaussian hypothesis one can also verify that the mean and variance of $X_{i,m}$ are at most $C' \exp(-c'2^{2m})$ for some $C', c' > 0$. If A is large enough, an application of the Chernoff bound (2.11) (or more precisely, the refinement in Exercise 2.1.3) then gives (after some computation)

$$\mathbf{P}(|S_{n,m}| \geq 2^{-m-1}An) \leq C'2^{-m} \exp(-c'An)$$

(say) for some $C', c' > 0$, and the claim follows. $\square$

Exercise 2.1.6. Show that the hypothesis that A is sufficiently large can be replaced by the hypothesis that $A > 0$ is independent of n. *Hint:* There are several approaches available. One can adapt the above proof; one can modify the proof of the Chernoff inequality directly; or one can figure out a way to deduce the small A case from the large A case.

Exercise 2.1.7. Show that the sub-Gaussian hypothesis can be generalised to a sub-exponential tail hypothesis

$$\mathbf{P}(|X| \geq t) \leq C \exp(-ct^p)$$

provided that $p > 1$. Show that the result also extends to the case $0 < p \leq 1$, except with the exponent $\exp(-c_A n)$ replaced by $\exp(-c_A n^{p-\varepsilon})$ for some $\varepsilon > 0$. (I do not know if the ε loss can be removed, but it is easy to see that one cannot hope to do much better than this, just by considering the probability that X_1 (say) is already as large as An.)

2.1.3. Lipschitz combinations. In the preceding discussion, we had only considered the linear combination $X_1, \ldots, X_n$ of independent variables $X_1, \ldots, X_n$. Now we consider more general combinations $F(X)$, where we write $X := (X_1, \ldots, X_n)$ for short. Of course, to get any non-trivial results we must make some regularity hypotheses on F. It turns out that a particularly useful class of a regularity hypothesis here is a Lipschitz hypothesis—that small variations in X lead to small variations in $F(X)$. A simple example of this is *McDiarmid's inequality*:

Theorem 2.1.10 (McDiarmid's inequality). *Let $X_1, \ldots, X_n$ be independent random variables taking values in ranges $R_1, \ldots, R_n$, and let $F : R_1 \times \ldots \times R_n \to \mathbf{C}$ be a function with the property that if one freezes all but the ith coordinate of $F(x_1, \ldots, x_n)$ for some $1 \leq i \leq n$, then F only fluctuates by at most $c_i > 0$, thus*

$$|F(x_1, \ldots, x_{i-1}, x_i, x_{i+1}, \ldots, x_n)$$
$$- F(x_1, \ldots, x_{i-1}, x_i', x_{i+1}, \ldots, x_n)| \leq c_i$$

for all $x_j \in X_j$, $x_i' \in X_i$ for $1 \leq j \leq n$. Then for any $\lambda > 0$, one has

$$\mathbf{P}(|F(X) - \mathbf{E}F(X)| \geq \lambda\sigma) \leq C \exp(-c\lambda^2)$$

for some absolute constants $C, c > 0$, where $\sigma^2 := \sum_{i=1}^n c_i^2$.

Proof. We may assume that F is real. By symmetry, it suffices to show the one-sided estimate

$$(2.15) \qquad \mathbf{P}(F(X) - \mathbf{E}F(X) \geq \lambda\sigma^2) \leq C \exp(-c\lambda^2).$$

To compute this quantity, we again use the exponential moment method. Let $t > 0$ be a parameter to be chosen later, and consider the exponential moment

$$(2.16) \qquad \mathbf{E}\exp(tF(X)).$$

To compute this, let us condition $X_1, \ldots, X_{n-1}$ to be fixed, and look at the conditional expectation

$$\mathbf{E}(\exp(tF(X))|X_1, \ldots, X_{n-1}).$$

We can simplify this as

$$\mathbf{E}(\exp(tY)|X_1,\ldots,X_{n-1})\exp(t\mathbf{E}(F(X)|X_1,\ldots,X_{n-1}))$$

where

$$Y := F(X) - \mathbf{E}(F(X)|X_1,\ldots,X_{n-1}).$$

For $X_1,\ldots,X_{n-1}$ fixed, tY only fluctuates by at most tc_n and has mean zero. Applying (2.10), we conclude that

$$\mathbf{E}(\exp(tY)|X_1,\ldots,X_{n-1}) \leq \exp(O(t^2 c_n^2)).$$

Integrating out the conditioning, we see that we have upper bounded (2.16) by

$$\exp(O(t^2 c_n^2))\mathbf{E}\exp(t(\mathbf{E}(F(X)|X_1,\ldots,X_{n-1})).$$

We observe that $(\mathbf{E}(F(X)|X_1,\ldots,X_{n-1})$ is a function $F_{n-1}(X_1,\ldots,X_{n-1})$ of $X_1,\ldots,X_{n-1}$, where F_{n-1} obeys the same hypotheses as F (but for $n-1$ instead of n). We can then iterate the above computation n times and eventually upper bound (2.16) by

$$\exp(\sum_{i=1}^{n} O(t^2 c_i^2))\exp(t\mathbf{E}F(X)),$$

which we rearrange as

$$\mathbf{E}\exp(t(F(X) - \mathbf{E}F(X))) \leq \exp(O(t^2\sigma^2)),$$

and thus by Markov's inequality (1.13)

$$\mathbf{P}(F(X) - \mathbf{E}F(X) \geq \lambda\sigma) \leq \exp(O(t^2\sigma^2) - t\lambda\sigma).$$

Optimising in t then gives the claim.   □

Exercise 2.1.8. Show that McDiarmid's inequality implies Hoeffding's inequality (Exercise 2.1.4).

Remark 2.1.11. One can view McDiarmid's inequality as a *tensorisation* of Hoeffding's lemma, as it leverages the latter lemma for a single random variable to establish an analogous result for n random variables. It is possible to apply this tensorisation trick to random variables taking values in more sophisticated metric spaces than an interval $[a,b]$, leading to a class of concentration of measure inequalities known as *transportation cost-information inequalities*, which will not be discussed here.

The most powerful concentration of measure results, though, do not just exploit Lipschitz type behaviour in each individual variable, but *joint* Lipschitz behaviour. Let us first give a classical instance of this, in the special case when the $X_1,\ldots,X_n$ are Gaussian variables. A key property of Gaussian variables is that any linear combination of independent Gaussians is again an independent Gaussian:

Exercise 2.1.9. Let $X_1, \ldots, X_n$ be independent real Gaussian variables with $X_i = N(\mu_i, \sigma_i^2)_{\mathbf{R}}$, and let $c_1, \ldots, c_n$ be real constants. Show that $c_1 X_1 + \cdots + c_n X_n$ is a real Gaussian with mean $\sum_{i=1}^{n} c_i \mu_i$ and variance $\sum_{i=1}^{n} |c_i|^2 \sigma_i^2$.

Show that the same claims also hold with complex Gaussians and complex constants c_i.

Exercise 2.1.10 (Rotation invariance). Let $X = (X_1, \ldots, X_n)$ be an $\mathbf{R}^n$-valued random variable, where $X_1, \ldots, X_n \equiv N(0,1)_{\mathbf{R}}$ are iid real Gaussians. Show that for any orthogonal matrix $U \in O(n)$, $UX \equiv X$.

Show that the same claim holds for complex Gaussians (so X is now $\mathbf{C}^n$-valued), and with the orthogonal group $O(n)$ replaced by the unitary group $U(n)$.

Theorem 2.1.12 (Gaussian concentration inequality for Lipschitz functions). *Let $X_1, \ldots, X_n \equiv N(0,1)_{\mathbf{R}}$ be iid real Gaussian variables, and let $F : \mathbf{R}^n \to \mathbf{R}$ be a 1-Lipschitz function (i.e., $|F(x) - F(y)| \leq |x - y|$ for all $x, y \in \mathbf{R}^n$, where we use the Euclidean metric on $\mathbf{R}^n$). Then for any λ one has*

$$\mathbf{P}(|F(X) - \mathbf{E}F(X)| \geq \lambda) \leq C \exp(-c\lambda^2)$$

for some absolute constants $C, c > 0$.

Proof. We use the following elegant argument of Maurey and Pisier. By subtracting a constant from F, we may normalise $\mathbf{E}F(X) = 0$. By symmetry it then suffices to show the upper tail estimate

$$\mathbf{P}(F(X) \geq \lambda) \leq C \exp(-c\lambda^2).$$

By smoothing F slightly we may assume that F is smooth, since the general case then follows from a limiting argument. In particular, the Lipschitz bound on F now implies the gradient estimate

$$(2.17) \qquad\qquad |\nabla F(x)| \leq 1$$

for all $x \in \mathbf{R}^n$.

Once again, we use the exponential moment method. It will suffice to show that

$$\mathbf{E} \exp(tF(X)) \leq \exp(Ct^2)$$

for some constant $C > 0$ and all $t > 0$, as the claim follows from Markov's inequality (1.13) and optimisation in t as in previous arguments.

To exploit the Lipschitz nature of F, we will need to introduce a second copy of $F(X)$. Let Y be an independent copy of X. Since $\mathbf{E}F(Y) = 0$, we see from Jensen's inequality (Exercise 1.1.8) that

$$\mathbf{E} \exp(-tF(Y)) \geq 1$$

and thus (by independence of X and Y)

$$\mathbf{E}\exp(tF(X)) \leq \mathbf{E}\exp(t(F(X) - F(Y))).$$

It is tempting to use the fundamental theorem of calculus along a line segment,

$$F(X) - F(Y) = \int_0^1 \frac{d}{dt}F((1-t)Y + tX)\,dt,$$

to estimate $F(X) - F(Y)$, but it turns out for technical reasons to be better to use a circular arc instead,

$$F(X) - F(Y) = \int_0^{\pi/2} \frac{d}{d\theta}F(Y\cos\theta + X\sin\theta)\,d\theta.$$

The reason for this is that $X_\theta := Y\cos\theta + X\sin\theta$ is another Gaussian random variable equivalent to X, as is its derivative $X_\theta' := -Y\sin\theta + X\cos\theta$ (by Exercise 2.1.9); furthermore, and crucially, these two random variables are *independent* (by Exercise 2.1.10).

To exploit this, we first use Jensen's inequality (Exercise 1.1.8) to bound

$$\exp(t(F(X) - F(Y))) \leq \frac{2}{\pi}\int_0^{\pi/2} \exp\left(\frac{\pi t}{2}\frac{d}{d\theta}F(X_\theta)\right)\,d\theta.$$

Applying the chain rule and taking expectations, we have

$$\mathbf{E}\exp(t(F(X) - F(Y))) \leq \frac{2}{\pi}\int_0^{\pi/2} \mathbf{E}\exp\left(\frac{\pi t}{2}\nabla F(X_\theta)\cdot X_\theta'\right)\,d\theta.$$

Let us condition X_θ to be fixed, then $X_\theta' \equiv X$; applying Exercise 2.1.9 and (2.17), we conclude that $\frac{\pi t}{2}\nabla F(X_\theta)\cdot X_\theta'$ is normally distributed with standard deviation at most $\frac{\pi t}{2}$. As such we have

$$\mathbf{E}\exp\left(\frac{\pi t}{2}\nabla F(X_\theta)\cdot X_\theta'\right) \leq \exp(Ct^2)$$

for some absolute constant C; integrating out the conditioning on X_θ we obtain the claim. $\square$

Exercise 2.1.11. Show that Theorem 2.1.12 is equivalent to the inequality

$$\mathbf{P}(X \in A)\mathbf{P}(X \notin A_\lambda) \leq C\exp(-c\lambda^2)$$

holding for all $\lambda > 0$ and all measurable sets A, where $X = (X_1, \ldots, X_n)$ is an $\mathbf{R}^n$-valued random variable with iid Gaussian components $X_1, \ldots, X_n \equiv N(0,1)_{\mathbf{R}}$, and A_λ is the λ-neighbourhood of A.

Now we give a powerful concentration inequality of Talagrand, which we will rely heavily on later in this text.

Theorem 2.1.13 (Talagrand concentration inequality). *Let $K > 0$, and let $X_1, \ldots, X_n$ be independent complex variables with $|X_i| \leq K$ for all $1 \leq i \leq n$. Let $F : \mathbf{C}^n \to \mathbf{R}$ be a 1-Lipschitz convex function (where we identify $\mathbf{C}^n$ with $\mathbf{R}^{2n}$ for the purposes of defining "Lipschitz" and "convex"). Then for any λ one has*

$$(2.18) \qquad \mathbf{P}(|F(X) - \mathbf{M}F(X)| \geq \lambda K) \leq C \exp(-c\lambda^2)$$

and

$$(2.19) \qquad \mathbf{P}(|F(X) - \mathbf{E}F(X)| \geq \lambda K) \leq C \exp(-c\lambda^2)$$

for some absolute constants $C, c > 0$, where $\mathbf{M}F(X)$ is a median of $F(X)$.

We now prove the theorem, following the remarkable argument of Talagrand [**Ta1995**].

By dividing through by K we may normalise $K = 1$. X now takes values in the convex set $\Omega^n \subset \mathbf{C}^n$, where Ω is the unit disk in $\mathbf{C}$. It will suffice to establish the inequality

$$(2.20) \qquad \mathbf{E} \exp(cd(X, A)^2) \leq \frac{1}{\mathbf{P}(X \in A)}$$

for any convex set A in Ω^n and some absolute constant $c > 0$, where $d(X, A)$ is the Euclidean distance between X and A. Indeed, if one obtains this estimate, then one has

$$\mathbf{P}(F(X) \leq x)\mathbf{P}(F(X) \geq y) \leq \exp(-c|x - y|^2)$$

for any $y > x$ (as can be seen by applying (2.20) to the convex set $A := \{z \in \Omega^n : F(z) \leq x\}$). Applying this inequality of one of x, y equal to the median $\mathbf{M}F(X)$ of $F(X)$ yields (2.18), which in turn implies that

$$\mathbf{E}F(X) = \mathbf{M}F(X) + O(1),$$

which then gives (2.19).

We would like to establish (2.20) by induction on dimension n. In the case when $X_1, \ldots, X_n$ are Bernoulli variables, this can be done, see e.g., [**Ta2010b**, §1.5]. In the general case, it turns out that in order to close the induction properly, one must strengthen (2.20) by replacing the Euclidean distance $d(X, A)$ by an essentially larger quantity, which I will call the *combinatorial distance* $d_c(X, A)$ from X to A. For each vector $z = (z_1, \ldots, z_n) \in \mathbf{C}^n$ and $\omega = (\omega_1, \ldots, \omega_n) \in \{0, 1\}^n$, we say that ω *supports* z if z_i is non-zero only when ω_i is non-zero. Define the *combinatorial support* $U_A(X)$ of A relative to X to be all the vectors in $\{0, 1\}^n$ that support at least one vector in $A - X$. Define the *combinatorial hull* $V_A(X)$ of A relative to X to be the convex hull of $U_A(X)$, and then define *combinatorial distance* $d_c(X, A)$ to be the distance between $V_A(X)$ and the origin.

Lemma 2.1.14 (Combinatorial distance controls Euclidean distance). *Let A be a convex subset of Ω^n. Then $d(X, A) \leq 2d_c(X, A)$.*

Proof. Suppose $d_c(X, A) \leq r$. Then there exists a convex combination $t = (t_1, \ldots, t_n)$ of elements $\omega \in U_A(X) \subset \{0, 1\}^n$ which has magnitude at most r. For each such $\omega \in U_A(X)$, we can find a vector $z_\omega \in X - A$ supported by ω. As A, X both lie in Ω^n, every coefficient of z_ω has magnitude at most 2, and is thus bounded in magnitude by twice the corresponding coefficient of ω. If we then let z_t be the convex combination of the z_ω indicated by t, then the magnitude of each coefficient of z_t is bounded by twice the corresponding coefficient of t, and so $|z_t| \leq 2r$. On the other hand, as A is convex, z_t lies in $X - A$, and so $d(X, A) \leq 2r$. The claim follows. $\qquad\square$

Thus to show (2.20) it suffices (after a modification of the constant c) to show that

$$(2.21) \qquad\qquad \mathbf{E}\exp(cd_c(X, A)^2) \leq \frac{1}{\mathbf{P}(X \in A)}.$$

We first verify the one-dimensional case. In this case, $d_c(X, A)$ equals 1 when $X \notin A$, and 0 otherwise, and the claim follows from elementary calculus (for c small enough).

Now suppose that $n > 1$ and the claim has already been proven for $n-1$. We write $X = (X', X_n)$, and let $A_{X_n} := \{z' \in \Omega^{n-1} : (z', X_n) \in A\}$ be a slice of A. We also let $B := \{z' \in \Omega^{n-1} : (z', t) \in A \text{ for some } t \in \Omega\}$. We have the following basic inequality:

Lemma 2.1.15. *For any $0 \leq \lambda \leq 1$, we have*

$$d_c(X, A)^2 \leq (1 - \lambda)^2 + \lambda d_c(X', A_{X_n})^2 + (1 - \lambda)d_c(X', B)^2.$$

Proof. Observe that $U_A(X)$ contains both $U_{A_{X_n}}(X') \times \{0\}$ and $U_B(X') \times \{1\}$, and so by convexity, $V_A(X)$ contains $(\lambda t + (1 - \lambda)u, 1 - \lambda)$ whenever $t \in V_{A_{X_n}}(X')$ and $u \in V_B(X')$. The claim then follows from Pythagoras' theorem and the Cauchy-Schwarz inequality. $\qquad\square$

Let us now freeze X_n and consider the conditional expectation

$$\mathbf{E}(\exp(cd_c(X, A)^2)|X_n).$$

Using the above lemma (with some λ depending on X_n to be chosen later), we may bound the left-hand side of (2.21) by

$$e^{c(1-\lambda)^2}\mathbf{E}\big((e^{cd_c(X', A_{X_n})^2})^\lambda\big(e^{cd_c(X', B)^2}\big)^{1-\lambda}|X_n\big);$$

applying Hölder's inequality and the induction hypothesis (2.21), we can bound this by

$$e^{c(1-\lambda)^2} \frac{1}{\mathbf{P}(X' \in A_{X_n}|X_n)^\lambda \mathbf{P}(X' \in B|X_n)^{1-\lambda}}$$

which we can rearrange as

$$\frac{1}{\mathbf{P}(X' \in B)} e^{c(1-\lambda)^2} r^{-\lambda}$$

where $r := \mathbf{P}(X' \in A_{X_n}|X_n)/\mathbf{P}(X' \in B)$ (here we note that the event $X' \in B$ is independent of X_n). Note that $0 \le r \le 1$. We then apply the elementary inequality

$$\inf_{0 \le \lambda \le 1} e^{c(1-\lambda)^2} r^{-\lambda} \le 2 - r,$$

which can be verified by elementary calculus if c is small enough (in fact one can take $c = 1/4$). We conclude that

$$\mathbf{E}(\exp(cd_c(X, A)^2)|X_n) \le \frac{1}{\mathbf{P}(X' \in B)} \left(2 - \frac{\mathbf{P}(X' \in A_{X_n}|X_n)}{\mathbf{P}(X' \in B)}\right).$$

Taking expectations in n we conclude that

$$\mathbf{E}(\exp(cd_c(X, A)^2)) \le \frac{1}{\mathbf{P}(X' \in B)} \left(2 - \frac{\mathbf{P}(X \in A)}{\mathbf{P}(X' \in B)}\right).$$

Using the inequality $x(2 - x) \le 1$ with $x := \frac{\mathbf{P}(X \in A)}{\mathbf{P}(X' \in B)}$ we conclude (2.21) as desired.

The above argument was elementary, but rather "magical" in nature. Let us now give a somewhat different argument of Ledoux [**Le1995**], based on log-Sobolev inequalities, which gives the upper tail bound

(2.22) $$\mathbf{P}(F(X) - \mathbf{E}F(X) \ge \lambda K) \le C \exp(-c\lambda^2),$$

but curiously does not give the lower tail bound[9].

Once again we can normalise $K = 1$. By regularising F we may assume that F is smooth. The first step is to establish the following log-Sobolev inequality:

Lemma 2.1.16 (Log-Sobolev inequality). *Let* $F : \mathbf{C}^n \to \mathbf{R}$ *be a smooth convex function. Then*

$$\mathbf{E}F(X)e^{F(X)} \le (\mathbf{E}e^{F(X)})(\log \mathbf{E}e^{F(X)}) + C\mathbf{E}e^{F(X)}|\nabla F(X)|^2$$

for some absolute constant C *(independent of* n*).*

---

[9]The situation is not symmetric, due to the convexity hypothesis on F.

Remark 2.1.17. If one sets $f := e^{F/2}$ and normalises $\mathbf{E}f(X)^2 = 1$, this inequality becomes

$$\mathbf{E}|f(X)|^2 \log |f(X)|^2 \le 4C\mathbf{E}|\nabla f(X)|^2$$

which more closely resembles the classical log-Sobolev inequality (see [**Gr1975**] or [**Fe1969**]). The constant C here can in fact be taken to be 2; see [**Le1995**].

Proof. We first establish the 1-dimensional case. If we let Y be an independent copy of X, observe that the left-hand side can be rewritten as

$$\frac{1}{2}\mathbf{E}((F(X) - F(Y))(e^{F(X)} - e^{F(Y)})) + (\mathbf{E}F(X))(\mathbf{E}e^{F(X)}).$$

From Jensen's inequality (Exercise 1.1.8), $\mathbf{E}F(X) \le \log \mathbf{E}e^{F(X)}$, so it will suffice to show that

$$\mathbf{E}((F(X) - F(Y))(e^{F(X)} - e^{F(Y)})) \le 2C\mathbf{E}e^{F(X)}|\nabla F(X)|^2.$$

From convexity of F (and hence of e^F) and the bounded nature of X, Y, we have

$$F(X) - F(Y) = O(|\nabla F(X)|)$$

and

$$e^{F(X)} - e^{F(Y)} = O(|\nabla F(X)|e^{F(X)})$$

when $F(X) \ge F(Y)$, which leads to

$$((F(X) - F(Y))(e^{F(X)} - e^{F(Y)})) = O(e^{F(X)}|\nabla F(X)|^2)$$

in this case. Similarly, when $F(X) < F(Y)$ (swapping X and Y). The claim follows.

To show the general case, we induct on n (keeping care to ensure that the constant C does not change in this induction process). Write $X = (X', X_n)$, where $X' := (X_1, \ldots, X_{n-1})$. From induction hypothesis, we have

$$\mathbf{E}(F(X)e^{F(X)}|X_n) \le f(X_n)e^{f(X_n)} + C\mathbf{E}(e^{F(X)}|\nabla' F(X)|^2|X_n)$$

where ∇' is the $n-1$-dimensional gradient and $f(X_n) := \log \mathbf{E}(e^{F(X)}|X_n)$. Taking expectations, we conclude that

$$(2.23) \qquad \mathbf{E}F(X)e^{F(X)} \le \mathbf{E}f(X_n)e^{f(X_n)} + C\mathbf{E}e^{F(X)}|\nabla' F(X)|^2.$$

From the convexity of F and Hölder's inequality we see that f is also convex, and $\mathbf{E}e^{f(X_n)} = \mathbf{E}e^{F(X)}$. By the $n = 1$ case already established, we have

$$(2.24) \qquad \mathbf{E}f(X_n)e^{f(X_n)} \le (\mathbf{E}e^{F(X)})(\log \mathbf{E}e^{F(X)}) + C\mathbf{E}e^{f(X_n)}|f'(X_n)|^2.$$

Now, by the chain rule

$$e^{f(X_n)}|f'(X_n)|^2 = e^{-f(X_n)}|\mathbf{E}e^{F(X)}F_{x_n}(X)|^2$$

where F_{x_n} is the derivative of F in the x_n direction (where $(x_1, \dots, x_n)$ are the usual coordinates for $\mathbf{R}^n$). Applying Cauchy-Schwarz, we conclude

$$e^{f(X_n)} |f'(X_n)|^2 \leq \mathbf{E} e^{F(X)} |F_{x_n}(X)|^2.$$

Inserting this into (2.23), (2.24) we close the induction. $\qquad \square$

Now let F be convex and 1-Lipschitz. Applying the above lemma to tF for any $t > 0$, we conclude that

$$\mathbf{E} tF(X) e^{tF(X)} \leq (\mathbf{E} e^{tF(X)})(\log \mathbf{E} e^{tF(X)}) + Ct^2 \mathbf{E} e^{tF(X)};$$

setting $H(t) := \mathbf{E} e^{tF(X)}$, we can rewrite this as a differential inequality

$$tH'(t) \leq H(t) \log H(t) + Ct^2 H(t)$$

which we can rewrite as

$$\frac{d}{dt} \left(\frac{1}{t} \log H(t) \right) \leq C.$$

From Taylor expansion we see that

$$\frac{1}{t} \log H(t) \to \mathbf{E} F(X)$$

as $t \to 0$, and thus

$$\frac{1}{t} \log H(t) \leq \mathbf{E} F(X) + Ct$$

for any $t > 0$. In other words,

$$\mathbf{E} e^{tF(X)} \leq \exp(t \mathbf{E} F(X) + Ct^2).$$

By Markov's inequality (1.13), we conclude that

$$\mathbf{P}(F(X) - \mathbf{E} F(X) > \lambda) \leq \exp(Ct^2 - t\lambda);$$

optimising in t gives (2.22).

Remark 2.1.18. The same argument, starting with Gross's log-Sobolev inequality for the Gaussian measure, gives the upper tail component of Theorem 2.1.12, with no convexity hypothesis on F. The situation is now symmetric with respect to reflections $F \mapsto -F$, and so one obtains the lower tail component as well. The method of obtaining concentration inequalities from log-Sobolev inequalities (or related inequalities, such as Poincaré-type inequalities) by combining the latter with the exponential moment method is known as *Herbst's argument*, and can be used to establish a number of other functional inequalities of interest.

We now close with a simple corollary of the Talagrand concentration inequality, which will be extremely useful in the sequel.

Corollary 2.1.19 (Distance between random vector and a subspace). *Let*
$X_1, \ldots, X_n$ *be independent complex-valued random variables with mean zero
and variance 1, and bounded almost surely in magnitude by K. Let V be a
subspace of $\mathbf{C}^n$ of dimension d. Then for any $\lambda > 0$, one has*

$$\mathbf{P}(|d(X, V) - \sqrt{n - d}| \geq \lambda K) \leq C \exp(-c\lambda^2)$$

for some absolute constants $C, c > 0$.

Informally, this corollary asserts that the distance between a random
vector X and an arbitrary subspace V is typically equal to $\sqrt{n - \dim(V)} +
O(1)$.

Proof. The function $z \mapsto d(z, V)$ is convex and 1-Lipschitz. From Theorem
2.1.13, one has

$$\mathbf{P}(|d(X, V) - Md(X, V)| \geq \lambda K) \leq C \exp(-c\lambda^2).$$

To finish the argument, it then suffices to show that

$$Md(X, V) = \sqrt{n - d} + O(K).$$

We begin with a second moment calculation. Observe that

$$d(X, V)^2 = \|\pi(X)\|^2 = \sum_{1 \leq i,j \leq n} p_{ij} X_i \overline{X_j},$$

where π is the orthogonal projection matrix to the complement $V^\perp$ of V,
and p_{ij} are the components of π. Taking expectations, we obtain

$$(2.25) \qquad \mathbf{E}d(X, V)^2 = \sum_{i=1}^n p_{ii} = \mathrm{tr}(\pi) = n - d$$

where the latter follows by representing π in terms of an orthonormal basis
of $V^\perp$. This is close to what we need, but to finish the task we need to
obtain some concentration of $d(X, V)^2$ around its mean. For this, we write

$$d(X, V)^2 - \mathbf{E}d(X, V)^2 = \sum_{1 \leq i,j \leq n} p_{ij}(X_i \overline{X_j} - \delta_{ij})$$

where δ_{ij} is the Kronecker delta. The summands here are pairwise uncorrelated for $1 \leq i < j \leq n$, and the $i > j$ cases can be combined with the $i < j$
cases by symmetry. They are also pairwise independent (and hence pairwise
uncorrelated) for $1 \leq i = j \leq n$. Each summand also has a variance of
$O(K^2)$. We thus have the variance bound

$$\mathbf{Var}(d(X, V)^2) = O(K^2 \sum_{1 \leq i < j \leq n} |p_{ij}|^2) + O(K^2 \sum_{1 \leq i \leq n} |p_{ii}|^2) = O(K^2(n - d)),$$

where the latter bound comes from representing π in terms of an orthonormal
basis of $V^\perp$. From this, (2.25), and Chebyshev's inequality (1.26), we see

that the median of $d(X, V)^2$ is equal to $n - d + O(\sqrt{K^2(n - d)})$, which implies on taking square roots that the median of $d(X, V)$ is $\sqrt{n - d} + O(K)$, as desired. $\qquad\square$

2.2. The central limit theorem

Consider the sum $S_n := X_1 + \cdots + X_n$ of iid real random variables $X_1, \ldots, X_n \equiv X$ of finite mean μ and variance σ^2 for some $\sigma > 0$. Then the sum S_n has mean $n\mu$ and variance $n\sigma^2$, and so (by Chebyshev's inequality (1.26)) we expect S_n to usually have size $n\mu + O(\sqrt{n}\sigma)$. To put it another way, if we consider the *normalised sum*

$$(2.26) \qquad\qquad Z_n := \frac{S_n - n\mu}{\sqrt{n}\sigma},$$

then Z_n has been normalised to have mean zero and variance 1, and is thus usually of size $O(1)$.

In Section 2.1, we were able to establish various tail bounds on Z_n. For instance, from Chebyshev's inequality (1.26) one has

$$(2.27) \qquad\qquad \mathbf{P}(|Z_n| > \lambda) \leq \lambda^{-2},$$

and if the original distribution X was bounded or sub-Gaussian, we had the much stronger Chernoff bound

$$(2.28) \qquad\qquad \mathbf{P}(|Z_n| > \lambda) \leq C \exp(-c\lambda^2)$$

for some absolute constants $C, c > 0$; in other words, the Z_n are uniformly sub-Gaussian.

Now we look at the distribution of Z_n. The fundamental *central limit theorem* tells us the asymptotic behaviour of this distribution:

Theorem 2.2.1 (Central limit theorem). *Let $X_1, \ldots, X_n \equiv X$ be iid real random variables of finite mean μ and variance σ^2 for some $\sigma > 0$, and let Z_n be the normalised sum (2.26). Then as $n \to \infty$, Z_n converges in distribution to the standard normal distribution $N(0, 1)_{\mathbf{R}}$.*

Exercise 2.2.1. Show that Z_n does not converge in probability or in the almost sure sense (in the latter case, we view $X_1, X_2, \ldots$ as an infinite sequence of iid random variables). (*Hint:* The intuition here is that for two very different values $n_1 \ll n_2$ of n, the quantities Z_{n_1} and Z_{n_2} are almost independent of each other, since the bulk of the sum S_{n_2} is determined by those X_n with $n > n_1$. Now make this intuition precise.)

Exercise 2.2.2. Use Stirling's formula (Section 1.2) to verify the central limit theorem in the case when X is a Bernoulli distribution, taking the values 0 and 1 only. (This is a variant of Exercise 1.2.2 or Exercise 2.1.2. It is easy to see that once one does this, one can rescale and handle any other two-valued distribution also.)

Exercise 2.2.3. Use Exercise 2.1.9 to verify the central limit theorem in the case when X is Gaussian.

Note we are only discussing the case of real iid random variables. The case of complex random variables (or more generally, vector-valued random variables) is a little bit more complicated, and will be discussed later in this section.

The central limit theorem (and its variants, which we discuss below) are extremely useful tools in random matrix theory, in particular, through the control they give on random walks (which arise naturally from linear functionals of random matrices). But the central limit theorem can also be viewed as a "commutative" analogue of various spectral results in random matrix theory (in particular, we shall see in later sections that the *Wigner semicircle law* can be viewed in some sense as a "non-commutative" or "free" version of the central limit theorem). Because of this, the *techniques* used to prove the central limit theorem can often be adapted to be useful in random matrix theory. Because of this, we shall use this section to dwell on several different proofs of the central limit theorem, as this provides a convenient way to showcase some of the basic methods that we will encounter again (in a more sophisticated form) when dealing with random matrices.

2.2.1. Reductions. We first record some simple reductions one can make regarding the proof of the central limit theorem. Firstly, we observe *scale invariance*: if the central limit theorem holds for one random variable X, then it is easy to see that it also holds for $aX + b$ for any real a, b with $a \neq 0$. Because of this, one can *normalise* to the case when X has mean $\mu = 0$ and variance $\sigma^2 = 1$, in which case Z_n simplifies to

$$(2.29) \qquad\qquad Z_n = \frac{X_1 + \cdots + X_n}{\sqrt{n}}.$$

The other reduction we can make is *truncation*: to prove the central limit theorem for arbitrary random variables X of finite mean and variance, it suffices to verify the theorem for *bounded* random variables. To see this, we first need a basic linearity principle:

Exercise 2.2.4 (Linearity of convergence). Let V be a finite-dimensional real or complex vector space, X_n, Y_n be sequences of V-valued random variables (not necessarily independent), and let X, Y be another pair of V-valued random variables. Let c_n, d_n be scalars converging to c, d, respectively.

(i) If X_n converges in distribution to X, and Y_n converges in distribution to Y, and at least one of X, Y is deterministic, show that $c_n X_n + d_n Y_n$ converges in distribution to $cX + dY$.

(ii) If X_n converges in probability to X, and Y_n converges in probability to Y, show that $c_n X_n + d_n Y_n$ converges in probability to $cX + dY$.

(iii) If X_n converges almost surely to X, and Y_n converges almost surely Y, show that $c_n X_n + d_n Y_n$ converges almost surely to $cX + dY$.

Show that the first part of the exercise can fail if X, Y are not deterministic.

Now suppose that we have established the central limit theorem for bounded random variables, and want to extend to the unbounded case. Let X be an unbounded random variable, which we can normalise to have mean zero and unit variance. Let $N = N_n > 0$ be a truncation parameter depending on n which, as usual, we shall optimise later, and split $X = X_{\leq N} + X_{>N}$ in the usual fashion ($X_{\leq N} = X\mathbf{I}(|X| \leq N)$; $X_{>N} = X\mathbf{I}(|X| > N)$). Thus we have $S_n = S_{n,\leq N} + S_{n,>N}$ as usual.

Let $\mu_{\leq N}, \sigma^2_{\leq N}$ be the mean and variance of the bounded random variable $X_{\leq N}$. As we are assuming that the central limit theorem is already true in the bounded case, we know that if we fix N to be independent of n, then

$$Z_{n,\leq N} := \frac{S_{n,\leq N} - n\mu_{\leq N}}{\sqrt{n}\sigma_{\leq N}}$$

converges in distribution to $N(0,1)_{\mathbf{R}}$. By a diagonalisation argument, we conclude that there exists a sequence N_n going (slowly) to infinity with n, such that $Z_{n,\leq N_n}$ still converges in distribution to $N(0,1)_{\mathbf{R}}$.

For such a sequence, we see from dominated convergence that $\sigma_{\leq N_n}$ converges to $\sigma = 1$. As a consequence of this and Exercise 2.2.4, we see that

$$\frac{S_{n,\leq N_n} - n\mu_{\leq N_n}}{\sqrt{n}}$$

converges in distribution to $N(0,1)_{\mathbf{R}}$.

Meanwhile, from dominated convergence again, $\sigma_{>N_n}$ converges to 0. From this and (2.27) we see that

$$\frac{S_{n,>N_n} - n\mu_{>N_n}}{\sqrt{n}}$$

converges in distribution to 0. Finally, from linearity of expectation we have $\mu_{\leq N_n} + \mu_{>N_n} = \mu = 0$. Summing (using Exercise 2.2.4), we obtain the claim.

Remark 2.2.2. The truncation reduction is not needed for some proofs of the central limit (notably the Fourier-analytic proof), but is very convenient for some of the other proofs that we will give here, and will also be used at several places in later sections.

By applying the scaling reduction after the truncation reduction, we observe that to prove the central limit theorem, it suffices to do so for random variables X which are bounded *and* which have mean zero and unit variance. (Why is it important to perform the reductions in this order?)

2.2.2. The Fourier method. We now give the standard Fourier-analytic proof of the central limit theorem. Given any real random variable X, we introduce the *characteristic function* $F_X : \mathbf{R} \to \mathbf{C}$, defined by the formula

$$(2.30) \qquad\qquad F_X(t) := \mathbf{E}e^{itX}.$$

Equivalently, F_X is the Fourier transform of the probability measure μ_X.

Example 2.2.3. The signed Bernoulli distribution has characteristic function $F(t) = \cos(t)$.

Exercise 2.2.5. Show that the normal distribution $N(\mu, \sigma^2)_{\mathbf{R}}$ has characteristic function $F(t) = e^{it\mu} e^{-\sigma^2 t^2 / 2}$.

More generally, for a random variable X taking values in a real vector space $\mathbf{R}^d$, we define the characteristic function $F_X : \mathbf{R}^d \to \mathbf{C}$ by

$$(2.31) \qquad\qquad F_X(t) := \mathbf{E}e^{it \cdot X}$$

where $\cdot$ denotes the Euclidean inner product on $\mathbf{R}^d$. One can similarly define the characteristic function on complex vector spaces $\mathbf{C}^d$ by using the complex inner product

$$(z_1, \ldots, z_d) \cdot (w_1, \ldots, w_d) := \mathrm{Re}(z_1 \overline{w_1} + \cdots + z_d \overline{w_d})$$

(or equivalently, by identifying $\mathbf{C}^d$ with $\mathbf{R}^{2d}$ in the usual manner.)

More generally[10], one can define the characteristic function on any finite dimensional real or complex vector space V, by identifying V with $\mathbf{R}^d$ or $\mathbf{C}^d$.

The characteristic function is clearly bounded in magnitude by 1, and equals 1 at the origin. By the Lebesgue dominated convergence theorem, F_X is continuous in t.

Exercise 2.2.6 (Riemann-Lebesgue lemma). Show that if X is an absolutely continuous random variable taking values in $\mathbf{R}^d$ or $\mathbf{C}^d$, then $F_X(t) \to 0$ as $t \to \infty$. Show that the claim can fail when the absolute continuity hypothesis is dropped.

---

[10]Strictly speaking, one either has to select an inner product on V to do this, or else make the characteristic function defined on the *dual space* V^* instead of on V itself; see for instance [**Ta2010**, §1.12]. But we will not need to care about this subtlety in our applications.

Exercise 2.2.7. Show that the characteristic function F_X of a random variable X taking values in $\mathbf{R}^d$ or $\mathbf{C}^d$ is in fact uniformly continuous on its domain.

Let X be a real random variable. If we Taylor expand e^{itX} and formally interchange the series and expectation, we arrive at the heuristic identity

$$(2.32) \qquad\qquad F_X(t) = \sum_{k=0}^{\infty} \frac{(it)^k}{k!} \mathbf{E} X^k$$

which thus interprets the characteristic function of a real random variable X as a kind of generating function for the moments. One rigorous version of this identity is as follows.

Exercise 2.2.8 (Taylor expansion of characteristic function). Let X be a real random variable with finite k^{th} moment for some $k \geq 1$. Show that F_X is k times continuously differentiable, and one has the partial Taylor expansion

$$F_X(t) = \sum_{j=0}^{k} \frac{(it)^j}{j!} \mathbf{E} X^j + o(|t|^k)$$

where $o(|t|^k)$ is a quantity that goes to zero as $t \to 0$, times $|t|^k$. In particular, we have

$$\frac{d^j}{dt^j} F_X(t) = i^j \mathbf{E} X^j$$

for all $0 \leq j \leq k$.

Exercise 2.2.9. Establish (2.32) in the case that X is sub-Gaussian, and show that the series converges locally uniformly in t.

Note that the characteristic function depends only on the distribution of X: if $X \equiv Y$, then $F_X = F_Y$. The converse statement is true also: if $F_X = F_Y$, then $X \equiv Y$. This follows from a more general (and useful) fact, known as *Lévy's continuity theorem*.

Theorem 2.2.4 (Lévy continuity theorem, special case). *Let V be a finite-dimensional real or complex vector space, and let X_n be a sequence of V-valued random variables, and let X be an additional V-valued random variable. Then the following statements are equivalent:*

  (i) *F_{X_n} converges pointwise to F_X.*

  (ii) *X_n converges in distribution to X.*

Proof. Without loss of generality, we may take $V = \mathbf{R}^d$.

The implication of (i) from (ii) is immediate from (2.31) and the definition of convergence in distribution (see Definition 1.1.29), since the function $x \mapsto e^{it \cdot x}$ is bounded continuous.

Now suppose that (i) holds, and we wish to show that (ii) holds. By Exercise 1.1.25(iv), it suffices to show that

$$\mathbf{E}\varphi(X_n) \to \mathbf{E}\varphi(X)$$

whenever $\varphi : V \to \mathbf{R}$ is a continuous, compactly supported function. By approximating φ uniformly by Schwartz functions (e.g., using the Stone-Weierstrass theorem, see [**Ta2010**]), it suffices to show this for Schwartz functions φ. But then we have the Fourier inversion formula

$$\varphi(X_n) = \int_{\mathbf{R}^d} \hat{\varphi}(t) e^{it \cdot X_n} \, dt$$

where

$$\hat{\varphi}(t) := \frac{1}{(2\pi)^d} \int_{\mathbf{R}^d} \varphi(x) e^{-it \cdot x} \, dx$$

is a Schwartz function, and is, in particular, absolutely integrable (see e.g. [**Ta2010**, §1.12]). From the Fubini-Tonelli theorem, we thus have

$$(2.33) \qquad \mathbf{E}\varphi(X_n) = \int_{\mathbf{R}^d} \hat{\varphi}(t) F_{X_n}(t) \, dt$$

and similarly for X. The claim now follows from the Lebesgue dominated convergence theorem.   □

Remark 2.2.5. Setting $X_n := Y$ for all n, we see, in particular, the previous claim that $F_X = F_Y$ if and only if $X \equiv Y$. It is instructive to use the above proof as a guide to prove this claim directly.

Exercise 2.2.10 (Lévy's continuity theorem, full version). Let V be a finite-dimensional real or complex vector space, and let X_n be a sequence of V-valued random variables. Suppose that F_{X_n} converges pointwise to a limit F. Show that the following are equivalent:

(i) F is continuous at 0.

(ii) X_n is a tight sequence.

(iii) F is the characteristic function of a V-valued random variable X (possibly after extending the sample space).

(iv) X_n converges in distribution to some V-valued random variable X (possibly after extending the sample space).

Hint: To get from (ii) to the other conclusions, use Prokhorov's theorem (see Exercise 1.1.25) and Theorem 2.2.4. To get back to (ii) from (i), use (2.33) for a suitable Schwartz function φ. The other implications are easy once Theorem 2.2.4 is in hand.

Remark 2.2.6. Lévy's continuity theorem is very similar in spirit to *Weyl's criterion* in equidistribution theory (see e.g. [**KuNi2006**]).

Exercise 2.2.11 (Esséen concentration inequality). Let X be a random variable taking values in $\mathbf{R}^d$. Then for any $r > 0$, $\varepsilon > 0$, show that

$$(2.34) \qquad \sup_{x_0 \in \mathbf{R}^d} \mathbf{P}(|X - x_0| \leq r) \leq C_{d,\varepsilon} r^d \int_{t \in \mathbf{R}^d : |t| \leq \varepsilon/r} |F_X(t)| \, dt$$

for some constant $C_{d,\varepsilon}$ depending only on d and ε. (*Hint:* Use (2.33) for a suitable Schwartz function φ.) The left-hand side of (2.34) is known as the *small ball probability* of X at radius r.

In Fourier analysis, we learn that the Fourier transform is a particularly well-suited tool for studying convolutions. The probability theory analogue of this fact is that characteristic functions are a particularly well-suited tool for studying sums of independent random variables. More precisely, we have

Exercise 2.2.12 (Fourier identities). Let V be a finite-dimensional real or complex vector space, and let X, Y be independent random variables taking values in V. Then

$$(2.35) \qquad F_{X+Y}(t) = F_X(t) F_Y(t)$$

for all $t \in V$. Also, for any scalar c, one has

$$F_{cX}(t) = F_X(\bar{c}t)$$

and more generally, for any linear transformation $T : V \to V$, one has

$$F_{TX}(t) = F_X(T^* t).$$

Remark 2.2.7. Note that this identity (2.35), combined with Exercise 2.2.5 and Remark 2.2.5, gives a quick alternate proof of Exercise 2.1.9.

In particular, in the normalised setting (2.29), we have the simple relationship

$$(2.36) \qquad F_{Z_n}(t) = F_X(t/\sqrt{n})^n$$

that describes the characteristic function of Z_n in terms of that of X.

We now have enough machinery to give a quick proof of the central limit theorem:

Proof of Theorem 2.2.1. We may normalise X to have mean zero and variance 1. By Exercise 2.2.8, we thus have

$$F_X(t) = 1 - t^2/2 + o(|t|^2)$$

for sufficiently small t, or equivalently

$$F_X(t) = \exp(-t^2/2 + o(|t|^2))$$

for sufficiently small t. Applying (2.36), we conclude that

$$F_{Z_n}(t) \to \exp(-t^2/2)$$

as $n \to \infty$ for any fixed t. But by Exercise 2.2.5, $\exp(-t^2/2)$ is the characteristic function of the normal distribution $N(0,1)_{\mathbf{R}}$. The claim now follows from the Lévy continuity theorem. $\square$

Exercise 2.2.13 (Vector-valued central limit theorem). Let $\vec{X} = (X_1, \ldots, X_d)$ be a random variable taking values in $\mathbf{R}^d$ with finite second moment. Define the *covariance matrix* $\Sigma(\vec{X})$ to be the $d \times d$ matrix Σ whose ij^{th} entry is the covariance $\mathbf{E}(X_i - \mathbf{E}(X_i))(X_j - \mathbf{E}(X_j))$.

(i) Show that the covariance matrix is positive semi-definite real symmetric.

(ii) Conversely, given any positive definite real symmetric $d \times d$ matrix Σ and $\mu \in \mathbf{R}^d$, show that the *normal distribution* $N(\mu, \Sigma)_{\mathbf{R}^d}$, given by the absolutely continuous measure

$$\frac{1}{((2\pi)^d \det \Sigma)^{1/2}} e^{-(x-\mu) \cdot \Sigma^{-1}(x-\mu)/2} \, dx,$$

has mean μ and covariance matrix Σ, and has a characteristic function given by

$$F(t) = e^{i\mu \cdot t} e^{-t \cdot \Sigma t/2}.$$

How would one define the normal distribution $N(\mu, \Sigma)_{\mathbf{R}^d}$ if Σ degenerated to be merely positive semi-definite instead of positive definite?

(iii) If $\vec{S}_n := \vec{X}_1 + \cdots + \vec{X}_n$ is the sum of n iid copies of $\vec{X}$, show that $\frac{\vec{S}_n - n\mu}{\sqrt{n}}$ converges in distribution to $N(0, \Sigma(X))_{\mathbf{R}^d}$.

Exercise 2.2.14 (Complex central limit theorem). Let X be a complex random variable of mean $\mu \in \mathbf{C}$, whose real and imaginary parts have variance $\sigma^2/2$ and covariance 0. Let $X_1, \ldots, X_n \equiv X$ be iid copies of X. Show that as $n \to \infty$, the normalised sums (2.26) converge in distribution to the standard complex Gaussian $N(0,1)_{\mathbf{C}}$.

Exercise 2.2.15 (Lindeberg central limit theorem). Let $X_1, X_2, \ldots$ be a sequence of independent (but not necessarily identically distributed) real random variables, normalised to have mean zero and variance one. Assume the (strong) *Lindeberg condition*

$$\lim_{N \to \infty} \limsup_{n \to \infty} \frac{1}{n} \sum_{j=1}^{n} \mathbf{E}|X_{j,>N}|^2 = 0$$

where $X_{j,>N} := X_j \mathbf{I}(|X_j| > N)$ is the truncation of X_j to large values. Show that as $n \to \infty$, $\frac{X_1 + \cdots + X_n}{\sqrt{n}}$ converges in distribution to $N(0,1)_{\mathbf{R}}$. (*Hint:* Modify the truncation argument.)

A more sophisticated version of the Fourier-analytic method gives a more quantitative form of the central limit theorem, namely the *Berry-Esséen theorem.*

Theorem 2.2.8 (Berry-Esséen theorem). *Let X have mean zero, unit variance, and finite third moment. Let $Z_n := (X_1 + \cdots + X_n)/\sqrt{n}$, where $X_1, \ldots, X_n$ are iid copies of X. Then we have*

$$(2.37) \qquad \mathbf{P}(Z_n < a) = \mathbf{P}(G < a) + O\left(\frac{1}{\sqrt{n}}(\mathbf{E}|X|^3)\right)$$

uniformly for all $a \in \mathbf{R}$, where $G \equiv N(0,1)_\mathbf{R}$, and the implied constant is absolute.

Proof. (Optional) Write $\varepsilon := \mathbf{E}|X|^3/\sqrt{n}$; our task is to show that

$$\mathbf{P}(Z_n < a) = \mathbf{P}(G < a) + O(\varepsilon)$$

for all a. We may of course assume that $\varepsilon < 1$, as the claim is trivial otherwise.

Let $c > 0$ be a small absolute constant to be chosen later. Let $\eta : \mathbf{R} \to \mathbf{R}$ be a non-negative Schwartz function with total mass 1 whose Fourier transform is supported in $[-c, c]$, and let $\varphi : \mathbf{R} \to \mathbf{R}$ be the smoothed out version of $1_{(-\infty,0]}$, defined as

$$\varphi(x) := \int_\mathbf{R} 1_{(-\infty,0]}(x - \varepsilon y)\eta(y) \, dy.$$

Observe that φ is decreasing from 1 to 0.

We claim that it suffices to show that

$$(2.38) \qquad \mathbf{E}\varphi(Z_n - a) = \mathbf{E}\varphi(G - a) + O_\eta(\varepsilon)$$

for every a, where the subscript means that the implied constant depends on η. Indeed, suppose that (2.38) held. Define

$$(2.39) \qquad \rho := \sup_a |\mathbf{P}(Z_n < a) - \mathbf{P}(G < a)|,$$

thus our task is to show that $\rho = O(\varepsilon)$.

Let a be arbitrary, and let $K > 0$ be a large absolute constant to be chosen later. We write

$$\mathbf{P}(Z_n < a) \leq \mathbf{E}\varphi(Z_n - a - K\varepsilon) \\ + \mathbf{E}(1 - \varphi(Z_n - a - K\varepsilon))\mathbf{I}(Z_n < a)$$

and thus by (2.38)

$$\mathbf{P}(Z_n < a) \leq \mathbf{E}\varphi(G - a - K\varepsilon) \\ + \mathbf{E}(1 - \varphi(Z_n - a - K\varepsilon))\mathbf{I}(Z_n < a) + O_\eta(\varepsilon).$$

Meanwhile, from (2.39) and an integration by parts we see that

$$\mathbf{E}(1 - \varphi(Z_n - a - K\varepsilon))\mathbf{I}(Z_n < a) = \mathbf{E}(1 - \varphi(G - a - K\varepsilon))\mathbf{I}(G < a)$$
$$+ O((1 - \varphi(-K\varepsilon))\rho).$$

From the bounded density of G and the rapid decrease of η we have

$$\mathbf{E}\varphi(G - a - K\varepsilon) + \mathbf{E}(1 - \varphi(G - a - K\varepsilon))\mathbf{I}(G < a)$$
$$= \mathbf{P}(G < a) + O_{\eta,K}(\varepsilon).$$

Putting all this together, we see that

$$\mathbf{P}(Z_n < a) \le \mathbf{P}(G < a) + O_{\eta,K}(\varepsilon) + O((1 - \varphi(-K\varepsilon))\rho).$$

A similar argument gives a lower bound

$$\mathbf{P}(Z_n < a) \ge \mathbf{P}(G < a) - O_{\eta,K}(\varepsilon) - O(\varphi(K\varepsilon)\rho),$$

and so

$$|\mathbf{P}(Z_n < a) - \mathbf{P}(G < a)| \le O_{\eta,K}(\varepsilon) + O((1 - \varphi(-K\varepsilon))\rho) + O(\varphi(K\varepsilon)\rho).$$

Taking suprema over a, we obtain

$$\rho \le O_{\eta,K}(\varepsilon) + O((1 - \varphi(-K\varepsilon))\rho) + O(\varphi(K\varepsilon)\rho).$$

If K is large enough (depending on c), we can make $1 - \varphi(-K\varepsilon)$ and $\varphi(K\varepsilon)$ small, and thus absorb the latter two terms on the right-hand side into the left-hand side. This gives the desired bound $\rho = O(\varepsilon)$.

It remains to establish (2.38). Applying (2.33), it suffices to show that

$$(2.40) \qquad |\int_{\mathbf{R}} \hat{\varphi}(t)(F_{Z_n}(t) - F_G(t))\, dt| \le O(\varepsilon).$$

Now we estimate each of the various expressions. Standard Fourier-analytic computations show that

$$\hat{\varphi}(t) = \hat{1}_{(-\infty,a]}(t)\hat{\eta}(t/\varepsilon)$$

and that

$$\hat{1}_{(-\infty,a]}(t) = O(\frac{1}{1 + |t|}).$$

Since $\hat{\eta}$ was supported in $[-c, c]$, it suffices to show that

$$(2.41) \qquad \int_{|t| \le c/\varepsilon} \frac{|F_{Z_n}(t) - F_G(t)|}{1 + |t|}\, dt \le O(\varepsilon).$$

From Taylor expansion we have

$$e^{itX} = 1 + itX - \frac{t^2}{2}X^2 + O(|t|^3|X|^3)$$

for any t; taking expectations and using the definition of ε we have

$$F_X(t) = 1 - t^2/2 + O(\varepsilon\sqrt{n}|t|^3)$$

and, in particular,

$$F_X(t) = \exp(-t^2/2 + O(\varepsilon\sqrt{n}|t|^3))$$

if $|t| \leq c/\mathbf{E}|X|^3$ and c is small enough. Applying (2.36), we conclude that

$$F_{Z_n}(t) = \exp(-t^2/2 + O(\varepsilon|t|^3))$$

if $|t| \leq c\varepsilon$. Meanwhile, from Exercise 2.2.5 we have $F_G(t) = \exp(-t^2/2)$. Elementary calculus then gives us

$$|F_{Z_n}(t) - F_G(t)| \leq O(\varepsilon|t|^3 \exp(-t^2/4))$$

(say) if c is small enough. Inserting this bound into (2.41) we obtain the claim. $\qquad\square$

Exercise 2.2.16. Show that the error terms here are sharp (up to constants) when X is a signed Bernoulli random variable.

2.2.3. The moment method. The above Fourier-analytic proof of the central limit theorem is one of the quickest (and slickest) proofs available for this theorem, and is accordingly the "standard" proof given in probability textbooks. However, it relies quite heavily on the Fourier-analytic identities in Exercise 2.2.12, which in turn are extremely dependent on both the commutative nature of the situation (as it uses the identity $e^{A+B} = e^A e^B$) and on the independence of the situation (as it uses identities of the form $\mathbf{E}(e^A e^B) = (\mathbf{E}e^A)(\mathbf{E}e^B)$). When we turn to random matrix theory, we will often lose (or be forced to modify) one or both of these properties, which can make it much more difficult (though not always impossible[11]) to apply Fourier analysis. Because of this, it is also important to look for non-Fourier based methods to prove results such as the central limit theorem. These methods often lead to proofs that are lengthier and more technical than the Fourier proofs, but also tend to be more robust, and in particular, can often be extended to random matrix theory situations. Thus both the Fourier and non-Fourier proofs will be of importance in this text.

The most elementary (but still remarkably effective) method available in this regard is the *moment method*, which we have already used in Section 2.1. This method to understand the distribution of a random variable X via its moments X^k. In principle, this method is equivalent to the Fourier method, through the identity (2.32); but in practice, the moment method proofs tend to look somewhat different than the Fourier-analytic ones, and it is often more apparent how to modify them to non-independent or non-commutative settings.

---

[11]See, in particular, [**Ly2009**] for a recent successful application of Fourier-analytic methods to random matrix theory.

We first need an analogue of the Lévy continuity theorem. Here we encounter a technical issue: whereas the Fourier phases $x \mapsto e^{itx}$ were bounded, the moment functions $x \mapsto x^k$ become unbounded at infinity. However, one can deal with this issue as long as one has sufficient decay:

Theorem 2.2.9 (Carleman continuity theorem). *Let X_n be a sequence of uniformly sub-Gaussian real random variables, and let X be another sub-Gaussian random variable. Then the following statements are equivalent:*

(i) *For every $k = 0, 1, 2, \ldots$, $\mathbf{E}X_n^k$ converges pointwise to $\mathbf{E}X^k$.*

(ii) *X_n converges in distribution to X.*

Proof. We first show how (ii) implies (i). Let $N > 0$ be a truncation parameter, and let $\varphi : \mathbf{R} \to \mathbf{R}$ be a smooth function that equals 1 on $[-1, 1]$ and vanishes outside of $[-2, 2]$. Then for any k, the convergence in distribution implies that $\mathbf{E}X_n^k\varphi(X_n/N)$ converges to $\mathbf{E}X^k\varphi(X/N)$. On the other hand, from the uniform sub-Gaussian hypothesis, one can make $\mathbf{E}X_n^k(1 - \varphi(X_n/N))$ and $\mathbf{E}X^k(1 - \varphi(X/N))$ arbitrarily small for fixed k by making N large enough. Summing, and then letting N go to infinity, we obtain (i).

Conversely, suppose (i) is true. From the uniform sub-Gaussian hypothesis, the X_n have $(k + 1)^{\text{st}}$ moment bounded by $(Ck)^{k/2}$ for all $k \geq 1$ and some C independent of k (see Exercise 1.1.4). From Taylor's theorem with remainder (and Stirling's formula, Section 1.2) we conclude

$$F_{X_n}(t) = \sum_{j=0}^{k} \frac{(it)^j}{j!}\mathbf{E}X_n^j + O((Ck)^{-k/2}|t|^{k+1})$$

uniformly in t and n. Similarly for X. Taking limits using (i) we see that

$$\limsup_{n \to \infty} |F_{X_n}(t) - F_X(t)| = O((Ck)^{-k/2}|t|^{k+1}).$$

Then letting $k \to \infty$, keeping t fixed, we see that $F_{X_n}(t)$ converges pointwise to $F_X(t)$ for each t, and the claim now follows from the Lévy continuity theorem. $\qquad\square$

Remark 2.2.10. One corollary of Theorem 2.2.9 is that the distribution of a sub-Gaussian random variable is uniquely determined by its moments (actually, this could already be deduced from Exercise 2.2.9 and Remark 2.2.5). The situation can fail for distributions with slower tails, for much the same reason that a smooth function is not determined by its derivatives at one point if that function is not analytic.

The Fourier inversion formula provides an easy way to recover the distribution from the characteristic function. Recovering a distribution from its moments is more difficult, and sometimes requires tools such as analytic continuation; this problem is known as the *inverse moment problem* and will not be discussed here.

To prove the central limit theorem, we know from the truncation method that we may assume, without loss of generality, that X is bounded (and, in particular, sub-Gaussian); we may also normalise X to have mean zero and unit variance. From the Chernoff bound (2.28) we know that the Z_n are uniformly sub-Gaussian; so by Theorem 2.2.9, it suffices to show that

$$\mathbf{E}Z_n^k \to \mathbf{E}G^k$$

for all $k = 0, 1, 2, \ldots$, where $G \equiv N(0, 1)_{\mathbf{R}}$ is a standard Gaussian variable.

The moments $\mathbf{E}G^k$ are easy to compute:

Exercise 2.2.17. Let k be a natural number, and let $G \equiv N(0, 1)_{\mathbf{R}}$. Show that $\mathbf{E}G^k$ vanishes when k is odd, and equal to $\frac{k!}{2^{k/2}(k/2)!}$ when k is even. (*Hint:* This can either be done directly by using the *Gamma function*, or by using Exercise 2.2.5 and Exercise 2.2.9.)

So now we need to compute $\mathbf{E}Z_n^k$. Using (2.29) and linearity of expectation, we can expand this as

$$\mathbf{E}Z_n^k = n^{-k/2} \sum_{1 \leq i_1, \ldots, i_k \leq n} \mathbf{E}X_{i_1} \ldots X_{i_k}.$$

To understand this expression, let us first look at some small values of k.

(i) For $k = 0$, this expression is trivially 1.

(ii) For $k = 1$, this expression is trivially 0, thanks to the mean zero hypothesis on X.

(iii) For $k = 2$, we can split this expression into the diagonal and off-diagonal components:

$$n^{-1} \sum_{1 \leq i \leq n} \mathbf{E}X_i^2 + n^{-1} \sum_{1 \leq i < j \leq n} \mathbf{E}2X_i X_j.$$

Each summand in the first sum is 1, as X has unit variance. Each summand in the second sum is 0, as the X_i have mean zero and are independent. So the second moment $\mathbf{E}Z_n^2$ is 1.

(iv) For $k = 3$, we have a similar expansion

$$n^{-3/2} \sum_{1 \leq i \leq n} \mathbf{E}X_i^3 + n^{-3/2} \sum_{1 \leq i < j \leq n} \mathbf{E}3X_i^2 X_j + 3X_i X_j^2$$
$$+ n^{-3/2} \sum_{1 \leq i < j < k \leq n} \mathbf{E}6X_i X_j X_k.$$

The summands in the latter two sums vanish because of the (joint) independence and mean zero hypotheses. The summands in the first sum need not vanish, but are $O(1)$, so the first term is $O(n^{-1/2})$, which is asymptotically negligible, so the third moment $\mathbf{E}Z_n^3$ goes to 0.

(v) For $k = 4$, the expansion becomes quite complicated:

$$n^{-2} \sum_{1 \le i \le n} \mathbf{E}X_i^4 + n^{-2} \sum_{1 \le i < j \le n} \mathbf{E}4X_i^3 X_j + 6X_i^2 X_j^2 + 4X_i X_j^3$$

$$+ n^{-2} \sum_{1 \le i < j < k \le n} \mathbf{E}12X_i^2 X_j X_k + 12X_i X_j^2 X_k + 12X_i X_j X_k^2$$

$$+ n^{-2} \sum_{1 \le i < j < k < l \le n} \mathbf{E}24X_i X_j X_k X_l.$$

Again, most terms vanish, except for the first sum, which is $O(n^{-1})$ and is asymptotically negligible, and the sum $n^{-2} \sum_{1 \le i < j \le n} \mathbf{E}6X_i^2 X_j^2$, which by the independence and unit variance assumptions works out to $n^{-2} 6\binom{n}{2} = 3 + o(1)$. Thus the fourth moment $\mathbf{E}Z_n^4$ goes to 3 (as it should).

Now we tackle the general case. Ordering the indices $i_1, \ldots, i_k$ as $j_1 < \cdots < j_m$ for some $1 \le m \le k$, with each j_r occurring with multiplicity $a_r \ge 1$ and using elementary enumerative combinatorics, we see that $\mathbf{E}Z_n^k$ is the sum of all terms of the form

$$(2.42) \qquad n^{-k/2} \sum_{1 \le j_1 < \cdots < j_m \le n} c_{k,a_1,\ldots,a_m} \mathbf{E}X_{j_1}^{a_1} \ldots X_{j_m}^{a_m}$$

where $1 \le m \le k$, $a_1, \ldots, a_m$ are positive integers adding up to k, and $c_{k,a_1,\ldots,a_m}$ is the multinomial coefficient

$$c_{k,a_1,\ldots,a_m} := \frac{k!}{a_1! \ldots a_m!}.$$

The total number of such terms depends only on k. More precisely, it is 2^{k-1} (exercise!), though we will not need this fact.

As we already saw from the small k examples, most of the terms vanish, and many of the other terms are negligible in the limit $n \to \infty$. Indeed, if any of the a_r are equal to 1, then every summand in (2.42) vanishes, by joint independence and the mean zero hypothesis. Thus, we may restrict attention to those expressions (2.42) for which all the a_r are at least 2. Since the a_r sum up to k, we conclude that m is at most $k/2$.

On the other hand, the total number of summands in (2.42) is clearly at most n^m (in fact it is $\binom{n}{m}$), and the summands are bounded (for fixed k) since X is bounded. Thus, if m is *strictly* less than $k/2$, then the expression in (2.42) is $O(n^{m-k/2})$ and goes to zero as $n \to \infty$. So, asymptotically, the only terms (2.42) which are still relevant are those for which m is *equal* to $k/2$. This already shows that $\mathbf{E}Z_n^k$ goes to zero when k is odd. When k is even, the only surviving term in the limit is now when $m = k/2$ and $a_1 = \cdots = a_m = 2$. But then by independence and unit variance, the expectation in (2.42) is 1, and so this term is equal to

$$n^{-k/2} \binom{n}{m} c_{k,2,\ldots,2} = \frac{1}{(k/2)!} \frac{k!}{2^{k/2}} + o(1),$$

and the main term is happily equal to the moment $\mathbf{E}G^k$ as computed in Exercise 2.2.17.

2.2.4. The Lindeberg swapping trick. The moment method proof of the central limit theorem that we just gave consisted of four steps:

(i) (Truncation and normalisation step) A reduction to the case when X was bounded with zero mean and unit variance.

(ii) (Inverse moment step) A reduction to a computation of asymptotic moments $\lim_{n \to \infty} \mathbf{E}Z_n^k$.

(iii) (Analytic step) Showing that most terms in the expansion of this asymptotic moment were zero, or went to zero as $n \to \infty$.

(iv) (Algebraic step) Using enumerative combinatorics to compute the remaining terms in the expansion.

In this particular case, the enumerative combinatorics was very classical and easy; it was basically asking for the number of ways one can place k balls in m boxes, so that the rth box contains a_r balls, and the answer is well known to be given by the multinomial $\frac{k!}{a_1!\ldots a_r!}$. By a small algebraic miracle, this result matched up nicely with the computation of the moments of the Gaussian $N(0,1)_{\mathbf{R}}$.

However, when we apply the moment method to more advanced problems, the enumerative combinatorics can become more non-trivial, requiring a fair amount of combinatorial and algebraic computation. The algebraic miracle that occurs at the end of the argument can then seem like a very fortunate but inexplicable coincidence, making the argument somehow unsatisfying despite being rigorous.

In [**Li1922**], Lindeberg observed that there was a very simple way to *decouple* the algebraic miracle from the analytic computations, so that all relevant algebraic identities only need to be verified in the special case of *Gaussian* random variables, in which everything is much easier to compute. This *Lindeberg swapping trick* (or *Lindeberg replacement trick*) will be very useful in the later theory of random matrices, so we pause to give it here in the simple context of the central limit theorem.

The basic idea is as follows. We repeat the truncation-and-normalisation and inverse moment steps in the preceding argument. Thus, $X_1, \ldots, X_n$ are iid copies of a bounded real random variable X of mean zero and unit variance, and we wish to show that $\mathbf{E}Z_n^k \to \mathbf{E}G^k$, where $G \equiv N(0,1)_{\mathbf{R}}$, where $k \geq 0$ is fixed.

Now let $Y_1, \ldots, Y_n$ be iid copies of the Gaussian itself: $Y_1, \ldots, Y_n \equiv N(0,1)_{\mathbf{R}}$. Because the sum of independent Gaussians is again a Gaussian (Exercise 2.1.9), we see that the random variable

$$W_n := \frac{Y_1 + \cdots + Y_n}{\sqrt{n}}$$

already has the same distribution as G: $W_n \equiv G$. Thus, it suffices to show that

$$\mathbf{E}Z_n^k = \mathbf{E}W_n^k + o(1).$$

Now we perform the analysis part of the moment method argument again. We can expand $\mathbf{E}Z_n^k$ into terms (2.42) as before, and discard all terms except for the $a_1 = \cdots = a_m = 2$ term as being $o(1)$. Similarly, we can expand $\mathbf{E}W_n^k$ into very similar terms (but with the X_i replaced by Y_i) and again discard all but the $a_1 = \cdots = a_m$ term.

But by hypothesis, the second moments of X and Y match: $\mathbf{E}X^2 = \mathbf{E}Y^2 = 1$. Thus, by joint independence, the $a_1 = \cdots = a_m = 2$ term (2.42) for X is exactly equal to that of Y. And the claim follows.

This is almost exactly the same proof as in the previous section, but note that we did *not* need to compute the multinomial coefficient $c_{k,a_1,\ldots,a_m}$, nor did we need to verify the miracle that this coefficient matched (up to normalising factors) to the moments of the Gaussian. Instead, we used the much more mundane "miracle" that the sum of independent Gaussians was again a Gaussian.

To put it another way, the Lindeberg replacement trick factors a universal limit theorem, such as the central limit theorem, into two components:

(i) a *universality* or *invariance* result, which shows that the distribution (or other statistics, such as moments) of some random variable $F(X_1, \ldots, X_n)$ is asymptotically unchanged in the limit $n \to \infty$ if each of the input variables X_i are replaced by a Gaussian substitute Y_i; and

(ii) the *Gaussian case*, which computes the asymptotic distribution (or other statistic) of $F(Y_1, \ldots, Y_n)$ in the case when $Y_1, \ldots, Y_n$ are all Gaussians.

The former type of result tends to be entirely analytic in nature (basically, one just needs to show that all error terms that show up when swapping X

with Y add up to $o(1)$), while the latter type of result tends to be entirely algebraic in nature (basically, one just needs to exploit the many pleasant algebraic properties of Gaussians). This decoupling of the analysis and algebra steps tends to simplify the argument both at a technical level and at a conceptual level, as each step then becomes easier to understand than the whole.

2.2.5. Individual swapping. In the above argument, we swapped all the original input variables $X_1, \ldots, X_n$ with Gaussians $Y_1, \ldots, Y_n$ *en masse*. There is also a variant of the Lindeberg trick in which the swapping is done *individually*. To illustrate the individual swapping method, let us use it to show the following weak version of the Berry-Esséen theorem:

Theorem 2.2.11 (Berry-Esséen theorem, weak form). *Let X have mean zero, unit variance, and finite third moment, and let φ be smooth with uniformly bounded derivatives up to third order. Let $Z_n := (X_1 + \cdots + X_n)/\sqrt{n}$, where $X_1, \ldots, X_n$ are iid copies of X. Then we have*

$$(2.43) \qquad \mathbf{E}\varphi(Z_n) = \mathbf{E}\varphi(G) + O(\frac{1}{\sqrt{n}}(\mathbf{E}|X|^3) \sup_{x \in \mathbf{R}} |\varphi'''(x)|)$$

where $G \equiv N(0,1)_{\mathbf{R}}$.

Proof. Let $Y_1, \ldots, Y_n$ and W_n be in the previous section. As $W_n \equiv G$, it suffices to show that

$$\mathbf{E}\varphi(Z_n) - \varphi(W_n) = o(1).$$

We telescope this (using linearity of expectation) as

$$\mathbf{E}\varphi(Z_n) - \varphi(W_n) = -\sum_{i=0}^{n-1} \mathbf{E}\varphi(Z_{n,i}) - \varphi(Z_{n,i+1})$$

where

$$Z_{n,i} := \frac{X_1 + \cdots + X_i + Y_{i+1} + \cdots + Y_n}{\sqrt{n}}$$

is a partially swapped version of Z_n. So it will suffice to show that

$$\mathbf{E}\varphi(Z_{n,i}) - \varphi(Z_{n,i+1}) = O((\mathbf{E}|X|^3) \sup_{x \in \mathbf{R}} |\varphi'''(x)|/n^{3/2})$$

uniformly for $0 \le i < n$.

We can write $Z_{n,i} = S_{n,i} + Y_{i+1}/\sqrt{n}$ and $Z_{n,i+1} = S_{n,i} + X_{i+1}/\sqrt{n}$, where

$$(2.44) \qquad S_{n,i} := \frac{X_1 + \cdots + X_i + Y_{i+2} + \cdots + Y_n}{\sqrt{n}}.$$

To exploit this, we use Taylor expansion with remainder to write

$$\varphi(Z_{n,i}) = \varphi(S_{n,i}) + \varphi'(S_{n,i})Y_{i+1}/\sqrt{n}$$
$$+ \frac{1}{2}\varphi''(S_{n,i})Y_{i+1}^2/n + O(|Y_{i+1}|^3/n^{3/2} \sup_{x\in\mathbf{R}} |\varphi'''(x)|)$$

and

$$\varphi(Z_{n,i+1}) = \varphi(S_{n,i}) + \varphi'(S_{n,i})X_{i+1}/\sqrt{n}$$
$$+ \frac{1}{2}\varphi''(S_{n,i})X_{i+1}^2/n + O(|X_{i+1}|^3/n^{3/2} \sup_{x\in\mathbf{R}} |\varphi'''(x)|)$$

where the implied constants depend on φ but not on n. Now, by construction, the moments of X_{i+1} and Y_{i+1} match to second order, thus

$$\mathbf{E}\varphi(Z_{n,i}) - \varphi(Z_{n,i+1}) = O(\mathbf{E}|Y_{i+1}|^3 \sup_{x\in\mathbf{R}} |\varphi'''(x)|/n^{3/2})$$
$$+ O(\mathbf{E}|X_{i+1}|^3 \sup_{x\in\mathbf{R}} |\varphi'''(x)|/n^{3/2}),$$

and the claim follows[12].   □

Remark 2.2.12. The above argument relied on Taylor expansion, and the hypothesis that the moments of X and Y matched to second order. It is not hard to see that if we assume more moments matching (e.g., $\mathbf{E}X^3 = \mathbf{E}Y^3 = 3$), and more smoothness on φ, we see that we can improve the $\frac{1}{\sqrt{n}}$ factor on the right-hand side. Thus we see that we expect swapping methods to become more powerful when more moments are matching. This is the guiding philosophy behind the *four moment theorem* [**TaVu2009b**], which (very) roughly speaking asserts that the spectral statistics of two random matrices are asymptotically indistinguishable if their coefficients have matching moments to fourth order. Unfortunately, due to reasons of space, we will not be able to cover the four moment theorem in detail in this text.

Theorem 2.2.11 is easily implied by Theorem 2.2.8 and an integration by parts. In the reverse direction, let us see what Theorem 2.2.11 tells us about the cumulative distribution function

$$\mathbf{P}(Z_n < a)$$

of Z_n. For any $\varepsilon > 0$, one can upper bound this expression by

$$\mathbf{E}\varphi(Z_n)$$

where φ is a smooth function equal to 1 on $(-\infty, a]$ that vanishes outside of $(-\infty, a + \varepsilon]$, and has third derivative $O(\varepsilon^{-3})$. By Theorem 2.2.11, we thus

---

[12]Note from Hölder's inequality that $\mathbf{E}|X|^3 \geq 1$.

have

$$\mathbf{P}(Z_n < a) \le \mathbf{E}\varphi(G) + O(\frac{1}{\sqrt{n}}(\mathbf{E}|X|^3)\varepsilon^{-3}).$$

On the other hand, as G has a bounded probability density function, we have

$$\mathbf{E}\varphi(G) = \mathbf{P}(G < a) + O(\varepsilon)$$

and so

$$\mathbf{P}(Z_n < a) \le \mathbf{P}(G < a) + O(\varepsilon) + O(\frac{1}{\sqrt{n}}(\mathbf{E}|X|^3)\varepsilon^{-3}).$$

A very similar argument gives the matching lower bound, thus

$$\mathbf{P}(Z_n < a) = \mathbf{P}(G < a) + O(\varepsilon) + O(\frac{1}{\sqrt{n}}(\mathbf{E}|X|^3)\varepsilon^{-3}).$$

Optimising in ε we conclude that

$$(2.45) \qquad \mathbf{P}(Z_n < a) = \mathbf{P}(G < a) + O(\frac{1}{\sqrt{n}}(\mathbf{E}|X|^3))^{1/4}.$$

Comparing this with Theorem 2.2.8 we see that the decay exponent in n in the error term has degraded by a factor of $1/4$. In our applications to random matrices, this type of degradation is acceptable, and so the swapping argument is a reasonable substitute for the Fourier-analytic one in this case. Also, this method is quite robust, and in particular, extends well to higher dimensions; we will return to this point in later sections, but see, for instance, [**TaVuKr2010**, Appendix D] for an example of a multidimensional Berry-Esséen theorem proven by this method.

On the other hand, there is another method that can recover this loss while still avoiding Fourier-analytic techniques; we turn to this topic next.

2.2.6. Stein's method. *Stein's method*, introduced by Charles Stein [**St1970**], is a powerful method to show convergence in distribution to a special distribution, such as the Gaussian. In several recent papers, this method has been used to control several expressions of interest in random matrix theory (e.g., the distribution of moments, or of the Stieltjes transform.) We will not use Stein's method in this text, but the method is of independent interest nonetheless.

The probability density function $\rho(x) := \frac{1}{\sqrt{2\pi}}e^{-x^2/2}$ of the standard normal distribution $N(0,1)_{\mathbf{R}}$ can be viewed as a solution to the ordinary differential equation

$$(2.46) \qquad \rho'(x) + x\rho(x) = 0.$$

One can take adjoints of this, and conclude (after an integration by parts) that ρ obeys the integral identity

$$\int_{\mathbf{R}} \rho(x)(f'(x) - xf(x))\, dx = 0$$

for any continuously differentiable f with both f and f' bounded (one can relax these assumptions somewhat). To put it another way, if $G \equiv N(0,1)$, then we have

(2.47) $\mathbf{E}f'(G) - Gf(G) = 0$

whenever f is continuously differentiable with f, f' both bounded.

It turns out that the converse is true: if X is a real random variable with the property that

$$\mathbf{E}f'(X) - Xf(X) = 0$$

whenever f is continuously differentiable with f, f' both bounded, then X is Gaussian. In fact, more is true, in the spirit of Theorem 2.2.4 and Theorem 2.2.9:

Theorem 2.2.13 (Stein continuity theorem). *Let X_n be a sequence of real random variables with uniformly bounded second moment, and let $G \equiv N(0,1)$. Then the following are equivalent:*

(i) $\mathbf{E}f'(X_n) - X_n f(X_n)$ *converges to zero whenever $f : \mathbf{R} \to \mathbf{R}$ is continuously differentiable with f, f' both bounded.*

(ii) X_n *converges in distribution to G.*

Proof. To show that (ii) implies (i), it is not difficult to use the uniform bounded second moment hypothesis and a truncation argument to show that $\mathbf{E}f'(X_n) - X_n f(X_n)$ converges to $\mathbf{E}f'(G) - Gf(G)$ when f is continuously differentiable with f, f' both bounded, and the claim then follows from (2.47).

Now we establish the converse. It suffices to show that

$$\mathbf{E}\varphi(X_n) - \mathbf{E}\varphi(G) \to 0$$

whenever $\varphi : \mathbf{R} \to \mathbf{R}$ is a bounded continuous function. We may normalise φ to be bounded in magnitude by 1.

Trivially, the function $\varphi(\cdot) - \mathbf{E}\varphi(G)$ has zero expectation when one substitutes G for the argument $\cdot$; thus

(2.48) $\dfrac{1}{\sqrt{2\pi}} \displaystyle\int_{-\infty}^{\infty} e^{-y^2/2}(\varphi(y) - \mathbf{E}\varphi(G))\, dy = 0.$

Comparing this with (2.47), one may thus hope to find a representation of the form

$$(2.49) \qquad \varphi(x) - \mathbf{E}\varphi(G) = f'(x) - xf(x)$$

for some continuously differentiable f with f, f' both bounded. This is a simple ODE and can be easily solved (by the method of integrating factors) to give a solution f, namely

$$(2.50) \qquad f(x) := e^{x^2/2} \int_{-\infty}^{x} e^{-y^2/2}(\varphi(y) - \mathbf{E}\varphi(G)) \, dy.$$

(One could dub f the *Stein transform* of φ, although this term does not seem to be in widespread use.) By the fundamental theorem of calculus, f is continuously differentiable and solves (2.49). Using (2.48), we may also write f as

$$(2.51) \qquad f(x) := -e^{x^2/2} \int_{x}^{\infty} e^{-y^2/2}(\varphi(y) - \mathbf{E}\varphi(G)) \, dy.$$

By completing the square, we see that $e^{-y^2/2} \leq e^{-x^2/2}e^{-x(y-x)}$. Inserting this into (2.50) and using the bounded nature of φ, we conclude that $f(x) = O_\varphi(1/|x|)$ for $x < -1$; inserting it instead into (2.51), we have $f(x) = O_\varphi(1/|x|)$ for $x > 1$. Finally, easy estimates give $f(x) = O_\varphi(1)$ for $|x| \leq 1$. Thus for all x we have

$$f(x) = O_\varphi(\frac{1}{1 + |x|})$$

which, when inserted back into (2.49), gives the boundedness of f' (and also of course gives the boundedness of f). In fact, if we rewrite (2.51) as

$$f(x) := -\int_{0}^{\infty} e^{-s^2/2}e^{-sx}(\varphi(x + s) - \mathbf{E}\varphi(G)) \, ds,$$

we see on differentiation under the integral sign (and using the Lipschitz nature of φ) that $f'(x) = O_\varphi(1/x)$ for $x > 1$; a similar manipulation (starting from (2.50)) applies for $x < -1$, and we in fact conclude that $f'(x) = O_\varphi(\frac{1}{1+|x|})$ for all x.

Applying (2.49) with $x = X_n$ and taking expectations, we have

$$\varphi(X_n) - \mathbf{E}\varphi(G) = f'(X_n) - X_n f(X_n).$$

By the hypothesis (i), the right-hand side goes to zero, hence the left-hand side does also, and the claim follows. □

The above theorem gave only a qualitative result (convergence in distribution), but the proof is quite quantitative, and can be used, in particular, to give Berry-Esséen type results. To illustrate this, we begin with a strengthening of Theorem 2.2.11 that reduces the number of derivatives of φ that need to be controlled:

Theorem 2.2.14 (Berry-Esséen theorem, less weak form). *Let X have mean zero, unit variance, and finite third moment, and let φ be smooth, bounded in magnitude by 1, and Lipschitz. Let $Z_n := (X_1 + \cdots + X_n)/\sqrt{n}$, where $X_1, \ldots, X_n$ are iid copies of X. Then we have*

$$(2.52) \qquad \mathbf{E}\varphi(Z_n) = \mathbf{E}\varphi(G) + O(\frac{1}{\sqrt{n}}(\mathbf{E}|X|^3)(1 + \sup_{x \in \mathbf{R}} |\varphi'(x)|))$$

where $G \equiv N(0,1)_{\mathbf{R}}$.

Proof. Set $A := 1 + \sup_{x \in \mathbf{R}} |\varphi'(x)|$.

Let f be the Stein transform (2.50) of φ, then by (2.49) we have

$$\mathbf{E}\varphi(Z_n) - \mathbf{E}\varphi(G) = \mathbf{E}f'(Z_n) - Z_n f(Z_n).$$

We expand $Z_n f(Z_n) = \frac{1}{\sqrt{n}} \sum_{i=1}^n X_i f(Z_n)$. For each i, we then split $Z_n = Z_{n;i} + \frac{1}{\sqrt{n}} X_i$, where $Z_{n;i} := (X_1 + \cdots + X_{i-1} + X_{i+1} + \cdots + X_n)/\sqrt{n}$ (cf. (2.44)). By the fundamental theorem of calculus, we have

$$\mathbf{E}X_i f(Z_n) = \mathbf{E}X_i f(Z_{n;i}) + \frac{1}{\sqrt{n}} X_i^2 f'(Z_{n;i} + \frac{t}{\sqrt{n}} X_i)$$

where t is uniformly distributed in $[0,1]$ and independent of all of the $X_1, \ldots, X_n$. Now observe that X_i and $Z_{n;i}$ are independent, and X_i has mean zero, so the first term on the right-hand side vanishes. Thus

$$(2.53) \qquad \mathbf{E}\varphi(Z_n) - \mathbf{E}\varphi(G) = \frac{1}{n} \sum_{i=1}^n \mathbf{E}f'(Z_n) - X_i^2 f'(Z_{n;i} + \frac{t}{\sqrt{n}} X_i).$$

Another application of independendence gives

$$\mathbf{E}f'(Z_{n;i}) = \mathbf{E}X_i^2 f'(Z_{n;i})$$

so we may rewrite (2.53) as

$$\frac{1}{n} \sum_{i=1}^n \mathbf{E}(f'(Z_n) - f'(Z_{n;i})) - X_i^2(f'(Z_{n;i} + \frac{t}{\sqrt{n}} X_i) - f'(Z_{n;i})).$$

Recall from the proof of Theorem 2.2.13 that $f(x) = O(1/(1 + |x|))$ and $f'(x) = O(A/(1 + |x|))$. By the product rule, this implies that $xf(x)$ has a Lipschitz constant of $O(A)$. Applying (2.49) and the definition of A, we conclude that f' has a Lipschitz constant of $O(A)$. Thus we can bound the previous expression as

$$\frac{1}{n} \sum_{i=1}^n \mathbf{E}\frac{1}{\sqrt{n}} O(A|X_i| + A|X_i|^3)$$

and the claim follows from Hölder's inequality. $\square$

This improvement already partially restores the exponent in (2.45) from
1/4 to 1/2. But one can do better still by pushing the arguments further.
Let us illustrate this in the model case when the X_i not only have bounded
third moment, but are in fact bounded:

Theorem 2.2.15 (Berry-Esséen theorem, bounded case). *Let X have mean
zero, unit variance, and be bounded by $O(1)$. Let $Z_n := (X_1 + \cdots + X_n)/\sqrt{n}$,
where $X_1, \ldots, X_n$ are iid copies of X. Then we have*

$$(2.54) \qquad \mathbf{P}(Z_n < a) = \mathbf{P}(G < a) + O(\frac{1}{\sqrt{n}})$$

whenever $a = O(1)$, where $G \equiv N(0,1)_{\mathbf{R}}$.

Proof. Write $\phi := 1_{(-\infty, a]}$, thus we seek to show that

$$\mathbf{E}\phi(Z_n) - \phi(G) = O(\frac{1}{\sqrt{n}}).$$

Let f be the Stein transform (2.50) of ϕ. The function ϕ is not continuous,
but it is not difficult to see (e.g., by a limiting argument) that we still have
the estimates $f(x) = O(1/(1+|x|))$ and $f'(x) = O(1)$ (in a weak sense), and
that xf has a Lipschitz norm of $O(1)$ (here we use the hypothesis $a = O(1)$).
A similar limiting argument gives

$$\mathbf{E}\phi(Z_n) - \phi(G) = \mathbf{E}f'(Z_n) - Z_n f(Z_n)$$

and by arguing as in the proof of Theorem 2.2.14, we can write the right-
hand side as

$$\frac{1}{n} \sum_{i=1}^{n} \mathbf{E}(f'(Z_n) - f'(Z_{n;i})) - X_i^2 (f'(Z_{n;i} + \frac{t}{\sqrt{n}}X_i) - f'(Z_{n;i})).$$

From (2.49), f' is equal to ϕ, plus a function with Lipschitz norm $O(1)$.
Thus, we can write the above expression as

$$\frac{1}{n} \sum_{i=1}^{n} \mathbf{E}(\phi(Z_n) - \phi(Z_{n;i})) - X_i^2 (\phi(Z_{n;i} + \frac{t}{\sqrt{n}}X_i) - \phi(Z_{n;i})) + O(1/\sqrt{n}).$$

The $\phi(Z_{n;i})$ terms cancel (due to the independence of X_i and $Z_{n;i}$, and the
normalised mean and variance of X_i), so we can simplify this as

$$\mathbf{E}\phi(Z_n) - \frac{1}{n} \sum_{i=1}^{n} \mathbf{E}X_i^2 \phi(Z_{n;i} + \frac{t}{\sqrt{n}}X_i)$$

and so we conclude that

$$\frac{1}{n} \sum_{i=1}^{n} \mathbf{E}X_i^2 \phi(Z_{n;i} + \frac{t}{\sqrt{n}}X_i) = \mathbf{E}\phi(G) + O(1/\sqrt{n}).$$

Since t and X_i are bounded, and ϕ is non-increasing, we have

$$\phi(Z_{n;i} + O(1/\sqrt{n})) \le \phi(Z_{n;i} + \frac{t}{\sqrt{n}}X_i) \le \phi(Z_{n;i} - O(1/\sqrt{n}));$$

applying the second inequality and using independence to once again eliminate the X_i^2 factor, we see that

$$\frac{1}{n}\sum_{i=1}^{n}\mathbf{E}\phi(Z_{n;i} - O(1/\sqrt{n})) \ge \mathbf{E}\phi(G) + O(1/\sqrt{n})$$

which implies (by another appeal to the non-increasing nature of ϕ and the bounded nature of X_i) that

$$\mathbf{E}\phi(Z_n - O(1/\sqrt{n})) \ge \mathbf{E}\phi(G) + O(1/\sqrt{n})$$

or, in other words, that

$$\mathbf{P}(Z_n \le a + O(1/\sqrt{n})) \ge \mathbf{P}(G \le a) + O(1/\sqrt{n}).$$

Similarly, using the lower bound inequalities, one has

$$\mathbf{P}(Z_n \le a - O(1/\sqrt{n})) \le \mathbf{P}(G \le a) + O(1/\sqrt{n}).$$

Moving a up and down by $O(1/\sqrt{n})$, and using the bounded density of G, we obtain the claim. $\square$

Actually, one can use Stein's method to obtain the full Berry-Esséen theorem, but the computations get somewhat technical, requiring an induction on n to deal with the contribution of the exceptionally large values of X_i; see [**BaHa1984**].

2.2.7. Predecessor comparison. Suppose one had never heard of the normal distribution, but one still suspected the existence of the central limit theorem; thus, one thought that the sequence Z_n of normalised distributions was converging in distribution to something, but was unsure what the limit was. Could one still work out what that limit was?

Certainly in the case of Bernoulli distributions, one could work explicitly using Stirling's formula (see Exercise 2.2.2), and the Fourier-analytic method would also eventually work. Let us now give a third way to (heuristically) derive the normal distribution as the limit of the central limit theorem. The idea is to compare Z_n with its predecessor Z_{n-1}, using the recursive formula

$$(2.55) \qquad\qquad Z_n = \frac{\sqrt{n-1}}{\sqrt{n}}Z_{n-1} + \frac{1}{\sqrt{n}}X_n$$

(normalising X_n to have mean zero and unit variance as usual; let us also truncate X_n to be bounded, for simplicity). Let us hypothesise that Z_n and Z_{n-1} are approximately the same distribution; let us also conjecture that this distribution is absolutely continuous, given as $\rho(x)\,dx$ for some smooth

$\rho(x)$. (If we secretly knew the central limit theorem, we would know that $\rho(x)$ is in fact $\frac{1}{\sqrt{2\pi}}e^{-x^2/2}$, but let us pretend that we did not yet know this fact.) Thus, for any test function φ, we expect

$$(2.56) \qquad \mathbf{E}\varphi(Z_n) \approx \mathbf{E}\varphi(Z_{n-1}) \approx \int_{\mathbf{R}} \varphi(x)\rho(x) \, dx.$$

Now let us try to combine this with (2.55). We assume φ to be smooth, and Taylor expand to third order:

$$\varphi(Z_n) = \varphi\left(\frac{\sqrt{n-1}}{\sqrt{n}}Z_{n-1}\right) + \frac{1}{\sqrt{n}}X_n\varphi'\left(\frac{\sqrt{n-1}}{\sqrt{n}}Z_{n-1}\right)$$
$$+ \frac{1}{2n}X_n^2\varphi''\left(\frac{\sqrt{n-1}}{\sqrt{n}}Z_{n-1}\right) + O(\frac{1}{n^{3/2}}).$$

Taking expectations, and using the independence of X_n and Z_{n-1}, together with the normalisations on X_n, we obtain

$$\mathbf{E}\varphi(Z_n) = \mathbf{E}\varphi\left(\frac{\sqrt{n-1}}{\sqrt{n}}Z_{n-1}\right) + \frac{1}{2n}\varphi''\left(\frac{\sqrt{n-1}}{\sqrt{n}}Z_{n-1}\right) + O(\frac{1}{n^{3/2}}).$$

Up to errors of $O(\frac{1}{n^{3/2}})$, one can approximate the second term here by $\frac{1}{2n}\varphi''(Z_{n-1})$. We then insert (2.56) and are led to the heuristic equation

$$\int_{\mathbf{R}} \varphi(x)\rho(x) \, dx \approx \int_{\mathbf{R}} \varphi\left(\frac{\sqrt{n-1}}{\sqrt{n}}x\right)\rho(x) + \frac{1}{2n}\varphi''(x)\rho(x) \, dx + O(\frac{1}{n^{3/2}}).$$

Changing variables for the first term on the right-hand side, and integrating by parts for the second term, we have

$$\int_{\mathbf{R}} \varphi(x)\rho(x) \, dx \approx \int_{\mathbf{R}} \varphi(x)\frac{\sqrt{n}}{\sqrt{n-1}}\rho\left(\frac{\sqrt{n}}{\sqrt{n-1}}x\right)$$
$$+ \frac{1}{2n}\varphi(x)\rho''(x) \, dx + O(\frac{1}{n^{3/2}}).$$

Since φ was an arbitrary test function, this suggests the heuristic equation

$$\rho(x) \approx \frac{\sqrt{n}}{\sqrt{n-1}}\rho\left(\frac{\sqrt{n}}{\sqrt{n-1}}x\right) + \frac{1}{2n}\rho''(x) + O(\frac{1}{n^{3/2}}).$$

Taylor expansion gives

$$\frac{\sqrt{n}}{\sqrt{n-1}}\rho\left(\frac{\sqrt{n}}{\sqrt{n-1}}x\right) = \rho(x) + \frac{1}{2n}\rho(x) + \frac{1}{2n}x\rho'(x) + O(\frac{1}{n^{3/2}}),$$

which leads us to the heuristic ODE

$$L\rho(x) = 0$$

where L is the *Ornstein-Uhlenbeck operator*

$$L\rho(x) := \rho(x) + x\rho'(x) + \rho''(x).$$

Observe that $L\rho$ is the total derivative of $x\rho(x) + \rho'(x)$; integrating from infinity, we thus get

$$x\rho(x) + \rho'(x) = 0$$

which is (2.46), and can be solved by standard ODE methods as $\rho(x) = ce^{-x^2/2}$ for some c; the requirement that probability density functions have total mass 1 then gives the constant c as $\frac{1}{\sqrt{2\pi}}$, as we knew it must.

The above argument was not rigorous, but one can make it so with a significant amount of PDE machinery. If we view n (or more precisely, $\log n$) as a time parameter, and view φ as depending on time, the above computations heuristically lead us eventually to the *Fokker-Planck equation*

$$\partial_t \rho(t, x) = L\rho$$

for the *Ornstein-Uhlenbeck process*, which is a linear parabolic equation that is fortunate enough that it can be solved exactly (indeed, it is not difficult to transform this equation to the linear heat equation by some straightforward changes of variable). Using the spectral theory of the Ornstein-Uhlenbeck operator L, one can show that solutions to this equation starting from an arbitrary probability distribution, are attracted to the Gaussian density function $\frac{1}{\sqrt{2\pi}}e^{-x^2/2}$, which as we saw is the steady state for this equation. The stable nature of this attraction can eventually be used to make the above heuristic analysis rigorous. However, this requires a substantial amount of technical effort (e.g., developing the theory of Sobolev spaces associated to L) and will not be attempted here. One can also proceed by relating the Fokker-Planck equation to the associated stochastic process, namely the Ornstein-Uhlenbeck process, but this requires one to first set up stochastic calculus, which we will not do here[13]. Stein's method, discussed above, can also be interpreted as a way of making the above computations rigorous (by not working with the density function ρ directly, but instead testing the random variable Z_n against various test functions φ).

This argument does, however, highlight two ideas which we will see again in later sections when studying random matrices. First, that it is profitable to study the distribution of some random object Z_n by comparing it with its predecessor Z_{n-1}, which one presumes to have almost the same distribution. Second, we see that it may potentially be helpful to approximate (in some weak sense) a discrete process (such as the iteration of the scheme (2.55)) with a continuous evolution (in this case, a Fokker-Planck equation) which can then be controlled using PDE methods.

---

[13]The various Taylor expansion calculations we have performed in this section, though, are closely related to stochastic calculus tools such as *Ito's lemma*.

2.3. The operator norm of random matrices

Now that we have developed the basic probabilistic tools that we will need, we now turn to the main subject of this text, namely the study of random matrices. There are many random matrix models (aka matrix ensembles) of interest—far too many to all be discussed here. We will thus focus on just a few simple models. First of all, we shall restrict attention to square matrices $M = (\xi_{ij})_{1 \leq i,j \leq n}$, where n is a (large) integer and the ξ_{ij} are real or complex random variables. (One can certainly study rectangular matrices as well, but for simplicity we will only look at the square case.) Then, we shall restrict to three main models:

(i) **Iid matrix ensembles**, in which the coefficients ξ_{ij} are iid random variables with a single distribution $\xi_{ij} \equiv \xi$. We will often normalise ξ to have mean zero and unit variance. Examples of iid models include the *Bernoulli ensemble* (aka *random sign matrices*) in which the ξ_{ij} are signed Bernoulli variables, the *real Gaussian matrix ensemble* in which $\xi_{ij} \equiv N(0,1)_{\mathbf{R}}$, and the *complex Gaussian matrix ensemble* in which $\xi_{ij} \equiv N(0,1)_{\mathbf{C}}$.

(ii) **Symmetric Wigner matrix ensembles**, in which the upper triangular coefficients ξ_{ij}, $j \geq i$ are jointly independent and real, but the lower triangular coefficients ξ_{ij}, $j < i$ are constrained to equal their transposes: $\xi_{ij} = \xi_{ji}$. Thus M by construction is always a real symmetric matrix. Typically, the strictly upper triangular coefficients will be iid, as will the diagonal coefficients, but the two classes of coefficients may have a different distribution. One example here is the *symmetric Bernoulli ensemble*, in which both the strictly upper triangular and the diagonal entries are signed Bernoulli variables; another important example is the *Gaussian Orthogonal Ensemble (GOE)*, in which the upper triangular entries have distribution $N(0,1)_{\mathbf{R}}$ and the diagonal entries have distribution $N(0,2)_{\mathbf{R}}$. (We will explain the reason for this discrepancy later.)

(iii) **Hermitian Wigner matrix ensembles**, in which the upper triangular coefficients are jointly independent, with the diagonal entries being real and the strictly upper triangular entries complex, and the lower triangular coefficients ξ_{ij}, $j < i$ are constrained to equal their adjoints: $\xi_{ij} = \overline{\xi_{ji}}$. Thus M by construction is always a Hermitian matrix. This class of ensembles contains the symmetric Wigner ensembles as a subclass. Another very important example is the *Gaussian Unitary Ensemble (GUE)*, in which all off-diagonal entries have distribution $N(0,1)_{\mathbf{C}}$, but the diagonal entries have distribution $N(0,1)_{\mathbf{R}}$.

Given a matrix ensemble M, there are many statistics of M that one may wish to consider, e.g., the eigenvalues or singular values of M, the trace and determinant, etc. In this section we will focus on a basic statistic, namely the *operator norm*

$$(2.57) \qquad \|M\|_{\mathrm{op}} := \sup_{x \in \mathbf{C}^n : |x|=1} |Mx|$$

of the matrix M. This is an interesting quantity in its own right, but also serves as a basic upper bound on many other quantities. (For instance, $\|M\|_{\mathrm{op}}$ is also the largest singular value $\sigma_1(M)$ of M and thus dominates the other singular values; similarly, all eigenvalues $\lambda_i(M)$ of M clearly have magnitude at most $\|M\|_{\mathrm{op}}$.) Because of this, it is particularly important to get good *upper tail bounds*,

$$\mathbf{P}(\|M\|_{\mathrm{op}} \geq \lambda) \leq \ldots,$$

on this quantity, for various thresholds λ. (Lower tail bounds are also of interest, of course; for instance, they give us confidence that the upper tail bounds are sharp.) Also, as we shall see, the problem of upper bounding $\|M\|_{\mathrm{op}}$ can be viewed as a non-commutative analogue[14] of upper bounding the quantity $|S_n|$ studied in Section 2.1.

An $n \times n$ matrix consisting entirely of 1s has an operator norm of exactly n, as can for instance be seen from the Cauchy-Schwarz inequality. More generally, any matrix whose entries are all uniformly $O(1)$ will have an operator norm of $O(n)$ (which can again be seen from Cauchy-Schwarz, or alternatively from *Schur's test* (see e.g. [**Ta2010**, §1.11]), or from a computation of the *Frobenius norm* (see (2.63))). However, this argument does not take advantage of possible cancellations in M. Indeed, from analogy with concentration of measure, when the entries of the matrix M are independent, bounded and have mean zero, we expect the operator norm to be of size $O(\sqrt{n})$ rather than $O(n)$. We shall see shortly that this intuition is indeed correct[15].

As mentioned before, there is an analogy here with the concentration of measure[16] phenomenon, and many of the tools used in the latter (e.g., the moment method) will also appear here. Similarly, just as many of the tools

---

[14]The analogue of the central limit theorem studied in Section 2.2 is the *Wigner semicircular law*, which will be studied in Section 2.4.

[15]One can see, though, that the mean zero hypothesis is important; from the triangle inequality we see that if we add the all-ones matrix (for instance) to a random matrix with mean zero, to obtain a random matrix whose coefficients all have mean 1, then at least one of the two random matrices necessarily has operator norm at least $n/2$.

[16]Indeed, we will be able to use some of the concentration inequalities from Section 2.1 directly to help control $\|M\|_{\mathrm{op}}$ and related quantities.

from concentration of measure could be adapted to help prove the central limit theorem, several of the tools seen here will be of use in deriving the semicircular law in Section 2.4.

The most advanced knowledge we have on the operator norm is given by the *Tracy-Widom law*, which not only tells us where the operator norm is concentrated in (it turns out, for instance, that for a Wigner matrix (with some additional technical assumptions), it is concentrated in the range $[2\sqrt{n} - O(n^{-1/6}), 2\sqrt{n} + O(n^{-1/6})]$), but what its distribution in that range is. While the methods in this section can eventually be pushed to establish this result, this is far from trivial, and will only be briefly discussed here. We will, however, discuss the Tracy-Widom law at several later points in the text.

2.3.1. The epsilon net argument. The slickest way to control $\|M\|_{\mathrm{op}}$ is via the moment method. But let us defer using this method for the moment, and work with a more "naive" way to control the operator norm, namely by working with the definition (2.57). From that definition, we see that we can view the upper tail event $\|M\|_{\mathrm{op}} > \lambda$ as a union of many simpler events:

$$(2.58) \qquad \mathbf{P}(\|M\|_{\mathrm{op}} > \lambda) \le \mathbf{P}(\bigvee_{x \in S} |Mx| > \lambda)$$

where $S := \{x \in \mathbf{C}^d : |x| = 1\}$ is the unit sphere in the complex space $\mathbf{C}^d$.

The point of doing this is that the event $|Mx| > \lambda$ is easier to control than the event $\|M\|_{\mathrm{op}} > \lambda$, and can in fact be handled by the concentration of measure estimates we already have. For instance:

Lemma 2.3.1. *Suppose that the coefficients ξ_{ij} of M are independent, have mean zero, and are uniformly bounded in magnitude by 1. Let x be a unit vector in $\mathbf{C}^n$. Then for sufficiently large A (larger than some absolute constant), one has*

$$\mathbf{P}(|Mx| \ge A\sqrt{n}) \le C \exp(-cAn)$$

for some absolute constants $C, c > 0$.

Proof. Let $X_1, \ldots, X_n$ be the n rows of M, then the column vector Mx has coefficients $X_i \cdot x$ for $i = 1, \ldots, n$. If we let $x_1, \ldots, x_n$ be the coefficients of x, so that $\sum_{j=1}^{n} |x_j|^2 = 1$, then $X_i \cdot x$ is just $\sum_{j=1}^{n} \xi_{ij} x_j$. Applying standard concentration of measure results (e.g., Exercise 2.1.4, Exercise 2.1.5, or Theorem 2.1.13), we see that each $X_i \cdot x$ is uniformly sub-Gaussian, thus

$$\mathbf{P}(|X_i \cdot x| \ge \lambda) \le C \exp(-c\lambda^2)$$

for some absolute constants $C, c > 0$. In particular, we have

$$\mathbf{E} e^{c|X_i \cdot x|^2} \le C$$

for some (slightly different) absolute constants $C, c > 0$. Multiplying these inequalities together for all i, we obtain

$$\mathbf{E}e^{c|Mx|^2} \leq C^n$$

and the claim then follows from Markov's inequality (1.14).   □

Thus (with the hypotheses of Lemma 2.3.1), we see that for each individual unit vector x, we have $|Mx| = O(\sqrt{n})$ with overwhelming probability. It is then tempting to apply the union bound and try to conclude that $\|M\|_{\mathrm{op}} = O(\sqrt{n})$ with overwhelming probability also. However, we encounter a difficulty: the unit sphere S is uncountable, and so we are taking the union over an uncountable number of events. Even though each event occurs with exponentially small probability, the union could well be everything.

Of course, it is extremely wasteful to apply the union bound to an uncountable union. One can pass to a countable union just by working with a countable dense subset of the unit sphere S instead of the sphere itself, since the map $x \mapsto |Mx|$ is continuous. Of course, this is still an infinite set and so we still cannot usefully apply the union bound. However, the map $x \mapsto |Mx|$ is not just continuous; it is *Lipschitz* continuous, with a Lipschitz constant of $\|M\|_{\mathrm{op}}$. Now, of course there is some circularity here because $\|M\|_{\mathrm{op}}$ is precisely the quantity we are trying to bound. Nevertheless, we can use this stronger continuity to refine the countable dense subset further, to a *finite* (but still quite dense) subset of S, at the slight cost of modifying the threshold λ by a constant factor. Namely:

Lemma 2.3.2. *Let Σ be a maximal $1/2$-net of the sphere S, i.e., a set of points in S that are separated from each other by a distance of at least $1/2$, and which is maximal with respect to set inclusion. Then for any $n \times n$ matrix M with complex coefficients, and any $\lambda > 0$, we have*

$$\mathbf{P}(\|M\|_{\mathrm{op}} > \lambda) \leq \mathbf{P}(\bigvee_{y \in \Sigma} |My| > \lambda/2).$$

Proof. By (2.57) (and compactness) we can find $x \in S$ such that

$$|Mx| = \|M\|_{\mathrm{op}}.$$

This point x need not lie in Σ. However, as Σ is a maximal $1/2$-net of S, we know that x lies within $1/2$ of some point y in Σ (since otherwise we could add x to Σ and contradict maximality). Since $|x - y| \leq 1/2$, we have

$$|M(x - y)| \leq \|M\|_{\mathrm{op}}/2.$$

By the triangle inequality we conclude that

$$|My| \geq \|M\|_{\mathrm{op}}/2.$$

In particular, if $\|M\|_{\mathrm{op}} > \lambda$, then $|My| > \lambda/2$ for some $y \in \Sigma$, and the claim follows. $\qquad\square$

Remark 2.3.3. Clearly, if one replaces the maximal $1/2$-net here with a maximal ε-net for some other $0 < \varepsilon < 1$ (defined in the obvious manner), then we get the same conclusion, but with $\lambda/2$ replaced by $\lambda(1 - \varepsilon)$.

Now that we have discretised the range of points y to be finite, the union bound becomes viable again. We first make the following basic observation:

Lemma 2.3.4 (Volume packing argument). *Let $0 < \varepsilon < 1$, and let Σ be an ε-net of the sphere S. Then Σ has cardinality at most $(C/\varepsilon)^n$ for some absolute constant $C > 0$.*

Proof. Consider the balls of radius $\varepsilon/2$ centred around each point in Σ; by hypothesis, these are disjoint. On the other hand, by the triangle inequality, they are all contained in the ball of radius $3/2$ centred at the origin. The volume of the latter ball is at most $(C/\varepsilon)^n$ the volume of any of the small balls, and the claim follows. $\qquad\square$

Exercise 2.3.1. Improve the bound $(C/\varepsilon)^n$ to C^n/ε^{n-1}. In the converse direction, if Σ is a *maximal* ε-net, show that Σ has cardinality *at least* c^n/ε^{n-1} for some absolute constant $c > 0$.

And now we get an upper tail estimate:

Corollary 2.3.5 (Upper tail estimate for iid ensembles). *Suppose that the coefficients ξ_{ij} of M are independent, have mean zero, and uniformly bounded in magnitude by 1. Then there exist absolute constants $C, c > 0$ such that*
$$\mathbf{P}(\|M\|_{\mathrm{op}} > A\sqrt{n}) \le C \exp(-cAn)$$
for all $A \ge C$. In particular, we have $\|M\|_{\mathrm{op}} = O(\sqrt{n})$ with overwhelming probability.

Proof. From Lemma 2.3.2 and the union bound, we have
$$\mathbf{P}(\|M\|_{\mathrm{op}} > A\sqrt{n}) \le \sum_{y \in \Sigma} \mathbf{P}(|My| > A\sqrt{n}/2)$$
where Σ is a maximal $1/2$-net of S. By Lemma 2.3.1, each of the probabilities $\mathbf{P}(|My| > A\sqrt{n}/2)$ is bounded by $C \exp(-cAn)$ if A is large enough. Meanwhile, from Lemma 2.3.4, Σ has cardinality $O(1)^n$. If A is large enough, the *entropy loss*[17] of $O(1)^n$ can be absorbed into the exponential gain of $\exp(-cAn)$ by modifying c slightly, and the claim follows. $\qquad\square$

---

[17]Roughly speaking, the *entropy* of a configuration is the logarithm of the number of possible states that configuration can be in. When applying the union bound to control all possible configurations at once, one often loses a factor proportional to the number of such states; this factor is sometimes referred to as the *entropy factor* or *entropy loss* in one's argument.

Exercise 2.3.2. If Σ is a maximal $1/4$-net instead of a maximal $1/2$-net, establish the following variant

$$\mathbf{P}(\|M\|_{\mathrm{op}} > \lambda) \leq \mathbf{P}(\bigvee_{x,y \in \Sigma} |x^*My| > \lambda/4)$$

of Lemma 2.3.2. Use this to provide an alternate proof of Corollary 2.3.5.

The above result was for matrices with independent entries, but it easily extends to the Wigner case:

Corollary 2.3.6 (Upper tail estimate for Wigner ensembles). *Suppose that the coefficients ξ_{ij} of M are independent for $j \geq i$, mean zero, and uniformly bounded in magnitude by 1, and let $\xi_{ij} := \overline{\xi_{ji}}$ for $j < i$. Then there exist absolute constants $C, c > 0$ such that*

$$\mathbf{P}(\|M\|_{\mathrm{op}} > A\sqrt{n}) \leq C \exp(-cAn)$$

for all $A \geq C$. In particular, we have $\|M\|_{\mathrm{op}} = O(\sqrt{n})$ with overwhelming probability.

Proof. From Corollary 2.3.5, the claim already holds for the upper-triangular portion of M, as well as for the strict lower-triangular portion of M. The claim then follows from the triangle inequality (adjusting the constants C, c appropriately). $\square$

Exercise 2.3.3. Generalise Corollary 2.3.5 and Corollary 2.3.6 to the case where the coefficients ξ_{ij} have uniform sub-Gaussian tails, rather than being uniformly bounded in magnitude by 1.

Remark 2.3.7. What we have just seen is a simple example of an *epsilon net argument*, which is useful when controlling a supremum of random variables $\sup_{x \in S} X_x$ such as (2.57), where each individual random variable X_x is known to obey a large deviation inequality (in this case, Lemma 2.3.1). The idea is to use metric arguments (e.g., the triangle inequality, see Lemma 2.3.2) to refine the set of parameters S to take the supremum over to an ε-net $\Sigma = \Sigma_\varepsilon$ for some suitable ε, and then apply the union bound. One takes a loss based on the cardinality of the ε-net (which is basically the *covering number* of the original parameter space at scale ε), but one can hope that the bounds from the large deviation inequality are strong enough (and the metric entropy bounds sufficiently accurate) to overcome this entropy loss.

There is of course the question of what scale ε to use. In this simple example, the scale $\varepsilon = 1/2$ sufficed. In other contexts, one has to choose the scale ε more carefully. In more complicated examples with no natural preferred scale, it often makes sense to take a large range of scales (e.g., $\varepsilon = 2^{-j}$ for $j = 1, \ldots, J$) and *chain* them together by using telescoping series such as $X_x = X_{x_1} + \sum_{j=1}^{J} X_{x_{j+1}} - X_{x_j}$ (where x_j is the nearest point in Σ_j to

x for $j = 1, \ldots, J$, and x_{J+1} is x by convention) to estimate the supremum, the point being that one can hope to exploit cancellations between adjacent elements of the sequence X_{x_j}. This is known as the method of *chaining*. There is an even more powerful refinement of this method, known as the method of *generic chaining*, which has a large number of applications; see [**Ta2005**] for a beautiful and systematic treatment of the subject. However, we will not use this method in this text.

2.3.2. A symmetrisation argument (optional). We pause here to record an elegant *symmetrisation argument* that exploits convexity to allow us to reduce, without loss of generality, to the symmetric case $M \equiv -M$, albeit at the cost of losing a factor of 2. We will not use this type of argument directly in this text, but it is often used elsewhere in the literature.

Let M be any random matrix with mean zero, and let $\tilde{M}$ be an independent copy of M. Then, conditioning on M, we have

$$\mathbf{E}(M - \tilde{M}|M) = M.$$

As the operator norm $M \mapsto \|M\|_{\mathrm{op}}$ is convex, we can then apply Jensen's inequality (Exercise 1.1.8) to conclude that

$$\mathbf{E}(\|M - \tilde{M}\|_{\mathrm{op}}|M) \geq \|M\|_{\mathrm{op}}.$$

Undoing the conditioning over M, we conclude that

$$(2.59) \qquad\qquad \mathbf{E}\|M - \tilde{M}\|_{\mathrm{op}} \geq \mathbf{E}\|M\|_{\mathrm{op}}.$$

Thus, to upper bound the expected operator norm of M, it suffices to upper bound the expected operator norm of $M - \tilde{M}$. The point is that even if M is not symmetric ($M \not\equiv -M$), $M - \tilde{M}$ is automatically symmetric.

One can modify (2.59) in a few ways, given some more hypotheses on M. Suppose now that $M = (\xi_{ij})_{1 \leq i,j \leq n}$ is a matrix with independent entries, thus $M - \tilde{M}$ has coefficients $\xi_{ij} - \tilde{\xi}_{ij}$ where $\tilde{\xi}_{ij}$ is an independent copy of ξ_{ij}. Introduce a random sign matrix $E = (\varepsilon_{ij})_{1 \leq i,j \leq n}$ which is (jointly) independent of $M, \tilde{M}$. Observe that as the distribution of $\xi_{ij} - \tilde{\xi}_{ij}$ is symmetric, that

$$(\xi_{ij} - \tilde{\xi}_{ij}) \equiv (\xi_{ij} - \tilde{\xi}_{ij})\varepsilon_{ij},$$

and thus

$$(M - \tilde{M}) \equiv (M - \tilde{M}) \cdot E$$

where $A \cdot B := (a_{ij}b_{ij})_{1 \leq i,j \leq n}$ is the *Hadamard product* of $A = (a_{ij})_{1 \leq i,j \leq n}$ and $B = (b_{ij})_{1 \leq i,j \leq n}$. We conclude from (2.59) that

$$\mathbf{E}\|M\|_{\mathrm{op}} \leq \mathbf{E}\|(M - \tilde{M}) \cdot E\|_{\mathrm{op}}.$$

By the distributive law and the triangle inequality we have

$$\|(M - \tilde{M}) \cdot E\|_{\mathrm{op}} \leq \|M \cdot E\|_{\mathrm{op}} + \|\tilde{M} \cdot E\|_{\mathrm{op}}.$$

But as $M \cdot E \equiv \tilde{M} \cdot E$, the quantities $\|M \cdot E\|_{\mathrm{op}}$ and $\|\tilde{M} \cdot E\|_{\mathrm{op}}$ have the same expectation. We conclude the *symmetrisation inequality*

$$(2.60) \qquad\qquad \mathbf{E}\|M\|_{\mathrm{op}} \le 2\mathbf{E}\|M \cdot E\|_{\mathrm{op}}.$$

Thus, if one does not mind losing a factor of two, one has the freedom to randomise the sign of each entry of M independently (assuming that the entries were already independent). Thus, in proving Corollary 2.3.5, one could have reduced to the case when the ξ_{ij} were symmetric, though in this case this would not have made the argument that much simpler.

Sometimes it is preferable to multiply the coefficients by a Gaussian rather than by a random sign. Again, let $M = (\xi_{ij})_{1 \le i,j \le n}$ have independent entries with mean zero. Let $G = (g_{ij})_{1 \le i,j \le n}$ be a real Gaussian matrix independent of M, thus the $g_{ij} \equiv N(0,1)_{\mathbf{R}}$ are iid. We can split $G = E \cdot |G|$, where $E := (\mathrm{sgn}(g_{ij}))_{1 \le i,j \le n}$ and $|G| = (|g_{ij}|)_{1 \le i,j \le n}$. Note that E, M, $|G|$ are independent, and E is a random sign matrix. In particular, (2.60) holds. We now use

Exercise 2.3.4. If $g \equiv N(0,1)_{\mathbf{R}}$, show that $\mathbf{E}|g| = \sqrt{\frac{2}{\pi}}$.

From this exercise we see that

$$\mathbf{E}(M \cdot E \cdot |G| | M, E) = \sqrt{\frac{2}{\pi}} M \cdot E$$

and hence by Jensen's inequality (Exercise 1.1.8) again

$$\mathbf{E}(\|M \cdot E \cdot |G|\|_{\mathrm{op}} | M, E) \ge \sqrt{\frac{2}{\pi}} \|M \cdot E\|_{\mathrm{op}}.$$

Undoing the conditional expectation in M, E and applying (2.60) we conclude the *Gaussian symmetrisation inequality*

$$(2.61) \qquad\qquad \mathbf{E}\|M\|_{\mathrm{op}} \le \sqrt{2\pi}\mathbf{E}\|M \cdot G\|_{\mathrm{op}}.$$

Thus, for instance, when proving Corollary 2.3.5, one could have inserted a random Gaussian in front of each coefficient. This would have made the proof of Lemma 2.3.1 marginally simpler (as one could compute directly with Gaussians, and reduce the number of appeals to concentration of measure results) but in this case the improvement is negligible. In other situations though it can be quite helpful to have the additional random sign or random Gaussian factor present. For instance, we have the following result of Latala [**La2005**]:

Theorem 2.3.8. *Let $M = (\xi_{ij})_{1 \leq i,j \leq n}$ be a matrix with independent mean zero entries, obeying the second moment bounds*

$$\sup_i \sum_{j=1}^n \mathbf{E}|\xi_{ij}|^2 \leq K^2 n,$$

$$\sup_j \sum_{i=1}^n \mathbf{E}|\xi_{ij}|^2 \leq K^2 n,$$

and the fourth moment bound

$$\sum_{i=1}^n \sum_{j=1}^n \mathbf{E}|\xi_{ij}|^4 \leq K^4 n^2$$

for some $K > 0$. Then $\mathbf{E}\|M\|_{\mathrm{op}} = O(K\sqrt{n})$.

Proof (Sketch only). Using (2.61) one can replace ξ_{ij} by $\xi_{ij} \cdot g_{ij}$ without much penalty. One then runs the epsilon-net argument with an explicit net, and uses concentration of measure results for Gaussians (such as Theorem 2.1.12) to obtain the analogue of Lemma 2.3.1. The details are rather intricate, and we refer the interested reader to [**La2005**]. □

As a corollary of Theorem 2.3.8, we see that if we have an iid matrix (or Wigner matrix) of mean zero whose entries have a fourth moment of $O(1)$, then the expected operator norm is $O(\sqrt{n})$. The fourth moment hypothesis is sharp. To see this, we make the trivial observation that the operator norm of a matrix $M = (\xi_{ij})_{1 \leq i,j \leq n}$ bounds the magnitude of any of its coefficients, thus

$$\sup_{1 \leq i,j \leq n} |\xi_{ij}| \leq \|M\|_{\mathrm{op}}$$

or, equivalently, that

$$\mathbf{P}(\|M\|_{\mathrm{op}} \leq \lambda) \leq \mathbf{P}(\bigvee_{1 \leq i,j \leq n} |\xi_{ij}| \leq \lambda).$$

In the iid case $\xi_{ij} \equiv \xi$, and setting $\lambda = A\sqrt{n}$ for some fixed A independent of n, we thus have

$$(2.62) \qquad \mathbf{P}(\|M\|_{\mathrm{op}} \leq A\sqrt{n}) \leq \mathbf{P}(|\xi| \leq A\sqrt{n})^{n^2}.$$

With the fourth moment hypothesis, one has from dominated convergence that

$$\mathbf{P}(|\xi| \leq A\sqrt{n}) \geq 1 - o_A(1/n^2),$$

and so the right-hand side of (2.62) is asymptotically trivial. But with weaker hypotheses than the fourth moment hypothesis, the rate of convergence of $\mathbf{P}(|\xi| \leq A\sqrt{n})$ to 1 can be slower, and one can easily build examples for which the right-hand side of (2.62) is $o_A(1)$ for every A, which forces $\|M\|_{\mathrm{op}}$ to typically be much larger than $\sqrt{n}$ on the average.

Remark 2.3.9. The symmetrisation inequalities remain valid with the operator norm replaced by any other convex norm on the space of matrices. The results are also just as valid for rectangular matrices as for square ones.

2.3.3. Concentration of measure. Consider a random matrix M of the type considered in Corollary 2.3.5 (e.g., a random sign matrix). We now know that the operator norm $\|M\|_{\mathrm{op}}$ is of size $O(\sqrt{n})$ with overwhelming probability. But there is much more that can be said. For instance, by taking advantage of the convexity and Lipschitz properties of $\|M\|_{\mathrm{op}}$, we have the following quick application of Talagrand's inequality (Theorem 2.1.13):

Proposition 2.3.10. *Let M be as in Corollary 2.3.5. Then for any $\lambda > 0$, one has*
$$\mathbf{P}(|\|M\|_{\mathrm{op}} - \mathbf{M}\|M\|_{\mathrm{op}}| \geq \lambda) \leq C \exp(-c\lambda^2)$$
for some absolute constants $C, c > 0$, where $\mathbf{M}\|M\|_{\mathrm{op}}$ is a median value for $\|M\|_{\mathrm{op}}$. The same result also holds with $\mathbf{M}\|M\|_{\mathrm{op}}$ replaced by the expectation $\mathbf{E}\|M\|_{\mathrm{op}}$.

Proof. We view $\|M\|_{\mathrm{op}}$ as a function $F((\xi_{ij})_{1 \leq i,j \leq n})$ of the independent complex variables ξ_{ij}, thus F is a function from $\mathbf{C}^{n^2}$ to $\mathbf{R}$. The convexity of the operator norm tells us that F is convex. The triangle inequality, together with the elementary bound

(2.63) $$\|M\|_{\mathrm{op}} \leq \|M\|_F$$

(easily proven by Cauchy-Schwarz), where

(2.64) $$\|M\|_F := (\sum_{i=1}^{n} \sum_{j=1}^{n} |\xi_{ij}|^2)^{1/2}$$

is the *Frobenius norm* (also known as the *Hilbert-Schmidt norm* or *2-Schatten norm*), tells us that F is Lipschitz with constant 1. The claim then follows directly from Talagrand's inequality (Theorem 2.1.13). $\qquad\square$

Exercise 2.3.5. Establish a similar result for the matrices in Corollary 2.3.6.

From Corollary 2.3.5 we know that the median or expectation of $\|M\|_{\mathrm{op}}$ is of size $O(\sqrt{n})$; we now know that $\|M\|_{\mathrm{op}}$ concentrates around this median to width at most $O(1)$. (This turns out to be non-optimal; the Tracy-Widom law actually gives a concentration of $O(n^{-1/6})$, under some additional assumptions on M. Nevertheless, this level of concentration is already non-trivial.)

However, this argument does not tell us much about what the median or expected value of $\|M\|_{\mathrm{op}}$ actually *is*. For this, we will need to use other methods, such as the moment method which we turn to next.

Remark 2.3.11. Talagrand's inequality, as formulated in Theorem 2.1.13, relies heavily on convexity. Because of this, we cannot apply this argument directly to non-convex matrix statistics, such as singular values $\sigma_j(M)$ other than the largest singular value $\sigma_1(M)$. Nevertheless, one can still use this inequality to obtain good concentration results, by using the convexity of related quantities, such as the partial sums $\sum_{j=1}^{J} \sigma_j(M)$; see [**Me2004**]. Other approaches include the use of alternate large deviation inequalities, such as those arising from log-Sobolev inequalities (see e.g., [**Gu2009**]), or by using more abstract versions of Talagrand's inequality (see [**AlKrVu2002**], [**GuZe2000**]).

2.3.4. The moment method. We now bring the moment method to bear on the problem, starting with the easy moments and working one's way up to the more sophisticated moments. It turns out that it is easier to work first with the case when M is symmetric or Hermitian; we will discuss the non-symmetric case near the end of this section.

The starting point for the moment method is the observation that for symmetric or Hermitian M, the operator norm $\|M\|_{\mathrm{op}}$ is equal to the ℓ^{∞} norm

$$(2.65) \qquad \|M\|_{\mathrm{op}} = \max_{1 \le i \le n} |\lambda_i|$$

of the eigenvalues $\lambda_1, \ldots, \lambda_n \in \mathbf{R}$ of M. On the other hand, we have the standard linear algebra identity

$$\mathrm{tr}(M) = \sum_{i=1}^{n} \lambda_i$$

and more generally

$$\mathrm{tr}(M^k) = \sum_{i=1}^{n} \lambda_i^k.$$

In particular, if $k = 2, 4, \ldots$ is an even integer, then $\mathrm{tr}(M^k)^{1/k}$ is just the ℓ^k norm of these eigenvalues, and we have the inequalities

$$(2.66) \qquad \|M\|_{\mathrm{op}}^k \le \mathrm{tr}(M^k) \le n\|M\|_{\mathrm{op}}^k.$$

To put this another way, knowledge of the kth moment $\mathrm{tr}(M^k)$ controls the operator norm up to a multiplicative factor of $n^{1/k}$. Taking larger and larger k, we should thus obtain more accurate control on the operator norm[18].

---

[18]This is also the philosophy underlying the *power method* in numerical linear algebra.

Remark 2.3.12. In most cases, one expects the eigenvalues to be reasonably uniformly distributed, in which case the upper bound in (2.66) is closer to the truth than the lower bound. One scenario in which this can be rigorously established is if it is known that the eigenvalues of M all come with a high multiplicity. This is often the case for matrices associated with group actions (particularly those which are *quasirandom* in the sense of Gowers [**Go2008**]). However, this is usually not the case with most random matrix ensembles, and we must instead proceed by increasing k as described above.

Let's see how this method works in practice. The simplest case is that of the second moment $\text{tr}(M^2)$, which in the Hermitian case works out to

$$\text{tr}(M^2) = \sum_{i=1}^{n} \sum_{j=1}^{n} |\xi_{ij}|^2 = \|M\|_F^2.$$

Note that (2.63) is just the $k = 2$ case of the lower inequality in (2.66), at least in the Hermitian case.

The expression $\sum_{i=1}^{n} \sum_{j=1}^{n} |\xi_{ij}|^2$ is easy to compute in practice. For instance, for the symmetric Bernoulli ensemble, this expression is exactly equal to n^2. More generally, if we have a Wigner matrix in which all off-diagonal entries have mean zero and unit variance, and the diagonal entries have mean zero and bounded variance (this is the case for instance for GOE), then the off-diagonal entries have mean 1, and by the law of large numbers[19] we see that this expression is almost surely asymptotic to n^2.

From the weak law of large numbers, we see, in particular, that one has

(2.67)
$$\sum_{i=1}^{n} \sum_{j=1}^{n} |\xi_{ij}|^2 = (1 + o(1))n^2$$

asymptotically almost surely.

Exercise 2.3.6. If the ξ_{ij} have uniformly sub-exponential tail, show that we in fact have (2.67) with overwhelming probability.

Applying (2.66), we obtain the bounds

(2.68) $(1 + o(1))\sqrt{n} \leq \|M\|_{\text{op}} \leq (1 + o(1))n$

asymptotically almost surely. This is already enough to show that the median of $\|M\|_{\text{op}}$ is at least $(1+o(1))\sqrt{n}$, which complements (up to constants) the upper bound of $O(\sqrt{n})$ obtained from the epsilon net argument. But the upper bound here is terrible; we will need to move to higher moments to improve it.

---

[19]There is of course a dependence between the upper triangular and lower triangular entries, but this is easy to deal with by folding the sum into twice the upper triangular portion (plus the diagonal portion, which is lower order).

Accordingly, we now turn to the fourth moment. For simplicity let us assume that all entries ξ_{ij} have zero mean and unit variance. To control moments beyond the second moment, we will also assume that all entries are bounded in magnitude by some K. We expand

$$\operatorname{tr}(M^4) = \sum_{1 \le i_1, i_2, i_3, i_4 \le n} \xi_{i_1 i_2} \xi_{i_2 i_3} \xi_{i_3 i_4} \xi_{i_4 i_1}.$$

To understand this expression, we take expectations:

$$\mathbf{E} \operatorname{tr}(M^4) = \sum_{1 \le i_1, i_2, i_3, i_4 \le n} \mathbf{E} \xi_{i_1 i_2} \xi_{i_2 i_3} \xi_{i_3 i_4} \xi_{i_4 i_1}.$$

One can view this sum graphically, as a sum over length four cycles in the vertex set $\{1, \ldots, n\}$; note that the four edges $\{i_1, i_2\}, \{i_2, i_3\}, \{i_3, i_4\}, \{i_4, i_1\}$ are allowed to be degenerate if two adjacent ξ_i are equal. The value of each term

(2.69) $$\mathbf{E} \xi_{i_1 i_2} \xi_{i_2 i_3} \xi_{i_3 i_4} \xi_{i_4 i_1}$$

in this sum depends on what the cycle does.

First, there is the case when all of the four edges $\{i_1, i_2\}, \{i_2, i_3\}, \{i_3, i_4\}$, $\{i_4, i_1\}$ are distinct. Then the four factors $\xi_{i_1 i_2}, \ldots, \xi_{i_4 i_1}$ are independent; since we are assuming them to have mean zero, the term (2.69) vanishes. Indeed, the same argument shows that the only terms that do not vanish are those in which each edge is repeated at least twice. A short combinatorial case check then shows that, up to cyclic permutations of the i_1, i_2, i_3, i_4 indices there are now only a few types of cycles in which the term (2.69) does not automatically vanish:

(i) $i_1 = i_3$, but i_2, i_4 are distinct from each other and from i_1.

(ii) $i_1 = i_3$ and $i_2 = i_4$.

(iii) $i_1 = i_2 = i_3$, but i_4 is distinct from i_1.

(iv) $i_1 = i_2 = i_3 = i_4$.

In the first case, the independence and unit variance assumptions tell us that (2.69) is 1, and there are $O(n^3)$ such terms, so the total contribution here to $\mathbf{E} \operatorname{tr}(M^4)$ is at most $O(n^3)$. In the second case, the unit variance and bound by K tells us that the term is $O(K^2)$, and there are $O(n^2)$ such terms, so the contribution here is $O(n^2 K^2)$. Similarly, the contribution of the third type of cycle is $O(n^2)$, and the fourth type of cycle is $O(nK^2)$, so we can put it all together to get

$$\mathbf{E} \operatorname{tr}(M^4) \le O(n^3) + O(n^2 K^2).$$

In particular, if we make the hypothesis $K = O(\sqrt{n})$, then we have

$$\mathbf{E} \operatorname{tr}(M^4) \le O(n^3),$$

and thus by Markov's inequality (1.13) we see that for any $\varepsilon > 0$, $\operatorname{tr}(M^4) \leq O_\varepsilon(n^3)$ with probability at least $1 - \varepsilon$. Applying (2.66), this leads to the upper bound

$$\|M\|_{\mathrm{op}} \leq O_\varepsilon(n^{3/4})$$

with probability at least $1 - \varepsilon$; a similar argument shows that for any fixed $\varepsilon > 0$, one has

$$\|M\|_{\mathrm{op}} \leq n^{3/4+\varepsilon}$$

with high probability. This is better than the upper bound obtained from the second moment method, but still non-optimal.

Exercise 2.3.7. If $K = o(\sqrt{n})$, use the above argument to show that

$$(\mathbf{E}\|M\|_{\mathrm{op}}^4)^{1/4} \geq (2^{1/4} + o(1))\sqrt{n},$$

which in some sense improves upon (2.68) by a factor of $2^{1/4}$. In particular, if $K = O(1)$, conclude that the median of $\|M\|_{\mathrm{op}}$ is at least $(2^{1/4} + o(1))\sqrt{n}$.

Now let us take a quick look at the sixth moment, again with the running assumption of a Wigner matrix in which all entries have mean zero, unit variance, and bounded in magnitude by K. We have

$$\mathbf{E}\operatorname{tr}(M^6) = \sum_{1 \leq i_1,\ldots,i_6 \leq n} \mathbf{E}\xi_{i_1 i_2} \cdots \xi_{i_5 i_6} \xi_{i_6 i_1},$$

a sum over cycles of length 6 in $\{1, \ldots, n\}$. Again, most of the summands here vanish; the only ones which do not are those cycles in which each edge occurs at least twice (so in particular, there are at most three distinct edges).

Classifying all the types of cycles that could occur here is somewhat tedious, but it is clear that there are going to be $O(1)$ different types of cycles. But we can organise things by the multiplicity of each edge, leaving us with four classes of cycles to deal with:

(i) Cycles in which there are three distinct edges, each occurring two times.

(ii) Cycles in which there are two distinct edges, one occurring twice and one occurring four times.

(iii) Cycles in which there are two distinct edges, each occurring three times[20].

(iv) Cycles in which a single edge occurs six times.

It is not hard to see that summands coming from the first type of cycle give a contribution of 1, and there are $O(n^4)$ of these (because such cycles span at most four vertices). Similarly, the second and third types of cycles

---

[20] Actually, this case ends up being impossible, due to a "bridges of Königsberg" type of obstruction, but we will retain it for this discussion.

give a contribution of $O(K^2)$ per summand, and there are $O(n^3)$ summands; finally, the fourth type of cycle gives a contribution of $O(K^4)$, with $O(n^2)$ summands. Putting this together we see that

$$\mathbf{E}\,\mathrm{tr}(M^6) \le O(n^4) + O(n^3 K^2) + O(n^2 K^4);$$

so, in particular, if we assume $K = O(\sqrt{n})$ as before, we have

$$\mathbf{E}\,\mathrm{tr}(M^6) \le O(n^4)$$

and if we then use (2.66) as before we see that

$$\|M\|_{\mathrm{op}} \le O_\varepsilon(n^{2/3})$$

with probability $1 - \varepsilon$, for any $\varepsilon > 0$; so we are continuing to make progress towards what we suspect (from the epsilon net argument) to be the correct bound of $n^{1/2}$.

Exercise 2.3.8. If $K = o(\sqrt{n})$, use the above argument to show that

$$(\mathbf{E}\|M\|_{\mathrm{op}}^6)^{1/6} \ge (5^{1/6} + o(1))\sqrt{n}.$$

In particular, if $K = O(1)$, conclude that the median of $\|M\|_{\mathrm{op}}$ is at least $(5^{1/6} + o(1))\sqrt{n}$. Thus this is a (slight) improvement over Exercise 2.3.7.

Let us now consider the general k^{th} moment computation under the same hypotheses as before, with k an even integer, and make some modest attempt to track the dependency of the constants on k. Again, we have

$$(2.70) \qquad \mathbf{E}\,\mathrm{tr}(M^k) = \sum_{1 \le i_1, \ldots, i_k \le n} \mathbf{E}\xi_{i_1 i_2} \ldots \xi_{i_k i_1},$$

which is a sum over cycles of length k. Again, the only non-vanishing expectations are those for which each edge occurs twice; in particular, there are at most $k/2$ edges, and thus at most $k/2 + 1$ vertices.

We divide the cycles into various classes, depending on which edges are equal to each other. (More formally, a class is an equivalence relation $\sim$ on a set of k labels, say $\{1, \ldots, k\}$ in which each equivalence class contains at least two elements, and a cycle of k edges $\{i_1, i_2\}, \ldots, \{i_k, i_1\}$ lies in the class associated to $\sim$ when we have that $\{i_j, i_{j+1}\} = \{i_{j'}, i_{j'+1}\}$ iff $j \sim j'$, where we adopt the cyclic notation $i_{k+1} := i_1$.)

How many different classes could there be? We have to assign up to $k/2$ labels to k edges, so a crude upper bound here is $(k/2)^k$.

Now consider a given class of cycle. It has j edges $e_1, \ldots, e_j$ for some $1 \le j \le k/2$, with multiplicities $a_1, \ldots, a_j$, where $a_1, \ldots, a_j$ are at least 2 and add up to k. The j edges span at most $j+1$ vertices; indeed, in addition to the first vertex i_1, one can specify all the other vertices by looking at the first appearance of each of the j edges $e_1, \ldots, e_j$ in the path from i_1 to i_k, and recording the final vertex of each such edge. From this, we see that the total number of cycles in this particular class is at most n^{j+1}. On the other hand, because each ξ_{ij} has mean zero, unit variance, and is bounded by K, the ath moment of this coefficient is at most K^{a-2} for any $a \geq 2$. Thus each summand in (2.70) coming from a cycle in this class has magnitude at most

$$K^{a_1-2} \ldots K^{a_j-2} = K^{a_1+\cdots+a_j-2j} = K^{k-2j}.$$

Thus the total contribution of this class to (2.70) is $n^{j+1}K^{k-2j}$, which we can upper bound by

$$\max(n^{\frac{k}{2}+1}, n^2 K^{k-2}) = n^{k/2+1} \max(1, K/\sqrt{n})^{k-2}.$$

Summing up over all classes, we obtain the (somewhat crude) bound

$$\mathbf{E}\,\mathrm{tr}(M^k) \leq (k/2)^k n^{k/2+1} \max(1, K/\sqrt{n})^{k-2}$$

and thus by (2.66),

$$\mathbf{E}\|M\|_{\mathrm{op}}^k \leq (k/2)^k n^{k/2+1} \max(1, K/\sqrt{n})^{k-2}$$

and so by Markov's inequality (1.13) we have

$$\mathbf{P}(\|M\|_{\mathrm{op}} \geq \lambda) \leq \lambda^{-k}(k/2)^k n^{k/2+1} \max(1, K/\sqrt{n})^{k-2}$$

for all $\lambda > 0$. This, for instance, places the median of $\|M\|_{\mathrm{op}}$ at

$$O(n^{1/k} k \sqrt{n} \max(1, K/\sqrt{n})).$$

We can optimise this in k by choosing k to be comparable to $\log n$, and so we obtain an upper bound of $O(\sqrt{n} \log n \max(1, K/\sqrt{n}))$ for the median; indeed, a slight tweaking of the constants tells us that $\|M\|_{\mathrm{op}} = O(\sqrt{n} \log n \max(1, K/\sqrt{n}))$ with high probability.

The same argument works if the entries have at most unit variance rather than unit variance, thus we have shown

Proposition 2.3.13 (Weak upper bound). *Let M be a random Hermitian matrix, with the upper triangular entries ξ_{ij}, $i \leq j$ being independent with mean zero and variance at most 1, and bounded in magnitude by K. Then $\|M\|_{\mathrm{op}} = O(\sqrt{n} \log n \max(1, K/\sqrt{n}))$ with high probability.*

When $K \leq \sqrt{n}$, this gives an upper bound of $O(\sqrt{n} \log n)$, which is still off by a logarithmic factor from the expected bound of $O(\sqrt{n})$. We will remove this logarithmic loss later in this section.

2.3.5. Computing the moment to top order. Now let us consider the case when $K = o(\sqrt{n})$, and each entry has variance exactly 1. We have an upper bound

$$\mathbf{E}\,\mathrm{tr}(M^k) \leq (k/2)^k n^{k/2+1};$$

let us try to get a more precise answer here (as in Exercises 2.3.7, 2.3.8). Recall that each class of cycles contributed a bound of $n^{j+1}K^{k-2j}$ to this expression. If $K = o(\sqrt{n})$, we see that such expressions are $o_k(n^{k/2+1})$ whenever $j < k/2$, where the $o_k()$ notation means that the decay rate as $n \to \infty$ can depend on k. So the total contribution of all such classes is $o_k(n^{k/2+1})$.

Now we consider the remaining classes with $j = k/2$. For such classes, each equivalence class of edges contains exactly two representatives, thus each edge is repeated exactly once. The contribution of each such cycle to (2.70) is exactly 1, thanks to the unit variance and independence hypothesis. Thus, the total contribution of these classes to $\mathbf{E} \operatorname{tr}(M^k)$ is equal to a purely combinatorial quantity, namely the number of cycles of length k on $\{1, \ldots, n\}$ in which each edge is repeated exactly once, yielding $k/2$ unique edges. We are thus faced with the enumerative combinatorics problem of bounding this quantity as precisely as possible.

With $k/2$ edges, there are at most $k/2 + 1$ vertices traversed by the cycle. If there are fewer than $k/2 + 1$ vertices traversed, then there are at most $O_k(n^{k/2}) = o_k(n^{k/2+1})$ cycles of this type, since one can specify such cycles by identifying up to $k/2$ vertices in $\{1, \ldots, n\}$ and then matching those coordinates with the k vertices of the cycle. So we set aside these cycles, and only consider those cycles which traverse exactly $k/2 + 1$ vertices. Let us call such cycles (i.e., cycles of length k with each edge repeated exactly once, and traversing exactly $k/2 + 1$ vertices) *non-crossing cycles* of length k in $\{1, \ldots, n\}$. Our remaining task is then to count the number of non-crossing cycles.

Example 2.3.14. Let a, b, c, d be distinct elements of $\{1, \ldots, n\}$. Then $(i_1, \ldots, i_6) = (a, b, c, d, c, b)$ is a non-crossing cycle of length 6, as is (a, b, a, c, a, d). Any cyclic permutation of a non-crossing cycle is again a non-crossing cycle.

Exercise 2.3.9. Show that a cycle of length k is non-crossing if and only if there exists a tree[21] in $\{1, \ldots, n\}$ of $k/2$ edges and $k/2 + 1$ vertices, such that the cycle lies in the tree and traverses each edge in the tree exactly twice.

Exercise 2.3.10. Let $i_1, \ldots, i_k$ be a cycle of length k. Arrange the integers $1, \ldots, k$ around a circle. Whenever $1 \le a < b \le k$ are such that $i_a = i_b$, with no c between a and b for which $i_a = i_c = i_b$, draw a line segment between a and b. Show that the cycle is non-crossing if and only if the number of line segments is exactly $k/2 - 1$, and the line segments do not cross each other. This may help explain the terminology "non-crossing".

---

[21] In graph theory, a *tree* is a finite collection of vertices and (undirected) edges between vertices, which do not contain any cycles.

Now we can complete the count. If k is a positive even integer, define a *Dyck word*[22] of length k to be the number of words consisting of left and right parentheses (,) of length k, such that when one reads from left to right, there are always at least as many left parentheses as right parentheses (or in other words, the parentheses define a valid nesting). For instance, the only Dyck word of length 2 is (), the two Dyck words of length 4 are (()) and ()(), and the five Dyck words of length 6 are

$$()()(), (())(), ()(()), (())(), ((())),$$

and so forth.

Lemma 2.3.15. *The number of non-crossing cycles of length k in $\{1, \ldots, n\}$ is equal to $C_{k/2}n(n-1)\ldots(n-k/2)$, where $C_{k/2}$ is the number of Dyck words of length k. (The number $C_{k/2}$ is also known as the $(k/2)^{\text{th}}$ Catalan number.)*

Proof. We will give a *bijective proof*. Namely, we will find a way to store a non-crossing cycle as a Dyck word, together with an (ordered) sequence of $k/2 + 1$ distinct elements from $\{1, \ldots, n\}$, in such a way that any such pair of a Dyck word and ordered sequence generates exactly one non-crossing cycle. This will clearly give the claim.

So, let us take a non-crossing cycle $i_1, \ldots, i_k$. We imagine traversing this cycle from i_1 to i_2, then from i_2 to i_3, and so forth until we finally return to i_1 from i_k. On each leg of this journey, say from i_j to i_{j+1}, we either use an edge that we have not seen before, or else we are using an edge for the second time. Let us say that the leg from i_j to i_{j+1} is an *innovative* leg if it is in the first category, and a *returning* leg otherwise. Thus there are $k/2$ innovative legs and $k/2$ returning legs. Clearly, it is only the innovative legs that can bring us to vertices that we have not seen before. Since we have to visit $k/2 + 1$ distinct vertices (including the vertex i_1 we start at), we conclude that each innovative leg must take us to a new vertex. We thus record, in order, each of the new vertices we visit, starting at i_1 and adding another vertex for each innovative leg; this is an ordered sequence of $k/2 + 1$ distinct elements of $\{1, \ldots, n\}$. Next, traversing the cycle again, we write a (whenever we traverse an innovative leg, and a) otherwise. This is clearly a Dyck word. For instance, using the examples in Example 2.3.14, the non-crossing cycle (a, b, c, d, c, b) gives us the ordered sequence (a, b, c, d) and the Dyck word $((()))$, while (a, b, a, c, a, d) gives us the ordered sequence (a, b, c, d) and the Dyck word ()()().

We have seen that every non-crossing cycle gives rise to an ordered sequence and a Dyck word. A little thought shows that the cycle can be

---

[22]Dyck words are also closely related to *Dyck paths* in enumerative combinatorics.

uniquely reconstructed from this ordered sequence and Dyck word (the key point being that whenever one is performing a returning leg from a vertex v, one is forced to return along the unique innovative leg that discovered v). A slight variant of this thought also shows that every Dyck word of length k and ordered sequence of $k/2+1$ distinct elements gives rise to a non-crossing cycle. This gives the required bijection, and the claim follows. $\qquad\square$

Next, we recall the classical formula for the Catalan number:

Exercise 2.3.11. Establish the recurrence

$$C_{n+1} = \sum_{i=0}^{n} C_i C_{n-i}$$

for any $n \geq 1$ (with the convention $C_0 = 1$), and use this to deduce that

(2.71) $$C_{k/2} := \frac{k!}{(k/2+1)!(k/2)!}$$

for all $k = 2, 4, 6, \ldots$.

Exercise 2.3.12. Let k be a positive even integer. Given a string of $k/2$ left parentheses and $k/2$ right parentheses which is *not* a Dyck word, define the *reflection* of this string by taking the first right parenthesis which does not have a matching left parenthesis, and then reversing all the parentheses after that right parenthesis. Thus, for instance, the reflection of ())(() is ())))(. Show that there is a bijection between non-Dyck words with $k/2$ left parentheses and $k/2$ right parentheses, and arbitrary words with $k/2 - 1$ left parentheses and $k/2+1$ right parentheses. Use this to give an alternate proof of (2.71).

Note that $n(n-1)\ldots(n-k/2) = (1 + o_k(1))n^{k/2+1}$. Putting all the above computations together, we conclude

Theorem 2.3.16 (Moment computation). *Let M be a real symmetric random matrix, with the upper triangular elements ξ_{ij}, $i \leq j$ jointly independent with mean zero and variance one, and bounded in magnitude by $o(\sqrt{n})$. Let k be a positive even integer. Then we have*

$$\mathbf{E}\operatorname{tr}(M^k) = (C_{k/2} + o_k(1))n^{k/2+1}$$

where $C_{k/2}$ is given by (2.71).

Remark 2.3.17. An inspection of the proof also shows that if we allow the ξ_{ij} to have variance at most one, rather than equal to one, we obtain the upper bound

$$\mathbf{E}\operatorname{tr}(M^k) \leq (C_{k/2} + o_k(1))n^{k/2+1}.$$

Exercise 2.3.13. Show that Theorem 2.3.16 also holds for Hermitian random matrices. (*Hint:* The main point is that with non-crossing cycles, each non-innovative leg goes in the reverse direction to the corresponding innovative leg—why?)

Remark 2.3.18. Theorem 2.3.16 can be compared with the formula

$$\mathbf{E} S^k = (C'_{k/2} + o_k(1)) n^{k/2}$$

derived in Section 2.1, where $S = X_1 + \cdots + X_n$ is the sum of n iid random variables of mean zero and variance one, and

$$C'_{k/2} := \frac{k!}{2^{k/2} (k/2)!}.$$

Exercise 2.3.10 shows that $C_{k/2}$ can be interpreted as the number of ways to join k points on the circle by $k/2 - 1$ non-crossing chords. In a similar vein, $C'_{k/2}$ can be interpreted as the number of ways to join k points on the circle by $k/2$ chords which are allowed to cross each other (except at the endpoints). Thus moments of Wigner-type matrices are in some sense the "non-crossing" version of moments of sums of random variables. We will discuss this phenomenon more when we turn to free probability in Section 2.5.

Combining Theorem 2.3.16 with (2.66) we obtain a lower bound

$$\mathbf{E} \| M \|_{\mathrm{op}}^k \geq (C_{k/2} + o_k(1)) n^{k/2}.$$

In the bounded case $K = O(1)$, we can combine this with Exercise 2.3.5 to conclude that the median (or mean) of $\| M \|_{\mathrm{op}}$ is at least $(C_{k/2}^{1/k} + o_k(1)) \sqrt{n}$. On the other hand, from Stirling's formula (Section 1.2) we see that $C_{k/2}^{1/k}$ converges to 2 as $k \to \infty$. Taking k to be a slowly growing function of n, we conclude

Proposition 2.3.19 (Lower Bai-Yin theorem). *Let M be a real symmetric random matrix, with the upper triangular elements ξ_{ij}, $i \leq j$ jointly independent with mean zero and variance one, and bounded in magnitude by $O(1)$. Then the median (or mean) of $\| M \|_{\mathrm{op}}$ is at least $(2 - o(1)) \sqrt{n}$.*

Remark 2.3.20. One can in fact obtain an exact asymptotic expansion of the moments $\mathbf{E} \operatorname{tr}(M^k)$ as a polynomial in n, known as the *genus expansion* of the moments. This expansion is, however, somewhat difficult to work with from a combinatorial perspective (except at top order) and will not be used here.

2.3.6. Removing the logarithm. The upper bound in Proposition 2.3.13 loses a logarithm in comparison to the lower bound coming from Theorem 2.3.16. We now discuss how to remove this logarithm.

Suppose that we could eliminate the $o_k(1)$ error in Theorem 2.3.16. Then from (2.66) we would have

$$\mathbf{E}\|M\|_{\mathrm{op}}^k \leq C_{k/2} n^{k/2+1}$$

and hence by Markov's inequality (1.13),

$$\mathbf{P}(\|M\|_{\mathrm{op}} > \lambda) \leq \lambda^{-k} C_{k/2} n^{k/2+1}.$$

Applying this with $\lambda = (2 + \varepsilon)\sqrt{n}$ for some fixed $\varepsilon > 0$, and setting k to be a large multiple of $\log n$, we see that $\|M\|_{\mathrm{op}} \leq (2 + O(\varepsilon))\sqrt{n}$ asymptotically almost surely, which on selecting ε to go to zero slowly in n gives in fact that $\|M\|_{\mathrm{op}} \leq (2 + o(1))\sqrt{n}$ asymptotically almost surely, thus complementing the lower bound in Proposition 2.3.19.

This argument was not rigorous because it did not address the $o_k(1)$ error. Without a more quantitative accounting of this error, one cannot set k as large as $\log n$ without losing control of the error terms; and indeed, a crude accounting of this nature will lose factors of k^k which are unacceptable. Nevertheless, by tightening the hypotheses a little bit and arguing more carefully, we can get a good bound, for k in the region of interest:

Theorem 2.3.21 (Improved moment bound). *Let M be a real symmetric random matrix, with the upper triangular elements ξ_{ij}, $i \leq j$ jointly independent with mean zero and variance one, and bounded in magnitude by $O(n^{0.49})$ (say). Let k be a positive even integer of size $k = O(\log^2 n)$ (say). Then we have*

$$\mathbf{E}\operatorname{tr}(M^k) = C_{k/2} n^{k/2+1} + O(k^{O(1)} 2^k n^{k/2+0.98})$$

where $C_{k/2}$ is given by (2.71). In particular, from the trivial bound $C_{k/2} \leq 2^k$ (which is obvious from the Dyck words definition) one has

(2.72) $$\mathbf{E}\operatorname{tr}(M^k) \leq (2 + o(1))^k n^{k/2+1}.$$

One can of course adjust the parameters $n^{0.49}$ and $\log^2 n$ in the above theorem, but we have tailored these parameters for our application to simplify the exposition slightly.

Proof. We may assume n large, as the claim is vacuous for bounded n.

We again expand using (2.70), and discard all the cycles in which there is an edge that only appears once. The contribution of the non-crossing cycles was already computed in the previous section to be

$$C_{k/2} n(n-1) \ldots (n - k/2),$$

which can easily be computed (e.g., by taking logarithms, or using Stirling's formula) to be $(C_{k/2} + o(1))n^{k/2+1}$. So the only task is to show that the net contribution of the remaining cycles is $O(k^{O(1)}2^k n^{k/2+0.98})$.

Consider one of these cycles $(i_1, \ldots, i_k)$; it has j distinct edges for some $1 \le j \le k/2$ (with each edge repeated at least once).

We order the j distinct edges $e_1, \ldots, e_j$ by their first appearance in the cycle. Let $a_1, \ldots, a_j$ be the multiplicities of these edges, thus the $a_1, \ldots, a_j$ are all at least 2 and add up to k. Observe from the moment hypotheses that the moment $\mathbf{E}|\xi_{ij}|^a$ is bounded by $O(n^{0.49})^{a-2}$ for $a \ge 2$. Since $a_1 + \cdots + a_j = k$, we conclude that the expression

$$\mathbf{E}\xi_{i_1 i_2} \cdots \xi_{i_k i_1}$$

in (2.70) has magnitude at most $O(n^{0.49})^{k-2j}$, and so the net contribution of the cycles that are not non-crossing is bounded in magnitude by

$$(2.73) \qquad \sum_{j=1}^{k/2} O(n^{0.49})^{k-2j} \sum_{a_1,\ldots,a_j} N_{a_1,\ldots,a_j}$$

where $a_1, \ldots, a_j$ range over integers that are at least 2 and which add up to k, and $N_{a_1,\ldots,a_j}$ is the number of cycles that are not non-crossing and have j distinct edges with multiplicity $a_1, \ldots, a_j$ (in order of appearance). It thus suffices to show that (2.73) is $O(k^{O(1)}2^k n^{k/2+0.98})$.

Next, we estimate $N_{a_1,\ldots,a_j}$ for a fixed $a_1, \ldots, a_j$. Given a cycle $(i_1, \ldots, i_k)$, we traverse its k legs (which each traverse one of the edges $e_1, \ldots, e_j$) one at a time and classify them into various categories:

(i) *High-multiplicity legs*, which use an edge e_i whose multiplicity a_i is larger than two.

(ii) *Fresh legs*, which use an edge e_i with $a_i = 2$ for the first time.

(iii) *Return legs*, which use an edge e_i with $a_i = 2$ that has already been traversed by a previous fresh leg.

We also subdivide fresh legs into *innovative* legs, which take one to a vertex one has not visited before, and *non-innovative* legs, which take one to a vertex that one has visited before.

At any given point in time when traversing this cycle, we define an *available* edge to be an edge e_i of multiplicity $a_i = 2$ that has already been traversed by its fresh leg, but not by its return leg. Thus, at any given point in time, one travels along either a high-multiplicity leg, a fresh leg (thus creating a new available edge), or one returns along an available edge (thus removing that edge from availability).

Call a return leg starting from a vertex v *forced* if, at the time one is performing that leg, there is only one available edge from v, and *unforced* otherwise (i.e., there are two or more available edges to choose from).

We suppose that there are $l := \#\{1 \leq i \leq j : a_i > 2\}$ high-multiplicity edges among the $e_1, \ldots, e_j$, leading to $j - l$ fresh legs and their $j - l$ return leg counterparts. In particular, the total number of high-multiplicity legs is

$$(2.74) \qquad \sum_{a_i > 2} a_i = k - 2(j - l).$$

Since $\sum_{a_i > 2} a_i \geq 3l$, we conclude the bound

$$(2.75) \qquad l \leq k - 2j.$$

We assume that there are m non-innovative legs among the $j - l$ fresh legs, leaving $j - l - m$ innovative legs. As the cycle is not non-crossing, we either have $j < k/2$ or $m > 0$.

Similarly, we assume that there are r unforced return legs among the $j - l$ total return legs. We have an important estimate:

Lemma 2.3.22 (Not too many unforced return legs). *We have*

$$r \leq 2(m + \sum_{a_i > 2} a_i).$$

In particular, from (2.74), (2.75), *we have*

$$r \leq O(k - 2j) + O(m).$$

Proof. Let v be a vertex visited by the cycle which is not the initial vertex i_1. Then the very first arrival at v comes from a fresh leg, which immediately becomes available. Each departure from v may create another available edge from v, but each subsequent arrival at v will delete an available leg from v, unless the arrival is along a non-innovative or high-multiplicity edge[23]. Finally, any returning leg that departs from v will also delete an available edge from v.

This has two consequences. First, if there are no non-innovative or high-multiplicity edges arriving at v, then whenever one arrives at v, there is at most one available edge from v, and so every return leg from v is forced (and there will be only one such return leg). If, instead, there are non-innovative or high-multiplicity edges arriving at v, then we see that the total number of return legs from v is at most one plus the number of such edges. In both cases, we conclude that the number of unforced return legs from v is bounded by twice the number of non-innovative or high-multiplicity edges arriving at v. Summing over v, one obtains the claim. $\square$

---

[23]Note that one can loop from v to itself and create an available edge, but this is along a non-innovative edge and so is not inconsistent with the previous statements.

Now we return to the task of counting $N_{a_1,\ldots,a_j}$, by recording various data associated to any given cycle $(i_1, \ldots, i_k)$ contributing to this number. First, fix m, r. We record the initial vertex i_1 of the cycle, for which there are n possibilities. Next, for each high-multiplicity edge e_i (in increasing order of i), we record all the a_i locations in the cycle where this edge is used; the total number of ways this can occur for each such edge can be bounded above by k^{a_i}, so the total entropy cost here is $k^{\sum_{a_i>2} a_i} = k^{k-2(j-l)}$. We also record the final endpoint of the first occurrence of the edge e_i for each such i; this list of l vertices in $\{1, \ldots, n\}$ has at most n^l possibilities.

For each innovative leg, we record the final endpoint of that leg, leading to an additional list of $j - l - m$ vertices with at most n^{j-l-m} possibilities.

For each non-innovative leg, we record the position of that leg, leading to a list of m numbers from $\{1, \ldots, k\}$, which has at most k^m possibilities.

For each unforced return leg, we record the position of the corresponding fresh leg, leading to a list of r numbers from $\{1, \ldots, k\}$, which has at most k^r possibilities.

Finally, we record a Dyck-like word of length k, in which we place a (whenever the leg is innovative, and) otherwise (the brackets need not match here). The total entropy cost here can be bounded above by 2^k.

We now observe that all this data (together with l, m, r) can be used to completely reconstruct the original cycle. Indeed, as one traverses the cycle, the data already tells us which edges are high-multiplicity, which ones are innovative, which ones are non-innovative, and which ones are return legs. In all edges in which one could possibly visit a new vertex, the location of that vertex has been recorded. For all unforced returns, the data tells us which fresh leg to backtrack upon to return to. Finally, for forced returns, there is only one available leg to backtrack to, and so one can reconstruct the entire cycle from this data.

As a consequence, for fixed l, m and r, there are at most

$$n k^{k-2(j-l)} n^l n^{j-l-m} k^m k^r 2^k$$

contributions to $N_{a_1,\ldots,a_j}$; using (2.75), (2.3.22) we can bound this by

$$k^{O(k-2j)+O(m)} n^{j-m+1} 2^k.$$

Summing over the possible values of m, r (recalling that we either have $j < k/2$ or $m > 0$, and also that $k = O(\log^2 n)$) we obtain

$$N_{a_1,\ldots,a_j} \leq k^{O(k-2j)+O(1)} n^{\max(j+1, k/2)} 2^k.$$

The expression (2.73) can then be bounded by

$$2^k \sum_{j=1}^{k/2} O(n^{0.49})^{k-2j} k^{O(k-2j)+O(1)} n^{\max(j+1,k/2)} \sum_{a_1,\ldots,a_j} 1.$$

When j is exactly $k/2$, then all the $a_1, \ldots, a_j$ must equal 2, and so the contribution of this case simplifies to $2^k k^{O(1)} n^{k/2}$. For $j < k/2$, the numbers $a_1 - 2, \ldots, a_j - 2$ are non-negative and add up to $k - 2j$, and so the total number of possible values for these numbers (for fixed j) can be bounded crudely by $j^{k-2j} \leq k^{k-2j}$ (for instance). Putting all this together, we can bound (2.73) by

$$2^k [k^{O(1)} n^{k/2} + \sum_{j=1}^{k/2-1} O(n^{0.49})^{k-2j} k^{O(k-2j)+O(1)} n^{j+1} k^{k-2j}],$$

which simplifies by the geometric series formula (and the hypothesis $k = O(\log^2 n)$) to

$$O(2^k k^{O(1)} n^{k/2+0.98}),$$

as required. $\qquad\qquad\qquad\qquad\qquad\qquad\qquad\qquad\qquad\qquad\qquad\qquad\square$

We can use this to conclude the following matching upper bound to Proposition 2.3.19, due to Bai and Yin [**BaYi1988**]:

Theorem 2.3.23 (Weak Bai-Yin theorem, upper bound). *Let $M = (\xi_{ij})_{1 \leq i,j \leq n}$ be a real symmetric matrix whose entries all have the same distribution ξ, with mean zero, variance one, and fourth moment $O(1)$. Then for every $\varepsilon > 0$ independent of n, one has $\|M\|_{\mathrm{op}} \leq (2+\varepsilon)\sqrt{n}$ asymptotically almost surely. In particular, $\|M\|_{\mathrm{op}} \leq (2 + o(1))\sqrt{n}$ asymptotically almost surely; as another consequence, the median of $\|M\|_{\mathrm{op}}$ is at most $(2+o(1))\sqrt{n}$. (If ξ is bounded, we see, in particular, from Proposition 2.3.19 that the median is in fact equal to $(2 + o(1))\sqrt{n}$.)*

The fourth moment hypothesis is best possible, as seen in the discussion after Theorem 2.3.8. We will discuss some generalisations and improvements of this theorem in other directions below.

Proof. To obtain Theorem 2.3.23 from Theorem 2.3.21 we use the truncation method. We split each ξ_{ij} as $\xi_{ij,\leq n^{0.49}} + \xi_{ij,>n^{0.49}}$ in the usual manner, and split $M = M_{\leq n^{0.49}} + M_{>n^{0.49}}$ accordingly. We would like to apply Theorem 2.3.21 to $M_{\leq n^{0.49}}$, but unfortunately the truncation causes some slight adjustment to the mean and variance of the $\xi_{ij,\leq n^{0.49}}$. The variance is not much of a problem; since ξ_{ij} had variance 1, it is clear that $\xi_{ij,\leq n^{0.49}}$ has variance at most 1, and it is easy to see that reducing the variance only serves to improve the bound (2.72). As for the mean, we use the mean zero nature of ξ_{ij} to write

$$\mathbf{E}\xi_{ij,\leq n^{0.49}} = -\mathbf{E}\xi_{ij,>n^{0.49}}.$$

To control the right-hand side, we use the trivial inequality $|\xi_{ij,\leq n^{0.49}}| \leq n^{-3\times 0.49}|\xi_{ij}|^4$ and the bounded fourth moment hypothesis to conclude that

$$\mathbf{E}\xi_{ij,\leq n^{0.49}} = O(n^{-1.47}).$$

Thus we can write $M_{\leq n^{0.49}} = \tilde{M}_{\leq n^{0.49}} + \mathbf{E}M_{\leq n^{0.49}}$, where $\tilde{M}_{\leq n^{0.49}}$ is the random matrix with coefficients

$$\tilde{\xi}_{ij,\leq n^{0.49}} := \xi_{ij,\leq n^{0.49}} - \mathbf{E}\xi_{ij,\leq n^{0.49}}$$

and $\mathbf{E}M_{\leq n^{0.49}}$ is a matrix whose entries have magnitude $O(n^{-1.47})$. In particular, by Schur's test this matrix has operator norm $O(n^{-0.47})$, and so by the triangle inequality

$$\|M\|_{\mathrm{op}} \leq \|\tilde{M}_{\leq n^{0.49}}\|_{\mathrm{op}} + \|M_{>n^{0.49}}\|_{\mathrm{op}} + O(n^{-0.47}).$$

The error term $O(n^{-0.47})$ is clearly negligible for n large, and it will suffice to show that

$$(2.76) \qquad\qquad \|\tilde{M}_{\leq n^{0.49}}\|_{\mathrm{op}} \leq (2 + \varepsilon/3)\sqrt{n}$$

and

$$(2.77) \qquad\qquad \|M_{>n^{0.49}}\|_{\mathrm{op}} \leq \frac{\varepsilon}{3}\sqrt{n}$$

asymptotically almost surely.

We first show (2.76). We can now apply Theorem 2.3.21 to conclude that

$$\mathbf{E}\|\tilde{M}_{\leq n^{0.49}}\|_{\mathrm{op}}^k \leq (2 + o(1))^k n^{k/2+1}$$

for any $k = O(\log^2 n)$. In particular, we see from Markov's inequality (1.13) that (2.76) holds with probability at most

$$\left(\frac{2 + o(1)}{2 + \varepsilon/3}\right)^k n.$$

Setting k to be a large enough multiple of $\log n$ (depending on ε), we thus see that this event (2.76) indeed holds asymptotically almost surely[24].

Now we turn to (2.77). The idea here is to exploit the sparseness of the matrix $M_{>n^{0.49}}$. First let us dispose of the event that one of the entries ξ_{ij} has magnitude larger than $\frac{\varepsilon}{3}\sqrt{n}$ (which would certainly cause (2.77) to fail). By the union bound, the probability of this event is at most

$$n^2\mathbf{P}\left(|\xi| \geq \frac{\varepsilon}{3}\sqrt{n}\right).$$

---

[24]Indeed, one can ensure it happens with overwhelming probability, by letting $k/\log n$ grow slowly to infinity.

By the fourth moment bound on ξ and dominated convergence, this expression goes to zero as $n \to \infty$. Thus, asymptotically almost surely, all entries are less than $\frac{\varepsilon}{3}\sqrt{n}$.

Now let us see how many non-zero entries there are in $M_{>n^{0.49}}$. By Markov's inequality (1.13) and the fourth moment hypothesis, each entry has a probability $O(n^{-4\times 0.49}) = O(n^{-1.96})$ of being non-zero; by the first moment method, we see that the expected number of entries is $O(n^{0.04})$. As this is much less than n, we expect it to be unlikely that any row or column has more than one entry. Indeed, from the union bound and independence, we see that the probability that any given row and column has at least two non-zero entries is at most

$$n^2 \times O(n^{-1.96})^2 = O(n^{-1.92})$$

and so by the union bound again, we see that with probability at least $1 - O(n^{-0.92})$ (and in particular, asymptotically almost surely), none of the rows or columns have more than one non-zero entry. As the entries have magnitude at most $\frac{\varepsilon}{3}\sqrt{n}$, the bound (2.77) now follows from Schur's test, and the claim follows. $\square$

We can upgrade the asymptotic almost sure bound to almost sure boundedness:

Theorem 2.3.24 (Strong Bai-Yin theorem, upper bound). *Let ξ be a real random variable with mean zero, variance 1, and finite fourth moment, and for all $1 \leq i \leq j$, let ξ_{ij} be an iid sequence with distribution ξ, and set $\xi_{ji} := \xi_{ij}$. Let $M_n := (\xi_{ij})_{1 \leq i,j \leq n}$ be the random matrix formed by the top left $n \times n$ block. Then almost surely one has $\limsup_{n \to \infty} \|M_n\|_{\mathrm{op}}/\sqrt{n} \leq 2$.*

Exercise 2.3.14. By combining the above results with Proposition 2.3.19 and Exercise 2.3.5, show that with the hypotheses of Theorem 2.3.24 with ξ bounded, one has $\lim_{n \to \infty} \|M_n\|_{\mathrm{op}}/\sqrt{n} = 2$ almost surely[25].

Proof. We first give ourselves an epsilon of room (cf. [**Ta2010**, §2.7]). It suffices to show that for each $\varepsilon > 0$, one has

$$(2.78) \qquad \limsup_{n \to \infty} \|M_n\|_{\mathrm{op}}/\sqrt{n} \leq 2 + \varepsilon$$

almost surely.

Next, we perform dyadic sparsification (as was done in the proof of the strong law of large numbers, Theorem 2.1.8). Observe that any minor of a matrix has its operator norm bounded by that of the larger matrix, and so $\|M_n\|_{\mathrm{op}}$ is increasing in n. Because of this, it will suffice to show (2.78) almost surely for n restricted to a *lacunary* sequence, such as $n = n_m :=$

---

[25]The same claim is true without the boundedness hypothesis; we will see this in Section 2.4.

$\lfloor(1+\varepsilon)^m\rfloor$ for $m = 1, 2, \ldots$, as the general case then follows by rounding n upwards to the nearest n_m (and adjusting ε a little bit as necessary).

Once we sparsified, it is now safe to apply the Borel-Cantelli lemma (Exercise 1.1.1), and it will suffice to show that

$$\sum_{m=1}^{\infty} \mathbf{P}(\|M_{n_m}\|_{\mathrm{op}} \geq (2 + \varepsilon)\sqrt{n_m}) < \infty.$$

To bound the probabilities $\mathbf{P}(\|M_{n_m}\|_{\mathrm{op}} \geq (2+\varepsilon)\sqrt{n_m})$, we inspect the proof of Theorem 2.3.23. Most of the contributions to this probability decay polynomially in n_m (i.e., are of the form $O(n_m^{-c})$ for some $c > 0$) and so are summable. The only contribution which can cause difficulty is the contribution of the event that one of the entries of M_{n_m} exceeds $\frac{\varepsilon}{3}\sqrt{n_m}$ in magnitude; this event was bounded by

$$n_m^2 \mathbf{P}(|\xi| \geq \frac{\varepsilon}{3}\sqrt{n_m}).$$

But if one sums over m using Fubini's theorem and the geometric series formula, we see that this expression is bounded by $O_\varepsilon(\mathbf{E}|\xi|^4)$, which is finite by hypothesis, and the claim follows. $\square$

Now we discuss some variants and generalisations of the Bai-Yin result.

First, we note that the results stated above require the diagonal and off-diagonal terms to have the same distribution. This is not the case for important ensembles such as the Gaussian Orthogonal Ensemble (GOE), in which the diagonal entries have twice as much variance as the off-diagonal ones. But this can easily be handled by considering the diagonal separately. For instance, consider a diagonal matrix $D = \mathrm{diag}(\xi_{11}, \ldots, \xi_{nn})$ where the $\xi_{ii} \equiv \xi$ are identically distributed with finite second moment. The operator norm of this matrix is just $\sup_{1 \leq i \leq n} |\xi_{ii}|$, and so by the union bound

$$\mathbf{P}(\|D\|_{\mathrm{op}} > \varepsilon\sqrt{n}) \leq n\mathbf{P}(|\xi| > \varepsilon\sqrt{n}).$$

From the finite second moment and dominated convergence, the right-hand side is $o_\varepsilon(1)$, and so we conclude that for every fixed $\varepsilon > 0$, $\|D\|_{\mathrm{op}} \leq \varepsilon\sqrt{n}$ asymptotically almost surely; diagonalising, we conclude that $\|D\|_{\mathrm{op}} = o(\sqrt{n})$ asymptotically almost surely. Because of this and the triangle inequality, we can modify the diagonal by any amount with identical distribution and bounded second moment (a similar argument also works for non-identical distributions if one has uniform control of some moment beyond the second, such as the fourth moment) while only affecting all operator norms by $o(\sqrt{n})$.

Exercise 2.3.15. Modify this observation to extend the weak and strong Bai-Yin theorems to the case where the diagonal entries are allowed to have different distribution than the off-diagonal terms, and need not be independent of each other or of the off-diagonal terms, but have uniformly bounded fourth moment.

Second, it is a routine matter to generalise the Bai-Yin result from real symmetric matrices to Hermitian matrices, basically for the same reasons that Exercise 2.3.13 works. We leave the details to the interested reader.

The Bai-Yin results also hold for iid random matrices, where $\xi_{ij} \equiv \xi$ has mean zero, unit variance, and bounded fourth moment; this is a result of Yin, Bai, and Krishnaiah [**YiBaKr1988**], building upon the earlier work of Geman [**Ge1980**]. Because of the lack of symmetry, the eigenvalues need not be real, and the bounds (2.66) no longer apply. However, there is a substitute, namely the bound

$$(2.79) \qquad \|M\|_{\mathrm{op}}^k \leq \mathrm{tr}((MM^*)^{k/2}) \leq n\|M\|_{\mathrm{op}}^k,$$

valid for any $n \times n$ matrix M with complex entries and every even positive integer k.

Exercise 2.3.16. Prove (2.79).

It is possible to adapt all of the above moment calculations for $\mathrm{tr}(M^k)$ in the symmetric or Hermitian cases to give analogous results for $\mathrm{tr}((MM^*)^{k/2})$ in the non-symmetric cases; we do not give the details here, but mention that the cycles now go back and forth along a bipartite graph with n vertices in each class, rather than in the complete graph on n vertices, although this ends up not affecting the enumerative combinatorics significantly. Another way of viewing this is through the simple observation that the operator norm of a non-symmetric matrix M is equal to the operator norm of the *augmented matrix*

$$(2.80) \qquad \tilde{M} := \begin{pmatrix} 0 & M \\ M^* & 0 \end{pmatrix},$$

which is a $2n \times 2n$ Hermitian matrix. Thus, one can to some extent identify an $n \times n$ iid matrix M with a $2n \times 2n$ Wigner-type matrix $\tilde{M}$, in which two $n \times n$ blocks of that matrix are set to zero.

Exercise 2.3.17. If M has singular values $\sigma_1, \ldots, \sigma_n$, show that $\tilde{M}$ has eigenvalues $\pm\sigma_1, \ldots, \pm\sigma_n$. This suggests that the theory of the singular values of an iid matrix should resemble to some extent the theory of eigenvalues of a Wigner matrix; we will see several examples of this phenomenon in later sections.

When one assumes more moment conditions on ξ than bounded fourth moment, one can obtain substantially more precise asymptotics on $\mathrm{tr}(M^k)$ than given by results such as Theorem 2.3.21, particularly if one also assumes that the underlying random variable ξ is symmetric (i.e., $\xi \equiv -\xi$). At a practical level, the advantage of symmetry is that it allows one to assume that the high-multiplicity edges in a cycle are traversed an *even* number of times; see the following exercise.

Exercise 2.3.18. Let X be a bounded real random variable. Show that X is symmetric if and only if $\mathbf{E}X^k = 0$ for all positive odd integers k.

Next, extend the previous result to the case when X is sub-Gaussian rather than bounded. (*Hint*: The slickest way to do this is via the characteristic function e^{itX} and analytic continuation; it is also instructive to find a "real-variable" proof that avoids the use of this function.)

By using these methods, it is in fact possible to show that under various hypotheses, $\|M\|_{\mathrm{op}}$ is concentrated in the range $[2\sqrt{n} - O(n^{-1/6}), 2\sqrt{n} + O(n^{-1/6})]$, and even to get a universal distribution for the normalised expression $(\|M\|_{\mathrm{op}} - 2\sqrt{n})n^{1/6}$, known as the *Tracy-Widom law*. See [**So1999**] for details. There have also been a number of subsequent variants and refinements of this result (as well as counterexamples when not enough moment hypotheses are assumed); see[26] [**So2004, SoFy2005, Ru2007, Pe2006, Vu2007, PeSo2007, Pe2009, Kh2009, TaVu2009c**].

2.4. The semicircular law

We can now turn attention to one of the centerpiece universality results in random matrix theory, namely the *Wigner semicircle law* for Wigner matrices. Recall from Section 2.3 that a *Wigner Hermitian matrix ensemble* is a random matrix ensemble $M_n = (\xi_{ij})_{1 \le i,j \le n}$ of Hermitian matrices (thus $\xi_{ij} = \overline{\xi_{ji}}$; this includes *real symmetric matrices* as an important special case), in which the upper-triangular entries ξ_{ij}, $i > j$ are iid complex random variables with mean zero and unit variance, and the diagonal entries ξ_{ii} are iid real variables, independent of the upper-triangular entries, with bounded mean and variance. Particular special cases of interest include the *Gaussian Orthogonal Ensemble (GOE)*, the *symmetric random sign matrices* (aka *symmetric Bernoulli ensemble*), and the *Gaussian Unitary Ensemble (GUE)*.

In Section 2.3 we saw that the operator norm of M_n was typically of size $O(\sqrt{n})$, so it is natural to work with the normalised matrix $\frac{1}{\sqrt{n}}M_n$. Accordingly, given any $n \times n$ Hermitian matrix M_n, we can form the (normalised)

---

[26]Similar results for some non-independent distributions are also available, see e.g., the paper [**DeGi2007**], which (like many of the other references cited above) builds upon the original work of Tracy and Widom [**TrWi2002**] that handled special ensembles such as GOE and GUE.

empirical spectral distribution (or *ESD* for short)

$$\mu_{\frac{1}{\sqrt{n}}M_n} := \frac{1}{n}\sum_{j=1}^{n}\delta_{\lambda_j(M_n)/\sqrt{n}},$$

of M_n, where $\lambda_1(M_n) \leq \cdots \leq \lambda_n(M_n)$ are the (necessarily real) eigenvalues of M_n, counting multiplicity. The ESD is a probability measure, which can be viewed as a distribution of the normalised eigenvalues of M_n.

When M_n is a random matrix ensemble, then the ESD $\mu_{\frac{1}{\sqrt{n}}M_n}$ is now a *random* measure; i.e., a random variable[27] taking values in the space $\mathrm{Pr}(\mathbf{R})$ of probability measures on the real line.

Now we consider the behaviour of the ESD of a sequence of Hermitian matrix ensembles M_n as $n \to \infty$. Recall from Section 1.1 that for any sequence of random variables in a σ-compact metrisable space, one can define notions of *convergence in probability* and *convergence almost surely*. Specialising these definitions to the case of random probability measures on $\mathbf{R}$, and to deterministic limits, we see that a sequence of random ESDs $\mu_{\frac{1}{\sqrt{n}}M_n}$ *converge in probability* (resp. *converge almost surely*) to a deterministic limit $\mu \in \mathrm{Pr}(\mathbf{R})$ (which, confusingly enough, is a deterministic probability measure!) if, for every test function $\varphi \in C_c(\mathbf{R})$, the quantities $\int_{\mathbf{R}} \varphi \, d\mu_{\frac{1}{\sqrt{n}}M_n}$ converge in probability (resp. converge almost surely) to $\int_{\mathbf{R}} \varphi \, d\mu$.

Remark 2.4.1. As usual, convergence almost surely implies convergence in probability, but not vice versa. In the special case of random probability measures, there is an even weaker notion of convergence, namely *convergence in expectation*, defined as follows. Given a random ESD $\mu_{\frac{1}{\sqrt{n}}M_n}$, one can form its *expectation* $\mathbf{E}\mu_{\frac{1}{\sqrt{n}}M_n} \in \mathrm{Pr}(\mathbf{R})$, defined via duality (the Riesz representation theorem) as

$$\int_{\mathbf{R}} \varphi \, d\mathbf{E}\mu_{\frac{1}{\sqrt{n}}M_n} := \mathbf{E}\int_{\mathbf{R}} \varphi \, d\mu_{\frac{1}{\sqrt{n}}M_n};$$

this probability measure can be viewed as the law of a *random* eigenvalue $\frac{1}{\sqrt{n}}\lambda_i(M_n)$ drawn from a random matrix M_n from the ensemble. We then say that the ESDs converge in expectation to a limit $\mu \in \mathrm{Pr}(\mathbf{R})$ if $\mathbf{E}\mu_{\frac{1}{\sqrt{n}}M_n}$ converges in the vague topology to μ, thus

$$\mathbf{E}\int_{\mathbf{R}} \varphi \, d\mu_{\frac{1}{\sqrt{n}}M_n} \to \int_{\mathbf{R}} \varphi \, d\mu$$

for all $\varphi \in C_c(\mathbf{R})$.

---

[27]Thus, the distribution of $\mu_{\frac{1}{\sqrt{n}}M_n}$ is a probability measure on probability measures!

In general, these notions of convergence are distinct from each other; but in practice, one often finds in random matrix theory that these notions are effectively equivalent to each other, thanks to the concentration of measure phenomenon.

Exercise 2.4.1. Let M_n be a sequence of $n \times n$ Hermitian matrix ensembles, and let μ be a continuous probability measure on $\mathbf{R}$.

(i) Show that $\mu_{\frac{1}{\sqrt{n}}M_n}$ converges almost surely to μ if and only if $\mu_{\frac{1}{\sqrt{n}}M_n}(-\infty, \lambda)$ converges almost surely to $\mu(-\infty, \lambda)$ for all $\lambda \in \mathbf{R}$.

(ii) Show that $\mu_{\frac{1}{\sqrt{n}}M_n}$ converges in probability to μ if and only if $\mu_{\frac{1}{\sqrt{n}}M_n}(-\infty, \lambda)$ converges in probability to $\mu(-\infty, \lambda)$ for all $\lambda \in \mathbf{R}$.

(iii) Show that $\mu_{\frac{1}{\sqrt{n}}M_n}$ converges in expectation to μ if and only if $\mathbf{E}\mu_{\frac{1}{\sqrt{n}}M_n}(-\infty, \lambda)$ converges to $\mu(-\infty, \lambda)$ for all $\lambda \in \mathbf{R}$.

We can now state the Wigner semicircular law.

Theorem 2.4.2 (Semicircular law). *Let M_n be the top left $n \times n$ minors of an infinite Wigner matrix $(\xi_{ij})_{i,j \geq 1}$. Then the ESDs $\mu_{\frac{1}{\sqrt{n}}M_n}$ converge almost surely (and hence also in probability and in expectation) to the* Wigner semicircular distribution

$$(2.81) \qquad \mu_{\mathrm{sc}} := \frac{1}{2\pi}(4 - |x|^2)^{1/2}_+ \, dx.$$

The semicircular law nicely complements the upper Bai-Yin theorem (Theorem 2.3.24), which asserts that (in the case when the entries have finite fourth moment, at least), the matrices $\frac{1}{\sqrt{n}}M_n$ almost surely have operator norm at most $2 + o(1)$. Note that the operator norm is the same thing as the largest magnitude of the eigenvalues. Because the semicircular distribution (2.81) is supported on the interval $[-2, 2]$ with positive density on the interior of this interval, Theorem 2.4.2 easily supplies the *lower Bai-Yin theorem*, that the operator norm of $\frac{1}{\sqrt{n}}M_n$ is almost surely *at least* $2 - o(1)$, and thus (in the finite fourth moment case) the norm is in fact *equal* to $2 + o(1)$. Indeed, we have just shown that the semcircular law provides an alternate proof of the lower Bai-Yin bound (Proposition 2.3.19).

As will become clearer in the Section 2.5, the semicircular law is the non-commutative (or *free probability*) analogue of the central limit theorem, with the semicircular distribution (2.81) taking on the role of the normal distribution. Of course, there is a striking difference between the two distributions, in that the former is compactly supported while the latter is merely sub-Gaussian. One reason for this is that the concentration of measure phenomenon is more powerful in the case of ESDs of Wigner matrices than it is for averages of iid variables; compare the concentration of measure results in Section 2.3 with those in Section 2.1.

There are several ways to prove (or at least to heuristically justify) the semicircular law. In this section we shall focus on the two most popular methods, the *moment method* and the *Stieltjes transform method*, together with a third (heuristic) method based on Dyson Brownian motion (see Section 3.1). In Section 2.5 we shall study the free probability approach, and in Section 2.6 we will study the determinantal processes method approach (although this method is initially only restricted to highly symmetric ensembles, such as GUE).

2.4.1. Preliminary reductions. Before we begin any of the proofs of the semicircular law, we make some simple observations which will reduce the difficulty of the arguments in the sequel.

The first observation is that the Cauchy interlacing law (Exercise 1.3.14) shows that the ESD of $\frac{1}{\sqrt{n}} M_n$ is very stable in n. Indeed, we see from the interlacing law that

$$\frac{n}{m} \mu_{\frac{1}{\sqrt{n}} M_n}(-\infty, \lambda/\sqrt{n}) - \frac{n-m}{m} \leq \mu_{\frac{1}{\sqrt{m}} M_m}(-\infty, \lambda/\sqrt{m})$$
$$\leq \frac{n}{m} \mu_{\frac{1}{\sqrt{n}} M_n}(-\infty, \lambda/\sqrt{n})$$

for any threshold λ and any $n > m > 0$.

Exercise 2.4.2. Using this observation, show that to establish the semicircular law (in any of the three senses of convergence), it suffices to do so for an arbitrary lacunary sequence $n_1, n_2, \ldots$ of n (thus $n_{j+1}/n_j \geq c$ for some $c > 1$ and all j).

The above lacunary reduction does not help one establish convergence in probability or expectation, but will be useful[28] when establishing almost sure convergence, as it significantly reduces the inefficiency of the union bound.

Next, we exploit the stability of the ESD with respect to perturbations, by taking advantage of the *Weilandt-Hoffmann inequality*

$$(2.82) \qquad \sum_{j=1}^{n} |\lambda_j(A+B) - \lambda_j(A)|^2 \leq \|B\|_F^2$$

for Hermitian matrices A, B, where $\|B\|_F := (\operatorname{tr} B^2)^{1/2}$ is the Frobenius norm (2.64) of B; see Exercise 1.3.6 or Exercise 1.3.4. We convert this inequality into an inequality about ESDs:

---

[28]Note that a similar lacunary reduction was also used to prove the strong law of large numbers, Theorem 2.1.8.

Lemma 2.4.3. *For any $n \times n$ Hermitian matrices A, B, any λ, and any $\varepsilon > 0$, we have*

$$\mu_{\frac{1}{\sqrt{n}}(A+B)}(-\infty, \lambda) \leq \mu_{\frac{1}{\sqrt{n}}(A)}(-\infty, \lambda + \varepsilon) + \frac{1}{\varepsilon^2 n^2} \|B\|_F^2$$

and similarly

$$\mu_{\frac{1}{\sqrt{n}}(A+B)}(-\infty, \lambda) \geq \mu_{\frac{1}{\sqrt{n}}(A)}(-\infty, \lambda - \varepsilon) - \frac{1}{\varepsilon^2 n^2} \|B\|_F^2.$$

Proof. We just prove the first inequality, as the second is similar (and also follows from the first, by reversing the sign of A, B).

Let $\lambda_i(A + B)$ be the largest eigenvalue of $A + B$ less than $\lambda\sqrt{n}$, and let $\lambda_j(A)$ be the largest eigenvalue of A less than $(\lambda + \varepsilon)\sqrt{n}$. Our task is to show that

$$i \leq j + \frac{1}{\varepsilon^2 n} \|B\|_F^2.$$

If $i \leq j$, then we are clearly done, so suppose that $i > j$. Then we have $|\lambda_l(A + B) - \lambda_l(A)| \geq \varepsilon\sqrt{n}$ for all $j < l \leq i$, and hence

$$\sum_{j=1}^{n} |\lambda_j(A + B) - \lambda_j(A)|^2 \geq \varepsilon^2(i - j)n.$$

The claim now follows from (2.82). $\qquad\qquad\qquad\qquad\qquad\qquad\square$

This has the following corollary:

Exercise 2.4.3 (Stability of ESD laws w.r.t. small perturbations). Let M_n be a sequence of random Hermitian matrix ensembles such that $\mu_{\frac{1}{\sqrt{n}}M_n}$ converges almost surely to a limit μ. Let N_n be another sequence of Hermitian random matrix ensembles such that $\frac{1}{n^2}\|N_n\|_F^2$ converges almost surely to zero. Show that $\mu_{\frac{1}{\sqrt{n}}(M_n+N_n)}$ converges almost surely to μ.

Show that the same claim holds if "almost surely" is replaced by "in probability" or "in expectation" throughout.

Informally, this exercise allows us to discard any portion of the matrix which is $o(n)$ in the Frobenius norm(2.64). For instance, the diagonal entries of M_n have a Frobenius norm of $O(\sqrt{n})$ almost surely, by the strong law of large numbers (Theorem 2.1.8). Hence, without loss of generality, we may set the diagonal equal to zero for the purposes of the semicircular law.

One can also remove any component of M_n that is of rank $o(n)$:

Exercise 2.4.4 (Stability of ESD laws w.r.t. small rank perturbations). Let M_n be a sequence of random Hermitian matrix ensembles such that $\mu_{\frac{1}{\sqrt{n}}M_n}$ converges almost surely to a limit μ. Let N_n be another sequence of random matrix ensembles such that $\frac{1}{n}\mathrm{rank}(N_n)$ converges almost surely to zero. Show that $\mu_{\frac{1}{\sqrt{n}}(M_n+N_n)}$ converges almost surely to μ. (*Hint:* Use the Weyl inequalities instead of the Wielandt-Hoffman inequality.)

Show that the same claim holds if "almost surely" is replaced by "in probability" or "in expectation" throughout.

In a similar vein, we may apply the truncation argument (much as was done for the central limit theorem in Section 2.2) to reduce the semicircular law to the bounded case:

Exercise 2.4.5. Show that in order to prove the semicircular law (in the almost sure sense), it suffices to do so under the additional hypothesis that the random variables are bounded; similarly, for the convergence in probability or in expectation senses.

Remark 2.4.4. These facts ultimately rely on the stability of eigenvalues with respect to perturbations. This stability is automatic in the Hermitian case, but for non-symmetric matrices, serious instabilities can occur due to the presence of *pseudospectrum*. We will discuss this phenomenon more in later sections (but see also [**Ta2009b**, §1.5]).

2.4.2. The moment method. We now prove the semicircular law via the *method of moments*, which we have already used several times in the previous sections. In order to use this method, it is convenient to use the preceding reductions to assume that the coefficients are bounded, the diagonal vanishes, and that n ranges over a lacunary sequence. We will implicitly assume these hypotheses throughout the rest of the section.

As we have already discussed the moment method extensively, much of the argument here will be delegated to exercises. A full treatment of these computations can be found in [**BaSi2010**].

The basic starting point is the observation that the moments of the ESD $\mu_{\frac{1}{\sqrt{n}}M_n}$ can be written as normalised traces of powers of M_n:

$$(2.83) \qquad \int_{\mathbf{R}} x^k \, d\mu_{\frac{1}{\sqrt{n}}M_n}(x) = \frac{1}{n}\operatorname{tr}(\frac{1}{\sqrt{n}}M_n)^k.$$

In particular, on taking expectations, we have

$$\int_{\mathbf{R}} x^k \, d\mathbf{E}\mu_{\frac{1}{\sqrt{n}}M_n}(x) = \mathbf{E}\frac{1}{n}\operatorname{tr}(\frac{1}{\sqrt{n}}M_n)^k.$$

From concentration of measure for the operator norm of a random matrix (Proposition 2.3.10), we see that the $\mathbf{E}\mu_{\frac{1}{\sqrt{n}}M_n}$ are uniformly sub-Gaussian; indeed, we have

$$\mathbf{E}\mu_{\frac{1}{\sqrt{n}}M_n}\{|x| \geq \lambda\} \leq Ce^{-c\lambda^2 n^2}$$

for $\lambda > C$, where C, c are absolute (so the decay in fact improves quite rapidly with n). From this and the Carleman continuity theorem (Theorem 2.2.9), we can now establish the circular law through computing the mean and variance of moments:

Exercise 2.4.6. (i) Show that to prove convergence in expectation to the semicircular law, it suffices to show that

$$(2.84) \qquad \mathbf{E}\frac{1}{n}\operatorname{tr}(\frac{1}{\sqrt{n}}M_n)^k = \int_{\mathbf{R}} x^k \, d\mu_{sc}(x) + o_k(1)$$

for $k = 1, 2, \ldots$, where $o_k(1)$ is an expression that goes to zero as $n \to \infty$ for fixed k (and fixed choice of coefficient distribution ξ).

(ii) Show that to prove convergence in probability to the semicircular law, it suffices to show (2.84) together with the variance bound

$$(2.85) \qquad \mathbf{Var}(\frac{1}{n}\operatorname{tr}(\frac{1}{\sqrt{n}}M_n)^k) = o_k(1)$$

for $k = 1, 2, \ldots$.

(iii) Show that to prove almost sure convergence to the semicircular law, it suffices to show (2.84) together with the variance bound

$$(2.86) \qquad \mathbf{Var}(\frac{1}{n}\operatorname{tr}(\frac{1}{\sqrt{n}}M_n)^k) = O_k(n^{-c_k})$$

for $k = 1, 2, \ldots$ and some $c_k > 0$. (Note here that it is useful to restrict n to a lacunary sequence!)

Ordinarily, computing second-moment quantities such as the left-hand side of (2.85) is harder than computing first-moment quantities such as (2.84). But one can obtain the required variance bounds from concentration of measure:

Exercise 2.4.7. (i) When k is a positive even integer, use Talagrand's inequality (Theorem 2.1.13) and convexity of the Schatten norm $\|A\|_{S^k} = (\operatorname{tr}(A^k))^{1/k}$ to establish (2.86) (and hence (2.85)) for such k.

(ii) For k odd, the formula $\|A\|_{S^k} = (\operatorname{tr}(A^k))^{1/k}$ still applies as long as A is positive definite. Applying this observation, the Bai-Yin theorem, and Talagrand's inequality to the S^k norms of $\frac{1}{\sqrt{n}}M_n + cI_n$ for a constant $c > 2$, establish (2.86) (and hence (2.85)) when k is odd also.

Remark 2.4.5. More generally, concentration of measure results (such as Talagrand's inequality, Theorem 2.1.13) can often be used to automatically upgrade convergence in expectation to convergence in probability or almost sure convergence. We will not attempt to formalise this principle here.

It is not difficult to establish (2.86), (2.85) through the moment method as well. Indeed, recall from Theorem 2.3.16 that we have the expected moment

(2.87) $$\mathbf{E}\frac{1}{n}\operatorname{tr}(\frac{1}{\sqrt{n}}M_n)^k = C_{k/2} + o_k(1)$$

for all $k = 1, 2, \ldots$, where the *Catalan number* $C_{k/2}$ is zero when k is odd, and is equal to

(2.88) $$C_{k/2} := \frac{k!}{(k/2+1)!(k/2)!}$$

for k even.

Exercise 2.4.8. By modifying the proof of Theorem 2.3.16, show that

(2.89) $$\mathbf{E}|\frac{1}{n}\operatorname{tr}(\frac{1}{\sqrt{n}}M_n)^k|^2 = C_{k/2}^2 + o_k(1)$$

and deduce (2.85). By refining the error analysis (e.g., using Theorem 2.3.21), also establish (2.86).

In view of the above computations, the establishment of the semicircular law now reduces to computing the moments of the semicircular distribution:

Exercise 2.4.9. Show that for any $k = 1, 2, 3, \ldots$, one has

$$\int_{\mathbf{R}} x^k \, d\mu_{\mathrm{sc}}(x) = C_{k/2}.$$

(*Hint:* Use a trigonometric substitution $x = 2\cos\theta$, and then express the integrand in terms of Fourier phases $e^{in\theta}$.)

This concludes the proof of the semicircular law (for any of the three modes of convergence).

Remark 2.4.6. In the spirit of the Lindeberg exchange method, observe that Exercise (2.4.9) is unnecessary if one already knows that the semicircular law holds for at least one ensemble of Wigner matrices (e.g., the GUE ensemble). Indeed, Exercise 2.4.9 can be *deduced* from such a piece of knowledge. In such a situation, it is not necessary to actually compute the main term $C_{k/2}$ on the right of (2.84); it would be sufficient to know that that limit is *universal*, in that it does not depend on the underlying distribution. In fact, it would even suffice to establish the slightly weaker statement

$$\mathbf{E}\frac{1}{n}\operatorname{tr}\left(\frac{1}{\sqrt{n}}M_n\right)^k = \mathbf{E}\frac{1}{n}\operatorname{tr}\left(\frac{1}{\sqrt{n}}M_n'\right)^k + o_k(1)$$

whenever M_n, M_n' are two ensembles of Wigner matrices arising from different underlying distributions (but still normalised to have mean zero, unit variance, and to be bounded (or at worst sub-Gaussian)). We will take advantage of this perspective later in this section.

Remark 2.4.7. The moment method also leads to good control on various *linear statistics* $\sum_{j=1}^{n} F(\lambda_j)$ of a Wigner matrix, and in particular, can be used to establish a central limit theorem for such statistics under some regularity conditions on F; see e.g. [**Jo1982**].

2.4.3. The Stieltjes transform method. The moment method was computationally intensive, but straightforward. As noted in Remark 2.4.6, even without doing much of the algebraic computation, it is clear that the moment method will show that some universal limit for Wigner matrices exists (or, at least, that the differences between the distributions of two different Wigner matrices converge to zero). But it is not easy to see from this method why the limit should be given by the semicircular law, as opposed to some other distribution (although one could eventually work this out from an inverse moment computation).

When studying the central limit theorem, we were able to use the Fourier method to control the distribution of random matrices in a cleaner way than in the moment method. Analogues of this method for random matrices exist, but require non-trivial formulae from non-commutative Fourier analysis, such as the *Harish-Chandra integration formula* (and also only work for highly symmetric ensembles, such as GUE or GOE), and will not be discussed in this text[29].

We now turn to another method, the *Stieltjes transform method*, pioneered in [**Pa1973**], developed further in [**Ba1993, Ba1993b**] and recently pushed yet further in [**ErScYa2008**], which uses complex-analytic methods rather than Fourier-analytic methods, and has turned out to be one of the most powerful and accurate tools in dealing with the ESD of random Hermitian matrices. Whereas the moment method started from the identity (2.83), the Stieltjes transform method proceeds from the identity

$$\int_{\mathbf{R}} \frac{1}{x-z} \, d\mu_{\frac{1}{\sqrt{n}}M_n}(x) = \frac{1}{n} \operatorname{tr}\left(\frac{1}{\sqrt{n}}M_n - zI\right)^{-1}$$

for any complex z not in the support of $\mu_{\frac{1}{\sqrt{n}}M_n}$. We refer to the expression on the left-hand side as the *Stieltjes transform*[30] of M_n or of $\mu_{\frac{1}{\sqrt{n}}M_n}$, and denote it by $s_{\mu_{\frac{1}{\sqrt{n}}M_n}}$ or as s_n for short. The expression $(\frac{1}{\sqrt{n}}M_n - zI)^{-1}$ is the normalised *resolvent* of M_n, and plays an important role in the spectral theory of that matrix. Indeed, in contrast to general-purpose methods such as the moment method, the Stieltjes transform method draws heavily on the

---

[29]Section 2.6, however, will contain some algebraic identities related in some ways to the non-commutative Fourier-analytic approach.

[30]This transform is also known as the *Cauchy transform*.

specific linear-algebraic structure of this problem, and in particular, on the rich structure of resolvents.

On the other hand, the Stieltjes transform can be viewed as a generating function of the moments via the Taylor series expansion

$$s_n(z) = -\frac{1}{z} - \frac{1}{z^2}\frac{1}{n^{3/2}}\operatorname{tr} M_n - \frac{1}{z^3}\frac{1}{n^2}\operatorname{tr} M_n^2 - \dots,$$

valid for z sufficiently large. This is somewhat (though not exactly) analogous to how the characteristic function $\mathbf{E}e^{itX}$ of a scalar random variable can be viewed as a generating function of the moments $\mathbf{E}X^k$.

Now let us study the Stieltjes transform method more systematically. Given any probability measure μ on the real line, we can form its *Stieltjes transform*

$$s_\mu(z) := \int_{\mathbf{R}} \frac{1}{x-z}\, d\mu(x)$$

for any z outside of the support of μ; in particular, the Stieltjes transform is well-defined on the upper and lower half-planes in the complex plane. Even without any further hypotheses on μ other than it is a probability measure, we can say a remarkable amount about how this transform behaves in z. Applying conjugations we obtain the symmetry

$$(2.90) \qquad \overline{s_\mu(z)} = s_\mu(\overline{z}),$$

so we may as well restrict attention to z in the upper half-plane (say). Next, from the trivial bound

$$|\frac{1}{x-z}| \le \frac{1}{|\operatorname{Im}(z)|}$$

one has the pointwise bound

$$(2.91) \qquad |s_\mu(z)| \le \frac{1}{|\operatorname{Im}(z)|}.$$

In a similar spirit, an easy application of dominated convergence gives the asymptotic

$$(2.92) \qquad s_\mu(z) = \frac{-1 + o_\mu(1)}{z}$$

where $o_\mu(1)$ is an expression that, for any fixed μ, goes to zero as z goes to infinity *non-tangentially* in the sense that $|\operatorname{Re}(z)|/|\operatorname{Im}(z)|$ is kept bounded, where the rate of convergence is allowed to depend on μ. From differentiation under the integral sign (or an application of *Morera's theorem* and Fubini's theorem) we see that $s_\mu(z)$ is *complex analytic* on the upper and lower half-planes; in particular, it is smooth away from the real axis. From the Cauchy integral formula (or differentiation under the integral sign) we in fact get some bounds for higher derivatives of the Stieltjes transform away from this axis:

$$(2.93) \qquad |\frac{d^j}{dz^j} s_\mu(z)| = O_j\left(\frac{1}{|\text{Im}(z)|^{j+1}}\right).$$

Informally, s_μ "behaves like a constant" at scales significantly less than the distance $|\text{Im}(z)|$ to the real axis; all the really interesting action here is going on near that axis.

The imaginary part of the Stieltjes transform is particularly interesting. Writing $z = a + ib$, we observe that

$$\text{Im}\frac{1}{x-z} = \frac{b}{(x-a)^2 + b^2} > 0$$

and so we see that

$$\text{Im}(s_\mu(z)) > 0$$

for z in the upper half-plane; thus s_μ is a complex-analytic map from the upper half-plane to itself, a type of function known as a *Herglotz function*[31].

One can also express the imaginary part of the Stieltjes transform as a convolution

$$(2.94) \qquad \text{Im}(s_\mu(a+ib)) = \pi\mu * P_b(a)$$

where P_b is the *Poisson kernel*

$$P_b(x) := \frac{1}{\pi}\frac{b}{x^2+b^2} = \frac{1}{b}P_1(\frac{x}{b}).$$

As is well known, these kernels form a family of *approximations to the identity*, and thus $\mu * P_b$ converges in the vague topology to μ (see e.g. [**Ta2010**, §1.13]). Thus we see that

$$\text{Im} s_\mu(\cdot + ib) \rightharpoonup \pi\mu$$

as $b \to 0^+$ in the vague topology, or equivalently (by (2.90)) that[32]

$$(2.95) \qquad \frac{s_\mu(\cdot + ib) - s_\mu(\cdot - ib)}{2\pi i} \rightharpoonup \mu$$

as $b \to 0^+$. Thus we see that a probability measure μ can be recovered in terms of the limiting behaviour of the Stieltjes transform on the real axis.

A variant of the above machinery gives us a criterion for convergence:

Exercise 2.4.10 (Stieltjes continuity theorem). Let μ_n be a sequence of random probability measures on the real line, and let μ be a deterministic probability measure.

---

[31]In fact, all complex-analytic maps from the upper half-plane to itself that obey the asymptotic (2.92) are of this form; this is a special case of the *Herglotz representation theorem*, which also gives a slightly more general description in the case when the asymptotic (2.92) is not assumed. A good reference for this material and its consequences is [**Ga2007**].

[32]The limiting formula (2.95) is closely related to the *Plemelj formula* in potential theory.

(i) μ_n converges almost surely to μ in the vague topology if and only if $s_{\mu_n}(z)$ converges almost surely to $s_\mu(z)$ for every z in the upper half-plane.

(ii) μ_n converges in probability to μ in the vague topology if and only if $s_{\mu_n}(z)$ converges in probability to $s_\mu(z)$ for every z in the upper half-plane.

(iii) μ_n converges in expectation to μ in the vague topology if and only if $\mathbf{E}s_{\mu_n}(z)$ converges to $s_\mu(z)$ for every z in the upper half-plane.

(*Hint:* The "only if" parts are fairly easy. For the "if" parts, take a test function $\phi \in C_c(\mathbf{R})$ and approximate $\int_{\mathbf{R}} \phi \, d\mu$ by $\int_{\mathbf{R}} \phi * P_b \, d\mu = \frac{1}{\pi} \operatorname{Im} \int_{\mathbf{R}} s_\mu(a + ib)\phi(a) \, da$. Then approximate this latter integral in turn by a Riemann sum, using (2.93).)

Thus, to prove the semicircular law, it suffices to show that for each z in the upper half-plane, the Stieltjes transform

$$s_n(z) = s_{\mu_{\frac{1}{\sqrt{n}} M_n}}(z) = \frac{1}{n} \operatorname{tr} \left(\frac{1}{\sqrt{n}} M_n - zI \right)^{-1}$$

converges almost surely (and thus in probability and in expectation) to the Stieltjes transform $s_{\mu_{sc}}(z)$ of the semicircular law.

It is not difficult to compute the Stieltjes transform $s_{\mu_{sc}}$ of the semicircular law, but let us hold off on that task for now, because we want to illustrate how the Stieltjes transform method can be used to *find* the semicircular law, even if one did not know this law in advance, by directly controlling $s_n(z)$. We will fix $z = a + ib$ to be a complex number not on the real line, and allow all implied constants in the discussion below to depend on a and b (we will focus here only on the behaviour as $n \to \infty$).

The main idea here is *predecessor comparison*: to compare the transform $s_n(z)$ of the $n \times n$ matrix M_n with the transform $s_{n-1}(z)$ of the top left $n-1 \times n-1$ minor M_{n-1}, or of other minors. For instance, we have the *Cauchy interlacing law* (Exercise 1.75), which asserts that the eigenvalues $\lambda_1(M_{n-1}), \ldots, \lambda_{n-1}(M_{n-1})$ of M_{n-1} intersperse that of $\lambda_1(M_n), \ldots, \lambda_n(M_n)$. This implies that for a complex number $a + ib$ with $b > 0$, the difference

$$\sum_{j=1}^{n-1} \frac{b}{(\lambda_j(M_{n-1})/\sqrt{n} - a)^2 + b^2} - \sum_{j=1}^{n} \frac{b}{(\lambda_j(M_n)/\sqrt{n} - a)^2 + b^2}$$

is an alternating sum of evaluations of the function $x \mapsto \frac{b}{(x-a)^2+b^2}$. The total variation of this function is $O(1)$ (recall that we are suppressing dependence of constants on a, b), and so the alternating sum above is $O(1)$. Writing this in terms of the Stieltjes transform, we conclude that

$$\sqrt{n(n-1)}s_{n-1}\left(\frac{\sqrt{n}}{\sqrt{n-1}}(a+ib)\right) - ns_n(a+ib) = O(1).$$

Applying (2.93) to approximate $s_{n-1}(\frac{\sqrt{n}}{\sqrt{n-1}}(a+ib))$ by $s_{n-1}(a+ib)$, we conclude that

$$(2.96) \qquad\qquad s_n(a+ib) = s_{n-1}(a+ib) + O(\frac{1}{n}).$$

So for fixed $z = a + ib$ away from the real axis, the Stieltjes transform $s_n(z)$ is quite stable in n.

This stability has the following important consequence. Observe that while the left-hand side of (2.96) depends on the $n \times n$ matrix M_n, the right-hand side depends only on the top left minor M_{n-1} of that matrix. In particular, it is *independent* of the nth row and column of M_n. This implies that this entire row and column has only a limited amount of *influence* on the Stieltjes transform $s_n(a + ib)$: no matter what value one assigns to this row and column (including possibly unbounded values, as long as one keeps the matrix Hermitian of course), the transform $s_n(a + ib)$ can only move by $O(\frac{|a|+|b|}{b^2 n})$.

By permuting the rows and columns, we obtain that in fact any row or column of M_n can influence $s_n(a + ib)$ by at most $O(\frac{1}{n})$. (This is closely related to the observation in Exercise 2.4.4 that low rank perturbations do not significantly affect the ESD.) On the other hand, the rows of (the upper triangular portion of) M_n are jointly independent. When M_n is a Wigner random matrix, we can then apply a standard concentration of measure result, such as McDiarmid's inequality (Theorem 2.1.10) to conclude concentration of s_n around its mean:

$$(2.97) \qquad \mathbf{P}(|s_n(a+ib) - \mathbf{E}s_n(a+ib)| \geq \lambda/\sqrt{n}) \leq Ce^{-c\lambda^2}$$

for all $\lambda > 0$ and some absolute constants $C, c > 0$. (This is not necessarily the strongest concentration result one can establish for the Stieltjes transform, but it will certainly suffice for our discussion here.) In particular, we see from the Borel-Cantelli lemma (Exercise 1.1.1) that for any fixed z away from the real line, $s_n(z) - \mathbf{E}s_n(z)$ converges almost surely (and thus also in probability) to zero. As a consequence, convergence of $s_n(z)$ in expectation automatically implies convergence in probability or almost sure convergence.

However, while concentration of measure tells us that $s_n(z)$ is close to its mean, it does not shed much light as to what this mean *is*. For this, we have to go beyond the Cauchy interlacing formula and deal with the resolvent $(\frac{1}{\sqrt{n}}M_n - zI_n)^{-1}$ more directly. First, we observe from the linearity of trace (and of expectation) that

$$\mathbf{E} s_n(z) = \frac{1}{n} \sum_{j=1}^{n} \mathbf{E}\left[\left(\frac{1}{\sqrt{n}} M_n - z I_n\right)^{-1}\right]_{jj}$$

where $[A]_{jj}$ denotes the jj component of a matrix A. Because M_n is a Wigner matrix, it is easy to see on permuting the rows and columns that all of the random variables $[(\frac{1}{\sqrt{n}} M_n - z I_n)^{-1}]_{jj}$ have the same distribution. Thus we may simplify the above formula as

(2.98) $$\mathbf{E} s_n(z) = \mathbf{E}\left[\left(\frac{1}{\sqrt{n}} M_n - z I_n\right)^{-1}\right]_{nn}.$$

So now we have to compute the last entry of an inverse of a matrix. There are of course a number of formulae for this, such as *Cramer's rule*. But it will be more convenient here to use a formula based instead on the *Schur complement*:

Exercise 2.4.11. Let A_n be a $n \times n$ matrix, let A_{n-1} be the top left $n-1 \times n-1$ minor, let a_{nn} be the bottom right entry of A_n, let $X \in \mathbf{C}^{n-1}$ be the right column of A_n with the bottom right entry removed, and let $Y^* \in (\mathbf{C}^{n-1})^*$ be the bottom row with the bottom right entry removed. In other words,

$$A_n = \begin{pmatrix} A_{n-1} & X \\ Y^* & a_{nn} \end{pmatrix}.$$

Assume that A_n and A_{n-1} are both invertible. Show that

$$[A_n^{-1}]_{nn} = \frac{1}{a_{nn} - Y^* A_{n-1}^{-1} X}.$$

(*Hint:* Solve the equation $A_n v = e_n$, where e_n is the nth basis vector, using the method of Schur complements (or from first principles).)

The point of this identity is that it describes (part of) the inverse of A_n in terms of the inverse of A_{n-1}, which will eventually provide a non-trivial recursive relationship between $s_n(z)$ and $s_{n-1}(z)$, which can then be played off against (2.96) to solve for $s_n(z)$ in the asymptotic limit $n \to \infty$.

In our situation, the matrix $\frac{1}{\sqrt{n}} M_n - z I_n$ and its minor $\frac{1}{\sqrt{n}} M_{n-1} - z I_{n-1}$ are automatically invertible. Inserting the above formula into (2.98) (and recalling that we normalised the diagonal of M_n to vanish), we conclude that

(2.99) $$\mathbf{E} s_n(z) = -\mathbf{E} \frac{1}{z + \frac{1}{n} X^* (\frac{1}{\sqrt{n}} M_{n-1} - z I_{n-1})^{-1} X - \frac{1}{\sqrt{n}} \xi_{nn}},$$

where $X \in \mathbf{C}^{n-1}$ is the top right column of M_n with the bottom entry ξ_{nn} removed.

One may be concerned that the denominator here could vanish. However, observe that z has imaginary part b if $z = a + ib$. Furthermore, from the spectral theorem we see that the imaginary part of $(\frac{1}{\sqrt{n}} M_{n-1} - z I_{n-1})^{-1}$ is positive definite, and so $X^*(\frac{1}{\sqrt{n}} M_{n-1} - z I_{n-1})^{-1} X$ has non-negative imaginary part. As a consequence, the magnitude of the denominator here is bounded below by $|b|$, and so its reciprocal is $O(1)$ (compare with (2.91)). So the reciprocal here is not going to cause any discontinuity, as we are considering b is fixed and non-zero.

Now we need to understand the expression $X^*(\frac{1}{\sqrt{n}} M_{n-1} - z I_{n-1})^{-1} X$. We write this as $X^* R X$, where R is the resolvent matrix $R := (\frac{1}{\sqrt{n}} M_{n-1} - z I_{n-1})^{-1}$. The distribution of the random matrix R could conceivably be quite complicated. However, the key point is that the vector X only involves the entries of M_n that do not lie in M_{n-1}, and so the random matrix R and the vector X are *independent*. Because of this, we can use the randomness of X to do most of the work in understanding the expression $X^* R X$, without having to know much about R at all.

To understand this, let us first condition R to be a *deterministic* matrix $R = (r_{ij})_{1 \leq i,j \leq n-1}$, and see what we can do with the expression $X^* R X$.

First, observe that R will not be arbitrary; indeed, from the spectral theorem we see that R will have operator norm at most $O(1)$. Meanwhile, from the Chernoff inequality (Theorem 2.1.3) or Hoeffding inequality (Exercise 2.1.4) we know that X has magnitude $O(\sqrt{n})$ with overwhelming probability. So we know that $X^* R X$ has magnitude $O(n)$ with overwhelming probability.

Furthermore, we can use concentration of measure as follows. Given any positive semi-definite matrix A of operator norm $O(1)$, the expression $(X^* A X)^{1/2} = \|A^{1/2} X\|$ is a Lipschitz function of X with operator norm $O(1)$. Applying Talagrand's inequality (Theorem 2.1.13) we see that this expression concentrates around its median:

$$\mathbf{P}(|(X^* A X)^{1/2} - \mathbf{M}(X^* A X)^{1/2}| \geq \lambda) \leq C e^{-c\lambda^2}$$

for any $\lambda > 0$. On the other hand, $\|A^{1/2} X\| = O(\|X\|)$ has magnitude $O(\sqrt{n})$ with overwhelming probability, so the median $\mathbf{M}(X^* A X)^{1/2}$ must be $O(\sqrt{n})$. Squaring, we conclude that

$$\mathbf{P}(|X^* A X - \mathbf{M} X^* A X| \geq \lambda \sqrt{n}) \leq C e^{-c\lambda^2}$$

(possibly after adjusting the absolute constants C, c). As usual, we may replace the median with the expectation:

$$\mathbf{P}(|X^* A X - \mathbf{E} X^* A X| \geq \lambda \sqrt{n}) \leq C e^{-c\lambda^2}.$$

This was for positive-definite matrices, but one can easily use the triangle inequality to generalise to self-adjoint matrices, and then to arbitrary matrices, of operator norm 1, and conclude that

$$(2.100) \qquad \mathbf{P}(|X^*RX - \mathbf{E}X^*RX| \geq \lambda\sqrt{n}) \leq Ce^{-c\lambda^2}$$

for any deterministic matrix R of operator norm $O(1)$.

But what is the expectation $\mathbf{E}X^*RX$? This can be expressed in components as

$$\mathbf{E}X^*RX = \sum_{i=1}^{n-1}\sum_{j=1}^{n-1} \mathbf{E}\overline{\xi_{in}} r_{ij} \xi_{jn}$$

where ξ_{in} are the entries of X, and r_{ij} are the entries of R. But the ξ_{in} are iid with mean zero and variance one, so the standard second moment computation shows that this expectation is nothing more than the trace

$$\mathrm{tr}(R) = \sum_{i=1}^{n-1} r_{ii}$$

of R. We have thus shown the concentration of the measure result

$$(2.101) \qquad \mathbf{P}(|X^*RX - \mathrm{tr}(R)| \geq \lambda\sqrt{n}) \leq Ce^{-c\lambda^2}$$

for any deterministic matrix R of operator norm $O(1)$, and any $\lambda > 0$. Informally, X^*RX is typically $\mathrm{tr}(R) + O(\sqrt{n})$.

The bound (2.101) was proven for deterministic matrices, but by using conditional expectation it also applies for any random matrix R, so long as that matrix is independent of X. In particular, we may apply it to our specific matrix of interest

$$R := \left(\frac{1}{\sqrt{n}} M_{n-1} - zI_{n-1}\right)^{-1}.$$

The trace of this matrix is essentially just the Stieltjes transform $s_{n-1}(z)$ at z. Actually, due to the normalisation factor being slightly off, we actually have

$$\mathrm{tr}(R) = n\frac{\sqrt{n}}{\sqrt{n-1}} s_{n-1}\left(\frac{\sqrt{n}}{\sqrt{n-1}} z\right),$$

but by using the smoothness (2.93) of the Stieltjes transform, together with the stability property (2.96) we can simplify this as

$$\mathrm{tr}(R) = n(s_n(z) + o(1)).$$

In particular, from (2.101) and (2.97), we see that

$$X^*RX = n(\mathbf{E}s_n(z) + o(1))$$

with overwhelming probability. Putting this back into (2.99), and recalling that the denominator is bounded away from zero, we have the remarkable *self-consistent equation*

$$(2.102) \qquad \mathbf{E}s_n(z) = -\frac{1}{z + \mathbf{E}s_n(z)} + o(1).$$

Note how this equation came by playing off two ways in which the spectral properties of a matrix M_n interacted with that of its minor M_{n-1}; first, via the Cauchy interlacing inequality, and second, via the Schur complement formula.

This equation already describes the behaviour of $\mathbf{E}s_n(z)$ quite well, but we will content ourselves with understanding the limiting behaviour as $n \to \infty$. From (2.93) and Fubini's theorem we know that the function $\mathbf{E}s_n$ is locally uniformly equicontinuous and locally uniformly bounded away from the real line. Applying the *Arzelá-Ascoli theorem*, we thus conclude that on a subsequence at least, $\mathbf{E}s_n$ converges locally uniformly to a limit s. This will be a Herglotz function (i.e., an analytic function mapping the upper half-plane to the upper half-plane), and taking limits in (2.102) (observing that the imaginary part of the denominator here is bounded away from zero) we end up with the exact equation

$$(2.103) \qquad s(z) = -\frac{1}{z + s(z)}.$$

We can of course solve this by the quadratic formula, obtaining

$$s(z) = -\frac{z \pm \sqrt{z^2 - 4}}{2} = -\frac{2}{z \pm \sqrt{z^2 - 4}}.$$

To figure out what branch of the square root one has to use here, we use (2.92), which easily implies[33] that

$$s(z) = \frac{-1 + o(1)}{z}$$

as z goes to infinity non-tangentially away from the real line. Also, we know that s has to be complex analytic (and in particular, continuous) away from the real line. From this and basic complex analysis, we conclude that

$$(2.104) \qquad s(z) = \frac{-z + \sqrt{z^2 - 4}}{2}$$

where $\sqrt{z^2 - 4}$ is the branch of the square root with a branch cut at $[-2, 2]$ and which equals z at infinity.

As there is only one possible subsequence limit of the $\mathbf{E}s_n$, we conclude that $\mathbf{E}s_n$ converges locally uniformly (and thus pointwise) to the function

---

[33]To justify this, one has to make the error term in (2.92) uniform in n, but this can be accomplished without difficulty using the Bai-Yin theorem (for instance).

(2.104), and thus (by the concentration of measure of $s_n(z)$) we see that for each z, $s_n(z)$ converges almost surely (and in probability) to $s(z)$.

Exercise 2.4.12. Find a direct proof (starting from (2.102), (2.92), and the smoothness of $\mathbf{E}s_n(z)$) that $\mathbf{E}s_n(z) = s(z) + o(1)$ for any fixed z, that avoids using the Arzelá-Ascoli theorem. (The basic point here is that one has to solve the approximate equation (2.102), using some robust version of the quadratic formula. The fact that $\mathbf{E}s_n$ is a Herglotz function will help eliminate various unwanted possibilities, such as one coming from the wrong branch of the square root.)

To finish computing the limiting ESD of Wigner matrices, we have to figure out what probability measure s comes from; but this is easily read off from (2.104) and (2.95):

$$(2.105) \qquad \frac{s(\cdot + ib) - s(\cdot - ib)}{2\pi i} \to \frac{1}{2\pi}(4 - x^2)_+^{1/2} \, dx = \mu_{\mathrm{sc}}$$

as $b \to 0$. Thus the semicircular law is the only possible measure which has Stieltjes transform s, and indeed a simple application of the Cauchy integral formula and (2.105) shows us that s is indeed the Stieltjes transform of μ_{sc}.

Putting all this together, we have completed the Stieltjes transform proof of the semicircular law.

Remark 2.4.8. In order to simplify the above exposition, we opted for a qualitative analysis of the semicircular law here, ignoring such questions as the rate of convergence to this law. However, an inspection of the above arguments reveals that it is easy to make all of the above analysis quite quantitative, with quite reasonable control on all terms[34]. In particular, it is not hard to use the above analysis to show that for $|\mathrm{Im}(z)| \geq n^{-c}$ for some small absolute constant $c > 0$, one has $s_n(z) = s(z) + O(n^{-c})$ with overwhelming probability. Combining this with a suitably quantitative version of the Stieltjes continuity theorem, this in turn gives a *polynomial rate* of convergence of the ESDs $\mu_{\frac{1}{\sqrt{n}}M_n}$ to the semicircular law μ_{sc}, in that one has

$$\mu_{\frac{1}{\sqrt{n}}M_n}(-\infty, \lambda) = \mu_{\mathrm{sc}}(-\infty, \lambda) + O(n^{-c})$$

with overwhelming probability for all $\lambda \in \mathbf{R}$.

A variant of this quantitative analysis can in fact get very good control on this ESD down to quite fine scales, namely to scales $\frac{\log^{O(1)} n}{n}$, which is only just a little bit larger than the mean spacing $O(1/n)$ of the normalised eigenvalues (recall that we have n normalised eigenvalues, constrained to lie

---

[34]One has to use Exercise 2.4.12 instead of the Arzelá-Ascoli theorem if one wants everything to be quantitative.

in the interval $[-2-o(1), 2+o(1)]$ by the Bai-Yin theorem). This was accomplished by Erdös, Schlein, and Yau [**ErScYa2008**][35] by using an additional observation, namely that the *eigenvectors* of a random matrix are very likely to be *delocalised* in the sense that their ℓ^2 energy is dispersed more or less evenly across its coefficients. Such delocalization has since proven to be a fundamentally important ingredient in the fine-scale spectral analysis of Wigner matrices, which is beyond the scope of this text.

2.4.4. Dyson Brownian motion and the Stieltjes transform (optional). In this section we explore how the Stieltjes transform interacts with the Dyson Brownian motion (which is presented in detail in Section 3.1). We let n be a large number, and let $M_n(t)$ be a Wiener process of Hermitian random matrices, with associated eigenvalues $\lambda_1(t), \ldots, \lambda_n(t)$, Stieltjes transforms

$$(2.106) \qquad s(t,z) := \frac{1}{n} \sum_{j=1}^{n} \frac{1}{\lambda_j(t)/\sqrt{n} - z}$$

and spectral measures

$$(2.107) \qquad \mu(t,z) := \frac{1}{n} \sum_{j=1}^{n} \delta_{\lambda_j(t)/\sqrt{n}}.$$

We now study how s, μ evolve in time in the asymptotic limit $n \to \infty$. Our computation will be only heuristic in nature.

Recall from Section 3.1 that the eigenvalues $\lambda_i = \lambda_i(t)$ undergo the Dyson Brownian motion

$$(2.108) \qquad d\lambda_i = dB_i + \sum_{j \neq i} \frac{dt}{\lambda_i - \lambda_j}.$$

Applying (2.106) and Taylor expansion (dropping all terms of higher order than dt, using the Ito heuristic $dB_i = O(dt^{1/2})$), we conclude that

$$ds = -\frac{1}{n^{3/2}} \sum_{i=1}^{n} \frac{dB_i}{(\lambda_i/\sqrt{n} - z)^2} - \frac{1}{n^2} \sum_{i=1}^{n} \frac{|dB_i|^2}{(\lambda_i/\sqrt{n} - z)^3}$$

$$-\frac{1}{n^{3/2}} \sum_{1 \leq i,j \leq n: i \neq j} \frac{dt}{(\lambda_i - \lambda_j)(\lambda_j/\sqrt{n} - z)^2}.$$

For z away from the real line, the term $\frac{1}{n^2} \sum_{i=1}^{n} \frac{|dB_i|^2}{(\lambda_i/\sqrt{n}-z)^3}$ is of size $O(dt/n)$ and can heuristically be ignored in the limit $n \to \infty$. Dropping this term,

---

[35] Strictly speaking, this paper assumed additional regularity hypotheses on the distribution ξ, but these conditions can be removed with the assistance of Talagrand's inequality, Theorem 2.1.13.

and then taking expectations to remove the Brownian motion term dB_i, we are led to

$$\mathbf{E}ds = -\mathbf{E}\frac{1}{n^{3/2}} \sum_{1 \leq i,j \leq n:i \neq j} \frac{dt}{(\lambda_i - \lambda_j)(\lambda_j/\sqrt{n} - z)^2}.$$

Performing the i summation using (2.106) we obtain

$$\mathbf{E}ds = -\mathbf{E}\frac{1}{n} \sum_{1 \leq j \leq n} \frac{s(\lambda_j/\sqrt{n})dt}{(\lambda_j/\sqrt{n} - z)^2}$$

where we adopt the convention that for real x, $s(x)$ is the average of $s(x+i0)$ and $s(x - i0)$. Using (2.107), this becomes

(2.109)
$$\mathbf{E}s_t = -\mathbf{E}\int_{\mathbf{R}} \frac{s(x)}{(x - z)^2} \, d\mu(x)$$

where the t subscript denotes differentiation in t. From (2.95) we heuristically have

$$s(x \pm i0) = s(x) \pm \pi i \mu(x)$$

(heuristically treating μ as a function rather than a measure) and on squaring one obtains

$$s(x \pm i0)^2 = (s(x)^2 - \pi^2 \mu^2(x)) \pm 2\pi i s(x)\mu(x).$$

From this the Cauchy integral formula around a slit in the real axis (using the bound (2.91) to ignore the contributions near infinity) we thus have

$$s^2(z) = \int_{\mathbf{R}} \frac{2s(x)}{x - z} \, d\mu(x)$$

and thus on differentiation in z, we have

$$2ss_z(z) = \int_{\mathbf{R}} \frac{2s(x)}{(x - z)^2} \, d\mu(x).$$

Comparing this with (2.109), we obtain

$$\mathbf{E}s_t + \mathbf{E}ss_z = 0.$$

From concentration of measure, we expect s to concentrate around its mean $\bar{s} := \mathbf{E}s$, and similarly s_z should concentrate around $\bar{s}_z$. In the limit $n \to \infty$, the expected Stieltjes transform $\bar{s}$ should thus obey the (complex) *Burgers' equation*

(2.110)
$$s_t + ss_z = 0.$$

To illustrate how this equation works in practice, let us give an informal derivation of the semicircular law. We consider the case when the Wiener process starts from $M(0) = 0$, thus $M_t \equiv \sqrt{t}G$ for a GUE matrix G. As such, we have the scaling symmetry

$$s(t, z) = \frac{1}{\sqrt{t}} s_{GUE} \left(\frac{z}{\sqrt{t}} \right)$$

where s_{GUE} is the asymptotic Stieltjes transform for GUE (which we secretly know to be given by (2.104), but let us pretend that we did not yet know this fact). Inserting this self-similar ansatz into (2.110) and setting $t = 1$, we conclude that

$$-\frac{1}{2} s_{GUE} - \frac{1}{2} z s'_{GUE} + s s'_{GUE} = 0;$$

multiplying by two and integrating, we conclude that

$$z s_{GUE} + s^2_{GUE} = C$$

for some constant C. But from the asymptotic (2.92) we see that C must equal -1. But then the above equation can be rearranged into (2.103), and so by repeating the arguments at the end of the previous section we can deduce the formula (2.104), which then gives the semicircular law by (2.95).

As is well known in PDE, one can solve Burgers' equation more generally by the *method of characteristics*. For reasons that will become clearer in Section 2.5, we now solve this equation by a slightly different (but ultimately equivalent) method. The idea is that rather than think of $s = s(t, z)$ as a function of z for fixed t, we think[36] of $z = z(t, s)$ as a function of s for fixed t. Note from (2.92) that we expect to be able to invert the relationship between s and z as long as z is large (and s is small).

To exploit this change of perspective, we think of s, z, t as all varying by infinitesimal amounts ds, dz, dt, respectively. Using (2.110) and the total derivative formula $ds = s_t dt + s_z dz$, we see that

$$ds = -s s_z dt + s_z dz.$$

If we hold s fixed (i.e., $ds = 0$), so that z is now just a function of t, and cancel out the s_z factor, we conclude that

$$\frac{dz}{dt} = s.$$

Integrating this, we see that

(2.111) $$z(t, s) = z(0, s) + ts.$$

This, in principle, gives a way to compute $s(t, z)$ from $s(0, z)$. First, we invert the relationship $s = s(0, z)$ to $z = z(0, s)$; then we add ts to $z(0, s)$; then we invert again to recover $s(t, z)$.

---

[36]This trick is sometimes known as the *hodograph transform*, especially if one views s as "velocity" and z as "position".

Since $M_t \equiv M_0 + \sqrt{t}G$, where G is a GUE matrix independent of M_0, we have thus given a formula to describe the Stieltjes transform of $M_0 + \sqrt{t}G$ in terms of the Stieltjes transform of M_0. This formula is a special case of a more general formula of Voiculescu for *free convolution*, with the operation of inverting the Stieltjes transform essentially being the famous *R-transform* of Voiculescu; we will discuss this more in the next section.

2.5. Free probability

In the foundations of modern probability, as laid out by Kolmogorov (and briefly reviewed in Section 1.1), the basic objects of study are constructed in the following order:

(i) First, one selects a *sample space* Ω, whose elements ω represent all the possible states that one's stochastic system could be in.

(ii) Then, one selects a *σ-algebra* $\mathcal{B}$ of *events* E (modeled by subsets of Ω), and assigns each of these events a *probability* $\mathbf{P}(E) \in [0,1]$ in a countably additive manner, so that the entire sample space has probability 1.

(iii) Finally, one builds (commutative) algebras of *random variables* X (such as complex-valued random variables, modeled by measurable functions from Ω to $\mathbf{C}$), and (assuming suitable integrability or moment conditions) one can assign *expectations* $\mathbf{E}X$ to each such random variable.

In *measure theory*, the underlying measure space Ω plays a prominent foundational role, with the measurable sets and measurable functions (the analogues of the events and the random variables) always being viewed as somehow being attached to that space. In probability theory, in contrast, it is the events and their probabilities that are viewed as being fundamental, with the sample space Ω being abstracted away as much as possible, and with the random variables and expectations being viewed as derived concepts. See Section 1.1 for further discussion of this philosophy.

However, it is possible to take the abstraction process one step further, and view the *algebra of random variables and their expectations* as being the foundational concept, and ignoring both the presence of the original sample space, the algebra of events, or the probability measure.

There are two reasons for wanting to shed (or abstract[37] away) these previously foundational structures. First, it allows one to more easily take certain types of limits, such as the large n limit $n \to \infty$ when considering

---

[37]This theme of using abstraction to facilitate the taking of the large n limit also shows up in the application of ergodic theory to combinatorics via the correspondence principle; see [**Ta2009**, §2.10] for further discussion.

$n \times n$ random matrices, because quantities built from the algebra of random variables and their expectations, such as the normalised moments of random matrices tend to be quite stable in the large n limit (as we have seen in previous sections), even as the sample space and event space varies with n.

Second, this abstract formalism allows one to generalise the classical, commutative theory of probability to the more general theory of *non-commutative probability*, which does not have a classical underlying sample space or event space, but is instead built upon a (possibly) *non-commutative* algebra of random variables (or "observables") and their expectations (or "traces"). This more general formalism not only encompasses classical probability, but also spectral theory (with matrices or operators taking the role of random variables, and the trace taking the role of expectation), random matrix theory (which can be viewed as a natural blend of classical probability and spectral theory), and quantum mechanics (with physical observables taking the role of random variables, and their expected value on a given quantum state being the expectation). It is also part of a more general "non-commutative way of thinking"[38] (of which *non-commutative geometry* and *quantum mechanics* are the most prominent examples), in which a space is understood primarily in terms of the ring or algebra of functions (or function-like objects, such as sections of bundles) placed on top of that space, and then the space itself is largely abstracted away in order to allow the algebraic structures to become less commutative. In short, the idea is to make *algebra* the foundation of the theory, as opposed to other possible choices of foundations such as sets, measures, categories, etc.

It turns out that non-commutative probability can be modeled using operator algebras such as C^*-*algebras, von Neumann algebras*, or algebras of bounded operators on a Hilbert space, with the latter being accomplished via the *Gelfand-Naimark-Segal construction*. We will discuss some of these models here, but just as probability theory seeks to abstract away its measure-theoretic models, the philosophy of non-commutative probability is also to downplay these operator algebraic models once some foundational issues are settled.

When one generalises the set of structures in one's theory, for instance from the commutative setting to the non-commutative setting, the notion of what it means for a structure to be "universal", "free", or "independent"

---

[38]Note that this foundational preference is to some extent a metamathematical one rather than a mathematical one; in many cases it is possible to rewrite the theory in a mathematically equivalent form so that some other mathematical structure becomes designated as the foundational one, much as probability theory can be equivalently formulated as the measure theory of probability measures. However, this does not negate the fact that a different choice of foundations can lead to a different way of thinking about the subject, and thus to ask a different set of questions and to discover a different set of proofs and solutions. Thus it is often of value to understand multiple foundational perspectives at once, to get a truly stereoscopic view of the subject.

can change. The most familiar example of this comes from group theory. If one restricts attention to the category of abelian groups, then the "freest" object one can generate from two generators e, f is the *free abelian group* of commutative words $e^n f^m$ with $n, m \in \mathbf{Z}$, which is isomorphic to the group $\mathbf{Z}^2$. If however one generalises to the non-commutative setting of arbitrary groups, then the "freest" object that can now be generated from two generators e, f is the *free group* $\mathbf{F}_2$ of non-commutative words $e^{n_1} f^{m_1} \ldots e^{n_k} f^{m_k}$ with $n_1, m_1, \ldots, n_k, m_k \in \mathbf{Z}$, which is a significantly larger extension of the free abelian group $\mathbf{Z}^2$.

Similarly, when generalising classical probability theory to non-commutative probability theory, the notion of what it means for two or more random variables to be independent changes. In the classical (commutative) setting, two (bounded, real-valued) random variables X, Y are *independent* if one has

$$\mathbf{E} f(X) g(Y) = 0$$

whenever $f, g : \mathbf{R} \to \mathbf{R}$ are well-behaved functions (such as polynomials) such that all of $\mathbf{E} f(X), \mathbf{E} g(Y)$ vanish. In the non-commutative setting, one can generalise the above definition to two *commuting* bounded self-adjoint variables; this concept is useful, for instance, in *quantum probability*, which is an abstraction of the theory of observables in quantum mechanics. But for two (bounded, self-adjoint) *non-commutative* random variables X, Y, the notion of classical independence no longer applies. As a substitute, one can instead consider the notion of being *freely independent* (or *free* for short), which means that

$$\mathbf{E} f_1(X) g_1(Y) \ldots f_k(X) g_k(Y) = 0$$

whenever $f_1, g_1, \ldots, f_k, g_k : \mathbf{R} \to \mathbf{R}$ are well-behaved functions such that all of $\mathbf{E} f_1(X), \mathbf{E} g_1(Y), \ldots, \mathbf{E} f_k(X), \mathbf{E} g_k(Y)$ vanish.

The concept of free independence was introduced by Voiculescu, and its study is now known as the subject of *free probability*. We will not attempt a systematic survey of this subject here; for this, we refer the reader to the surveys of Speicher [**Sp**] and of Biane [**Bi2003**]. Instead, we shall just discuss a small number of topics in this area to give the flavour of the subject only.

The significance of free probability to random matrix theory lies in the fundamental observation that random matrices which have independent entries in the classical sense, also tend to be independent[39] in the free probability sense, in the large n limit $n \to \infty$. Because of this, many tedious computations in random matrix theory, particularly those of an algebraic

---

[39]This is only possible because of the highly non-commutative nature of these matrices; as we shall see, it is not possible for non-trivial commuting independent random variables to be freely independent.

or enumerative combinatorial nature, can be done more quickly and systematically by using the framework of free probability, which by design is optimised for algebraic tasks rather than analytical ones.

Much as free groups are in some sense "maximally non-commutative", freely independent random variables are about as far from being commuting as possible. For instance, if X, Y are freely independent and of expectation zero, then $\mathbf{E}XYXY$ vanishes, but $\mathbf{E}XXYY$ instead factors as $(\mathbf{E}X^2)(\mathbf{E}Y^2)$. As a consequence, the behaviour of freely independent random variables can be quite different from the behaviour of their classically independent commuting counterparts. Nevertheless, there is a remarkably strong *analogy* between the two types of independence, in that results which are true in the classically independent case often have an interesting analogue in the freely independent setting. For instance, the central limit theorem (Section 2.2) for averages of classically independent random variables which, roughly speaking, asserts that such averages become Gaussian in the large n limit, has an analogue for averages of freely independent variables, the *free central limit theorem* which, roughly speaking, asserts that such averages become *semicircular* in the large n limit. One can then use this theorem to provide yet another proof of Wigner's semicircle law (Section 2.4).

Another important (and closely related) analogy is that while the distribution of sums of independent commutative random variables can be quickly computed via the characteristic function (i.e., the Fourier transform of the distribution), the distribution of sums of freely independent noncommutative random variables can be quickly computed using the *Stieltjes transform* instead (or with closely related objects, such as the *R-transform* of Voiculescu). This is strongly reminiscent of the appearance of the Stieltjes transform in random matrix theory, and indeed we will see many parallels between the use of the Stieltjes transform here and in Section 2.4.

As mentioned earlier, free probability is an excellent tool for computing various expressions of interest in random matrix theory, such as asymptotic values of normalised moments in the large n limit $n \to \infty$. Nevertheless, as it only covers the asymptotic regime in which n is sent to infinity while holding all other parameters fixed, there are some aspects of random matrix theory to which the tools of free probability are not sufficient by themselves to resolve (although it can be possible to combine free probability theory with other tools to then answer these questions). For instance, questions regarding the *rate* of convergence of normalised moments as $n \to \infty$ are not directly answered by free probability, though if free probability is combined with tools such as concentration of measure (Section 2.1), then such rate information can often be recovered. For similar reasons, free probability lets one understand the behaviour of kth moments as $n \to \infty$ for *fixed* k, but has more difficulty dealing with the situation in which k is allowed to grow slowly in n (e.g., $k = O(\log n)$). Because of this, free probability methods are effective at controlling the *bulk* of the spectrum of a random matrix, but have more difficulty with the *edges* of that spectrum as well as with fine-scale structure of the spectrum, although one can sometimes use free probability methods to understand operator norms (as studied in Section 2.3). Finally, free probability methods are most effective when dealing with matrices that are Hermitian with bounded operator norm, largely because the spectral theory of bounded self-adjoint operators in the infinite-dimensional setting of the large n limit is non-pathological[40]. For non-self-adjoint operators, free probability needs to be augmented with additional tools, most notably by bounds on least singular values, in order to recover the required stability for the various spectral data of random matrices to behave continuously with respect to the large n limit. We will return this latter point in Section 2.7.

2.5.1. Abstract probability theory. We will now slowly build up the foundations of non-commutative probability theory, which seeks to capture the abstract algebra of random variables and their expectations. The impatient reader who wants to move directly on to free probability theory may largely jump straight to the final definition at the end of this section, but it can be instructive to work with these foundations for a while to gain some intuition on how to handle non-commutative probability spaces.

To motivate the formalism of abstract (non-commutative) probability theory, let us first discuss the three key examples of non-commutative probability spaces, and then abstract away all features that are not shared in common by all three examples.

Example 2.5.1 (Random scalar variables). We begin with *classical probability theory*—the study of scalar random variables. In order to use the powerful tools of complex analysis (such as the Stieltjes transform), it is very convenient to allow our random variables to be complex valued. In order to meaningfully take expectations, we would like to require all of our random variables to also be absolutely integrable. But this requirement is not sufficient by itself to get good algebraic structure, because the product of two absolutely integrable random variables need not be absolutely integrable. As we want to have as much algebraic structure as possible, we will therefore restrict attention further, to the collection $L^{\infty-} := \bigcap_{k=1}^{\infty} L^k(\Omega)$ of random variables with all moments finite. This class is closed under multiplication, and all elements in this class have a finite trace (or expectation). One can of course restrict further, to the space $L^{\infty} = L^{\infty}(\Omega)$ of (essentially) bounded variables, but by doing so one loses important examples of random

---

[40]This is ultimately due to the stable nature of eigenvalues in the self-adjoint setting; see [**Ta2010b**, §1.5] for discussion.

variables, most notably Gaussians, so we will work instead[41] with the space $L^{\infty-}$.

The space $L^{\infty-}$ of complex-valued random variables with all moments finite now becomes an *algebra* over the complex numbers $\mathbf{C}$; i.e., it is a vector space over $\mathbf{C}$ that is also equipped with a bilinear multiplication operation $\cdot : L^{\infty-} \times L^{\infty-} \to L^{\infty-}$ that obeys the associative and distributive laws. It is also commutative, but we will suppress this property, as it is not shared by the other two examples we will be discussing. The deterministic scalar 1 then plays the role of the multiplicative unit in this algebra.

In addition to the usual algebraic operations, one can also take the *complex conjugate* or *adjoint* $X^* = \overline{X}$ of a complex-valued random variable X. This operation $* : L^{\infty-} \to L^{\infty-}$ interacts well with the other algebraic operations: it is in fact an *anti-automorphism* on $L^{\infty-}$, which means that it preserves addition $(X+Y)^* = X^* + Y^*$, reverses multiplication $(XY)^* = Y^* X^*$, is anti-homogeneous $((cX)^* = \overline{c}X^*$ for $c \in \mathbf{C})$, and it is invertible. In fact, it is its own inverse $((X^*)^* = X)$, and is thus an *involution*.

This package of properties can be summarised succinctly by stating that the space $L^{\infty-}$ of bounded complex-valued random variables is a (unital) *$*$-algebra*.

The expectation operator $\mathbf{E}$ can now be viewed as a map $\mathbf{E} : L^{\infty-} \to \mathbf{C}$. It obeys some obvious properties, such as being linear (i.e., $\mathbf{E}$ is a linear functional on L^∞). In fact, it is *$*$-linear*, which means that it is linear and also that $\mathbf{E}(X^*) = \overline{\mathbf{E}X}$ for all X. We also clearly have $\mathbf{E}1 = 1$. We will remark on some additional properties of expectation later.

Example 2.5.2 (Deterministic matrix variables). A second key example is that of (finite-dimensional) *spectral theory*—the theory of $n \times n$ complex-valued matrices $X \in M_n(\mathbf{C})$. (One can also consider infinite-dimensional spectral theory, of course, but for simplicity we only consider the finite-dimensional case in order to avoid having to deal with technicalities such as unbounded operators.) Like the space $L^{\infty-}$ considered in the previous example, $M_n(\mathbf{C})$ is a $*$-algebra, where the multiplication operation is of course given by matrix multiplication, the identity is the matrix identity $1 = I_n$, and the involution $X \mapsto X^*$ is given by the *matrix adjoint operation*. On the other hand, as is well known, this $*$-algebra is not commutative (for $n \geq 2$).

The analogue of the expectation operation here is the *normalised trace* $\tau(X) := \frac{1}{n} \operatorname{tr} X$. Thus $\tau : M_n(\mathbf{C}) \to \mathbf{C}$ is a $*$-linear functional on $M_n(\mathbf{C})$ that maps 1 to 1. The analogy between expectation and normalised trace is

---

[41] This will cost us some analytic structure—in particular, $L^{\infty-}$ will not be a Banach space, in contrast to L^∞—but as our focus is on the algebraic structure, this will be an acceptable price to pay.

particularly evident when comparing the moment method for scalar random variables (based on computation of the moments $\mathbf{E}X^k$) with the moment method in spectral theory (based on a computation of the moments $\tau(X^k)$).

Example 2.5.3 (Random matrix variables). *Random matrix theory* combines classical probability theory with finite-dimensional spectral theory, with the random variables of interest now being the random matrices $X \in L^{\infty-} \otimes M_n(\mathbf{C})$, all of whose entries have all moments finite. It is not hard to see that this is also a $*$-algebra with identity $1 = I_n$, which again will be non-commutative for $n \geq 2$. The normalised trace τ here is given by

$$\tau(X) := \mathbf{E}\frac{1}{n}\operatorname{tr} X,$$

thus one takes both the normalised matrix trace and the probabilistic expectation, in order to arrive at a deterministic scalar (i.e., a complex number). As before, we see that $\tau : L^{\infty-} \otimes M_n(\mathbf{C}) \to \mathbf{C}$ is a $*$-linear functional that maps 1 to 1. As we saw in Section 2.3, the moment method for random matrices is based on a computation of the moments $\tau(X^k) = \mathbf{E}\frac{1}{n}\operatorname{tr} X^k$.

Let us now simultaneously abstract the above three examples, but reserving the right to impose some additional axioms as needed:

Definition 2.5.4 (Non-commutative probability space, preliminary definition). A *non-commutative probability space* (or more accurately, a *potentially non-commutative* probability space) $(\mathcal{A}, \tau)$ will consist of a (potentially non-commutative) $*$-algebra $\mathcal{A}$ of (potentially non-commutative) *random variables* (or *observables*) with identity 1, together with a *trace* $\tau : \mathcal{A} \to \mathbf{C}$, which is a $*$-linear functional that maps 1 to 1. This trace will be required to obey a number of additional axioms which we will specify later in this section.

This definition is not yet complete, because we have not fully decided on what axioms to enforce for these spaces, but for now let us just say that the three examples $(L^{\infty-}, \mathbf{E})$, $(M_n(\mathbf{C}), \frac{1}{n}\operatorname{tr})$, $(L^{\infty-} \otimes M_n(\mathbf{C}), \mathbf{E}\frac{1}{n}\operatorname{tr})$ given above will obey these axioms and serve as model examples of non-commutative probability spaces. We mention that the requirement $\tau(1) = 1$ can be viewed as an abstraction of Kolmogorov's axiom that the sample space has probability 1.

To motivate the remaining axioms, let us try seeing how some basic concepts from the model examples carry over to the abstract setting.

First, we recall that every scalar random variable $X \in L^{\infty-}$ has a probability distribution μ_X, which is a probability measure on the complex plane $\mathbf{C}$; if X is self-adjoint (i.e., real-valued), so that $X = X^*$, then this distribution is supported on the real line $\mathbf{R}$. The condition that X lie in $L^{\infty-}$ ensures that this measure is rapidly decreasing, in the sense that $\int_{\mathbf{C}} |z|^k \, d\mu_X(x) < \infty$ for all k. The measure μ_X is related to the moments $\tau(X^k) = \mathbf{E}X^k$ by the formula

$$(2.112) \qquad\qquad \tau(X^k) = \int_{\mathbf{C}} z^k \, d\mu_X(z)$$

for $k = 0, 1, 2, \ldots$. In fact, one has the more general formula

$$(2.113) \qquad\qquad \tau(X^k(X^*)^l) = \int_{\mathbf{C}} z^k \overline{z}^l \, d\mu_X(z)$$

for $k, l = 0, 1, 2, \ldots$.

Similarly, every deterministic matrix $X \in M_n(\mathbf{C})$ has an *empirical spectral distribution* $\mu_X = \frac{1}{n} \sum_{i=1}^n \delta_{\lambda_i(X)}$, which is a probability measure on the complex plane $\mathbf{C}$. Again, if X is self-adjoint, then distribution is supported on the real line $\mathbf{R}$. This measure is related to the moments $\tau(X^k) = \frac{1}{n} \operatorname{tr} X^k$ by the same formula (2.112) as in the case of scalar random variables. Because n is finite, this measure is finitely supported (and in particular, is rapidly decreasing). As for (2.113), the spectral theorem tells us that this formula holds when X is normal (i.e., $XX^* = X^*X$), and in particular, if X is self-adjoint (of course, in this case (2.113) collapses to (2.112)), but is not true in general. Note that this subtlety does not appear in the case of scalar random variables because in this commutative setting, all elements are automatically normal.

Finally, for random matrices $X \in L^{\infty-} \otimes M_n(\mathbf{C})$, we can form the *expected empirical spectral distribution* $\mu_X = \mathbf{E}\frac{1}{n} \sum_{i=1}^n \delta_{\lambda_i(X)}$, which is again a rapidly decreasing probability measure on $\mathbf{C}$, which is supported on $\mathbf{R}$ if X is self-adjoint. This measure is again related to the moments $\tau(X^k) = \mathbf{E}\frac{1}{n} \operatorname{tr} X^k$ by the formula (2.112), and also by (2.113) if X is normal.

Now let us see whether we can set up such a spectral measure μ_X for an element X in an abstract non-commutative probability space $(\mathcal{A}, \tau)$. From the above examples, it is natural to try to define this measure through the formula (2.112), or equivalently (by linearity) through the formula

$$(2.114) \qquad\qquad \tau(P(X)) = \int_{\mathbf{C}} P(z) \, d\mu_X(z)$$

whenever $P : \mathbf{C} \to \mathbf{C}$ is a polynomial with complex coefficients (note that one can define $P(X)$ without difficulty as $\mathcal{A}$ is a $*$-algebra). In the normal case, one may hope to work with the more general formula

$$(2.115) \qquad\qquad \tau(P(X, X^*)) = \int_{\mathbf{C}} P(z, \overline{z}) \, d\mu_X(z)$$

whenever $P : \mathbf{C} \times \mathbf{C} \to \mathbf{C}$ is a polynomial of two complex variables (note that $P(X, X^*)$ can be defined unambiguously precisely when X is normal).

It is tempting to apply the *Riesz representation theorem* to (2.114) to define the desired measure μ_X, perhaps after first using the *Weierstrass approximation theorem* to pass from polynomials to continuous functions. However, there are multiple technical issues with this idea:

(i) In order for the polynomials to be dense in the continuous functions in the uniform topology on the support of μ_X, one needs the intended support $\sigma(X)$ of μ_X to be on the real line $\mathbf{R}$, or else one needs to work with the formula (2.115) rather than (2.114). Also, one also needs the intended support $\sigma(X)$ to be bounded for the Weierstrass approximation theorem to apply directly.

(ii) In order for the Riesz representation theorem to apply, the functional $P \mapsto \tau(P(X, X^*))$ (or $P \mapsto \tau(P(X))$) needs to be continuous in the uniform topology, thus one must be able to obtain a bound[42] of the form $|\tau(P(X, X^*))| \le C \sup_{z \in \sigma(X)} |P(z, \overline{z})|$ for some (preferably compact) set $\sigma(X)$.

(iii) In order to get a probability measure rather than a signed measure, one also needs some non-negativity: $\tau(P(X, X^*))$ needs to be non-negative whenever $P(z, \overline{z}) \ge 0$ for z in the intended support $\sigma(X)$.

To resolve the non-negativity issue, we impose an additional axiom on the non-commutative probability space $(\mathcal{A}, \tau)$:

Axiom 2.5.5 (Non-negativity). *For any $X \in \mathcal{A}$, we have $\tau(X^*X) \ge 0$. (Note that X^*X is self-adjoint, and so its trace $\tau(X^*X)$ is necessarily a real number.)*

In the language of *von Neumann algebras*, this axiom (together with the normalisation $\tau(1) = 1$) is essentially asserting that τ is a state. Note that this axiom is obeyed by all three model examples, and is also consistent with (2.115). It is the non-commutative analogue of the Kolmogorov axiom that all events have non-negative probability.

With this axiom, we can now define a positive semi-definite *inner product* $\langle, \rangle_{L^2(\tau)}$ on $\mathcal{A}$ by the formula

$$\langle X, Y \rangle_{L^2(\tau)} := \tau(X^*Y).$$

This obeys the usual axioms of an inner product, except that it is only positive semi-definite rather than positive definite. One can impose positive definiteness by adding an axiom that the trace τ is *faithful*, which means that $\tau(X^*X) = 0$ if and only if $X = 0$. However, we will not need the faithfulness axiom here.

---

[42]To get a probability measure, one in fact needs to have $C = 1$.

Without faithfulness, $\mathcal{A}$ is a semi-definite inner product space with semi-norm

$$\|X\|_{L^2(\tau)} := (\langle X, X \rangle_{L^2(\tau)})^{1/2} = \tau(X^*X)^{1/2}.$$

In particular, we have the *Cauchy-Schwarz inequality*

$$|\langle X, Y \rangle_{L^2(\tau)}| \leq \|X\|_{L^2(\tau)} \|Y\|_{L^2(\tau)}.$$

This leads to an important monotonicity:

Exercise 2.5.1 (Monotonicity). Let X be a self-adjoint element of a non-commutative probability space $(\mathcal{A}, \tau)$. Show that we have the monotonicity relationships

$$|\tau(X^{2k-1})|^{1/(2k-1)} \leq |\tau(X^{2k})|^{1/(2k)} \leq |\tau(X^{2k+2})|^{1/(2k+2)}$$

for any $k \geq 0$.

As a consequence, we can define the *spectral radius* $\rho(X)$ of a self-adjoint element X by the formula

(2.116) $$\rho(X) := \lim_{k \to \infty} |\tau(X^{2k})|^{1/(2k)},$$

in which case we obtain the inequality

(2.117) $$|\tau(X^k)| \leq \rho(X)^k$$

for any $k = 0, 1, 2, \ldots$. We then say that a self-adjoint element is *bounded* if its spectral radius is finite.

Example 2.5.6. In the case of random variables, the spectral radius is the essential supremum $\|X\|_{L^\infty}$, while for deterministic matrices, the spectral radius is the operator norm $\|X\|_{\mathrm{op}}$. For random matrices, the spectral radius is the essential supremum $\|\|X\|_{\mathrm{op}}\|_{L^\infty}$ of the operator norm.

Guided by the model examples, we expect that a bounded self-adjoint element X should have a spectral measure μ_X supported on the interval $[-\rho(X), \rho(X)]$. But how do we show this? It turns out that one can proceed by tapping the power of complex analysis, and introducing the *Stieltjes transform*

(2.118) $$s_X(z) := \tau((X - z)^{-1})$$

for complex numbers z. Now, this transform need not be defined for all z at present, because we do not know that $X - z$ is invertible in $\mathcal{A}$. However, we can avoid this problem by working *formally*. Indeed, we have the formal *Neumann series expansion*

$$(X - z)^{-1} = -\frac{1}{z} - \frac{X}{z^2} - \frac{X^2}{z^3} - \cdots$$

which leads to the formal Laurent series expansion

$$(2.119) \qquad s_X(z) = -\sum_{k=0}^{\infty} \frac{\tau(X^k)}{z^{k+1}}.$$

If X is bounded self-adjoint, then from (2.117) we see that this formal series actually converges in the region $|z| > \rho(X)$. We will thus *define* the Stieltjes transform $s_X(z)$ on the region $|z| > \rho(X)$ by this series expansion (2.119), and then extend to as much of the complex plane as we can by analytic continuation[43].

We now push the domain of definition of $s_X(z)$ into the disk $\{|z| \leq \rho(X)\}$. We need some preliminary lemmas.

Exercise 2.5.2. Let X be bounded self-adjoint. For any real number R, show that $\rho(R^2 + X^2) = R^2 + \rho(X)^2$. (*Hint:* Use (2.116), (2.117)).

Exercise 2.5.3. Let X be bounded normal. Show that

$$|\tau(X^k)| \leq \tau((X^*X)^k)^{1/2} \leq \rho(X^*X)^{k/2}.$$

Now let R be a large positive real number. The idea is to rewrite the (formal) Stieltjes transform $\tau((X - z)^{-1})$ using the formal identity

$$(2.120) \qquad (X - z)^{-1} = ((X + iR) - (z + iR))^{-1}$$

and take Neumann series again to arrive at the formal expansion

$$(2.121) \qquad s_X(z) = -\sum_{k=0}^{\infty} \frac{\tau((X + iR)^k)}{(z + iR)^{k+1}}.$$

From the previous two exercises we see that

$$|\tau((X + iR)^k)| \leq (R^2 + \rho(X)^2)^{k/2},$$

and so the above Laurent series converges for $|z + iR| > (R^2 + \rho(X)^2)^{1/2}$.

Exercise 2.5.4. Give a rigorous proof that the two series (2.119), (2.121) agree for z large enough.

We have thus extended $s_X(z)$ analytically to the region $\{z : |z + iR| > (R^2 + \rho(X)^2)^{1/2}\}$. Letting $R \to \infty$, we obtain an extension of $s_X(z)$ to the upper half-plane $\{z : \text{Im}(z) > 0\}$. A similar argument (shifting by $-iR$ instead of $+iR$) gives an extension to the lower half-plane, thus defining $s_X(z)$ analytically everywhere except on the interval $[-\rho(X), \rho(X)]$.

---

[43]There could in principle be some topological obstructions to this continuation, but we will soon see that the only place where singularities can occur is on the real interval $[-\rho(X), \rho(X)]$, and so no topological obstructions will appear. One can also work with the original definition (2.118) of the Stieltjes transform, but this requires imposing some additional analytic axioms on the non-commutative probability space, such as requiring that $\mathcal{A}$ be a C^*-algebra or a von Neumann algebra, and we will avoid discussing these topics here as they are not the main focus of free probability theory.

On the other hand, it is not possible to analytically extend $s_X(z)$ to the region $\{z : |z| > \rho(X) - \varepsilon\}$ for any $0 < \varepsilon < \rho(X)$. Indeed, if this were the case, then from the *Cauchy integral formula* (applied at infinity), we would have the identity

$$\tau(X^k) = -\frac{1}{2\pi i} \int_{|z|=R} s_X(z) z^k \, dz$$

for any $R > \rho(X) - \varepsilon$, which when combined with (2.116) implies that $\rho(X) \leq R$ for all such R, which is absurd. Thus the spectral radius $\rho(X)$ can also be interpreted as the radius of the smallest ball centred at the origin outside of which the Stieltjes transform can be analytically continued.

Now that we have the Stieltjes transform everywhere outside of $[-\rho(X), \rho(X)]$, we can use it to derive an important bound (which will soon be superceded by (2.114), but will play a key role in the proof of that stronger statement):

Proposition 2.5.7 (Boundedness). *Let X be bounded self-adjoint, and let $P : \mathbf{C} \to \mathbf{C}$ be a polynomial. Then*

$$|\tau(P(X))| \leq \sup_{x \in [-\rho(X), \rho(X)]} |P(x)|.$$

Proof (Sketch). We can of course assume that P is non-constant, as the claim is obvious otherwise. From Exercise 2.5.3 (replacing P with $P\overline{P}$, where $\overline{P}$ is the polynomial whose coefficients are the complex conjugate of that of P) we may reduce to the case when P has real coefficients, so that $P(X)$ is self-adjoint. Since X is bounded, it is not difficult (using (2.116), (2.117)) to show that $P(X)$ is bounded also (Exercise!).

As $P(X)$ is bounded self-adjoint, it has a Stieltjes transform defined outside of $[-\rho(P(X)), \rho(P(X))]$, which for large z is given by the formula

$$(2.122) \qquad s_{P(X)}(z) = -\sum_{k=0}^{\infty} \frac{\tau(P(X)^k)}{z^{k+1}}.$$

By the previous discussion, to establish the proposition it will suffice to show that the Stieltjes transform can be continued to the domain

$$\Omega := \mathbf{C} \setminus [- \sup_{x \in [-\rho(X), \rho(X)]} |P(x)|, \sup_{x \in [-\rho(X), \rho(X)]} |P(x)|].$$

For this, we observe the partial fractions decomposition

$$\frac{1}{P(w) - z} = \sum_{\zeta : P(\zeta) = z} \frac{P'(\zeta)^{-1}}{w - \zeta}$$

of $(P(w) - z)^{-1}$ into linear combinations of $(w - \zeta)^{-1}$, at least when the roots of $P - z$ are simple. Thus, formally, at least, we have the identity

$$s_{P(X)}(z) = \sum_{\zeta : P(\zeta) = z} \frac{1}{P'(\zeta)} s_X(\zeta).$$

One can verify this identity is consistent with (2.122) for z sufficiently large. (Exercise! *Hint:* First do the case when X is a scalar, then expand in Taylor series and compare coefficients, then use the agreement of the Taylor series to do the general case.)

If z is in the domain Ω, then all the roots ζ of $P(\zeta) = z$ lie outside the interval $[-\rho(X), \rho(X)]$. So we can use the above formula as a *definition* of $s_{P(X)}(z)$, at least for those $z \in \Omega$ for which the roots of $P - z$ are simple; but there are only finitely many exceptional z (arising from zeroes of P') and one can check (Exercise! *Hint:* use the analytic nature of s_X and the residue theorem to rewrite parts of $s_{P(X)}(z)$ as a contour integral.) that the singularities here are removable. It is easy to see (Exercise!) that $s_{P(X)}$ is holomorphic outside of these removable singularities, and the claim follows. □

Exercise 2.5.5. Fill in the steps marked (Exercise!) in the above proof.

From Proposition 2.5.7 and the *Weierstrass approximation theorem* (see e.g. [**Ta2010**, §1.10]), we see that the linear functional $P \mapsto \tau(P(X))$ can be uniquely extended to a bounded linear functional on $C([-\rho(X), \rho(X)])$, with an operator norm 1. Applying the *Riesz representation theorem* (see e.g. [**Ta2010**, §1.10]), we thus can find a unique *Radon measure* (or equivalently, Borel measure) μ_X on $[-\rho(X), \rho(X)]$ of *total variation* 1 obeying the identity (2.114) for all P. In particular, setting $P = 1$ see that μ_X has total mass 1; since it also has total variation 1, it must be a probability measure. We have thus shown the fundamental

Theorem 2.5.8 (Spectral theorem for bounded self-adjoint elements). *Let X be a bounded self-adjoint element of a non-commutative probability space $(\mathcal{A}, \tau)$. Then there exists a unique Borel probability measure μ_X on $[-\rho(X), \rho(X)]$ (known as the* spectral measure *of X) such that (2.114) holds for all polynomials $P : \mathbf{C} \to \mathbf{C}$.*

Remark 2.5.9. If one assumes some completeness properties of the non-commutative probability space, such as that $\mathcal{A}$ is a C^*-algebra or a von Neumann algebra, one can use this theorem to meaningfully define $F(X)$ for other functions $F : [-\rho(X), \rho(X)] \to \mathbf{C}$ than polynomials; specifically, one can do this for continuous functions F if $\mathcal{A}$ is a C^*-algebra, and for $L^\infty(\mu_X)$ functions F if $\mathcal{A}$ is a von Neumann algebra. Thus, for instance, we can start define absolute values $|X|$, or square roots $|X|^{1/2}$, etc. Such an assignment $F \mapsto F(X)$ is known as a *functional calculus*; it can be used, for instance, to go back and make rigorous sense of the formula (2.118). A functional calculus is a very convenient tool to have in operator algebra theory, and for that reason one often *completes* a non-commutative probability space into a C^*-algebra or von Neumann algebra, much as how it is often convenient to complete the rationals and work instead with the reals. However, we will proceed here instead by working with a (possibly incomplete) non-commutative probability space, and working primarily with *formal* expressions (e.g., formal power series in z) without trying to evaluate such expressions in some completed space. We can get away with this because we will be working exclusively in situations in which the spectrum of a random variable can be reconstructed exactly from its moments (which is, in particular, true in the case of bounded random variables). For unbounded random variables, one must usually instead use the full power of functional analysis and work with the spectral theory of unbounded operators on Hilbert spaces.

Exercise 2.5.6. Let X be a bounded self-adjoint element of a non-commutative probability space, and let μ_X be the spectral measure of X. Establish the formula

$$s_X(z) = \int_{[-\rho(X), \rho(X)]} \frac{1}{x - z} \, d\mu_X(x)$$

for all $z \in \mathbf{C} \backslash [-\rho(X), \rho(X)]$. Conclude that the support[44] of the spectral measure μ_X must contain at least one of the two points $-\rho(X), \rho(X)$.

Exercise 2.5.7. Let X be a bounded self-adjoint element of a non-commutative probability space with faithful trace. Show that $\rho(X) = 0$ if and only if $X = 0$.

Remark 2.5.10. It is possible to also obtain a spectral theorem for bounded normal elements along the lines of the above theorem (with μ_X now supported in a disk rather than in an interval, and with (2.114) replaced by (2.115)), but this is somewhat more complicated to show (basically, one needs to extend the self-adjoint spectral theorem to a pair of commuting self-adjoint elements, which is a little tricky to show by complex-analytic methods, as one has to use several complex variables).

The spectral theorem more or less completely describes the behaviour of a single (bounded self-adjoint) element X in a non-commutative probability space. As remarked above, it can also be extended to study multiple commuting self-adjoint elements. However, when one deals with multiple *non-commuting* elements, the spectral theorem becomes inadequate (and

---

[44]The *support* of a measure is the intersection of all the closed sets of full measure.

indeed, it appears that in general there is no usable substitute for this theorem). However, we can begin making a little bit of headway if we assume as a final (optional) axiom a very weak form of commutativity in the trace:

Axiom 2.5.11 (Trace). *For any two elements X, Y, we have $\tau(XY) = \tau(YX)$.*

Note that this axiom is obeyed by all three of our model examples. From this axiom, we can cyclically permute products in a trace, e.g., $\tau(XYZ) = \tau(YZX) = \tau(ZXY)$. However, we cannot take non-cyclic permutations; for instance, $\tau(XYZ)$ and $\tau(XZY)$ are distinct in general. This axiom is a trivial consequence of the commutative nature of the complex numbers in the classical setting, but can play a more non-trivial role in the non-commutative setting. It is, however possible, to develop a large part of free probability without this axiom, if one is willing instead to work in the category of von Neumann algebras. Thus, we shall leave it as an optional axiom:

Definition 2.5.12 (Non-commutative probability space, final definition). A *non-commutative probability space* $(\mathcal{A}, \tau)$ consists of a $*$-algebra $\mathcal{A}$ with identity 1, together with a $*$-linear functional $\tau : \mathcal{A} \to \mathbf{C}$, that maps 1 to 1 and obeys the non-negativity axiom. If τ obeys the trace axiom, we say that the non-commutative probability space is *tracial*. If τ obeys the faithfulness axiom, we say that the non-commutative probability space is *faithful*.

From this new axiom and the Cauchy-Schwarz inequality we can now get control on products of several non-commuting elements:

Exercise 2.5.8. Let $X_1, \ldots, X_k$ be bounded self-adjoint elements of a tracial non-commutative probability space $(\mathcal{A}, \tau)$. Show that
$$|\tau(X_1^{m_1} \ldots X_k^{m_k})| \leq \rho(X_1)^{m_1} \ldots \rho(X_k)^{m_k}$$
for any non-negative integers $m_1, \ldots, m_k$. (*Hint:* Induct on k, and use Cauchy-Schwarz to split up the product as evenly as possible, using cyclic permutations to reduce the complexity of the resulting expressions.)

Exercise 2.5.9. Let $\mathcal{A} \cap L^\infty(\tau)$ be those elements X in a tracial non-commutative probability space $(\mathcal{A}, \tau)$ whose real and imaginary parts $\mathrm{Re}(X) := \frac{X+X^*}{2}$, $\mathrm{Im}(X) := \frac{X-X^*}{2i}$ are bounded and self-adjoint; we refer to such elements simply as *bounded* elements. Show that this is a sub-$*$-algebra of $\mathcal{A}$.

This allows one to perform the following *Gelfand-Naimark-Segal (GNS) construction*. Recall that $\mathcal{A} \cap L^\infty(\tau)$ has a positive semi-definite inner product $\langle,\rangle_{L^2(\tau)}$. We can perform the Hilbert space completion of this inner product space (quotienting out by the elements of zero norm), leading to a complex Hilbert space $L^2(\tau)$ into which $\mathcal{A} \cap L^\infty(\tau)$ can be mapped as a dense subspace by an isometry[45] $\iota : \mathcal{A} \cap L^\infty(\tau) \to L^2(\tau)$.

The space $\mathcal{A} \cap L^\infty(\tau)$ acts on itself by multiplication, and thus also acts on the dense subspace $\iota(\mathcal{A} \cap L^\infty(\tau))$ of $L^2(\tau)$. We would like to extend this action to all of $L^2(\tau)$, but this requires an additional estimate:

Lemma 2.5.13. *Let* $(\mathcal{A}, \tau)$ *be a tracial non-commutative probability space. If* $X, Y \in \mathcal{A} \cap L^\infty(\tau)$ *with* X *self-adjoint, then*

$$\|XY\|_{L^2(\tau)} \leq \rho(X)\|Y\|_{L^2(\tau)}.$$

Proof. Squaring and cyclically permuting, it will suffice to show that

$$\tau(Y^* X^2 Y) \leq \rho(X)^2 \tau(Y^* Y).$$

Let $\varepsilon > 0$ be arbitrary. By Weierstrass approximation, we can find a polynomial P with real coefficients such that $x^2 + P(x)^2 = \rho(X)^2 + O(\varepsilon)$ on the interval $[-\rho(X), \rho(X)]$. By Proposition 2.5.7, we can thus write $X^2 + P(X)^2 = \rho(X)^2 + E$ where E is self-adjoint with $\rho(E) = O(\varepsilon)$. Multiplying on the left by Y^* and on the right by Y and taking traces, we obtain

$$\tau(Y^* X^2 Y) + \tau(Y^* P(X)^2 Y) \leq \rho(X)^2 \tau(Y^* Y) + \tau(Y^* E Y).$$

By non-negativity, $\tau(Y^* P(X)^2 Y) \geq 0$. By Exercise 2.5.8, we have $\tau(Y^* E Y) = O_Y(\varepsilon)$. Sending $\varepsilon \to 0$ we obtain the claim. $\square$

As a consequence, we see that the self-adjoint elements X of $\mathcal{A} \cap L^\infty(\tau)$ act in a bounded manner on all of $L^2(\tau)$, and so on taking real and imaginary parts, we see that the same is true for the non-self-adjoint elements too. Thus we can associate to each $X \in L^\infty(\tau)$ a bounded linear transformation $\overline{X} \in B(L^2(\tau))$ on the Hilbert space $L^2(\tau)$.

Exercise 2.5.10 (Gelfand-Naimark theorem). Show that the map $X \mapsto \overline{X}$ is a $*$-isomorphism from $\mathcal{A} \cap L^\infty(\tau)$ to a $*$-subalgebra of $B(L^2(\tau))$, and that one has the representation

$$\tau(X) = \langle e, \overline{X} e \rangle$$

for any $X \in L^\infty(\tau)$, where e is the unit vector $e := \iota(1)$.

Remark 2.5.14. The Gelfand-Naimark theorem required the tracial hypothesis only to deal with the error E in the proof of Lemma 2.5.13. One can also establish this theorem without this hypothesis, by assuming instead that the non-commutative space is a C^*-algebra; this provides a continuous functional calculus, so that we can replace P in the proof of Lemma 2.5.13

---

[45] This isometry is injective when $\mathcal{A}$ is faithful, but will have a non-trivial kernel otherwise.

by a continuous function and dispense with E altogether. This formulation of the Gelfand-Naimark theorem is the one which is usually seen in the literature.

The Gelfand-Naimark theorem identifies $\mathcal{A} \cap L^\infty(\tau)$ with a $*$-subalgebra of $B(L^2(\tau))$. The closure of this $*$-subalgebra in the weak operator topology[46] is then a *von Neumann algebra*, which we denote as $L^\infty(\tau)$. As a consequence, we see that non-commutative probability spaces are closely related to von Neumann algebras (equipped with a tracial state τ). However, we refrain from identifying the former completely with the latter, in order to allow ourselves the freedom to work with such spaces as $L^{\infty-}$, which is almost but not quite a von Neumann algebra. Instead, we use the looser (and more algebraic) definition in Definition 2.5.12.

2.5.2. Limits of non-commutative random variables. One benefit of working in an abstract setting is that it becomes easier to take certain types of limits. For instance, it is intuitively obvious that the cyclic groups $\mathbf{Z}/N\mathbf{Z}$ are "converging" in some sense to the integer group $\mathbf{Z}$. This convergence can be formalised by selecting a distinguished generator e of all groups involved (1 mod N in the case of $\mathbf{Z}/N\mathbf{Z}$, and 1 in the case of the integers $\mathbf{Z}$), and noting that the set of relations involving this generator in $\mathbf{Z}/N\mathbf{Z}$ (i.e., the relations $ne = 0$ when n is divisible by N) converge in a pointwise sense to the set of relations involving this generator in $\mathbf{Z}$ (i.e., the empty set). Here, to see the convergence, we viewed a group abstractly via the relations between its generators, rather than on a concrete realisation of a group as (say) residue classes modulo N.

We can similarly define convergence of random variables in non-commutative probability spaces as follows.

Definition 2.5.15 (Convergence). Let $(\mathcal{A}_n, \tau_n)$ be a sequence of non-commutative probability spaces, and let $(\mathcal{A}_\infty, \tau_\infty)$ be an additional non-commutative space. For each n, let $X_{n,1}, \ldots, X_{n,k}$ be a sequence of random variables in $\mathcal{A}_n$, and let $X_{\infty,1}, \ldots, X_{\infty,k}$ be a sequence of random variables in $\mathcal{A}_\infty$. We say that $X_{n,1}, \ldots, X_{n,k}$ converges *in the sense of moments* to $X_{\infty,1}, \ldots, X_{\infty,k}$ if we have

$$\tau_n(X_{n,i_1} \ldots X_{n,i_m}) \to \tau_\infty(X_{\infty,i_1} \ldots X_{\infty,i_m})$$

as $n \to \infty$ for any sequence $i_1, \ldots, i_m \in \{1, \ldots, k\}$. We say that $X_{n,1}, \ldots, X_{n,k}$ converge *in the sense of $*$-moments* to $X_{\infty,1}, \ldots, X_{\infty,k}$ if $X_{n,1}, \ldots, X_{n,k}$, $X_{n,1}^*, \ldots, X_{n,k}^*$ converges in the sense of moments to $X_{\infty,1}, \ldots, X_{\infty,k}$, $X_{\infty,1}^*, \ldots, X_{\infty,k}^*$.

---

[46]The *weak operator topology* on the space $B(H)$ of bounded operators on a Hilbert space is the weakest topology for which the coefficient maps $T \mapsto \langle Tu, v \rangle_H$ are continuous for each $u, v \in H$.

If $X_1, \ldots, X_k$ (viewed as a constant k-tuple in n) converges in the sense of moments (resp. $*$-moments) to $Y_1, \ldots, Y_k$, we say that $X_1, \ldots, X_k$ and $Y_1, \ldots, Y_k$ have *matching joint moments* (resp. *matching joint $*$-moments*).

Example 2.5.16. If X_n, Y_n converge in the sense of moments to X_∞, Y_∞, then we have, for instance, that

$$\tau_n(X_n Y_n^k X_n) \to \tau_\infty(X_\infty Y_\infty^k X_\infty)$$

as $n \to \infty$ for each k, while if they converge in the stronger sense of $*$-moments, then we obtain more limits, such as

$$\tau_n(X_n Y_n^k X_n^*) \to \tau_\infty(X_\infty Y_\infty^k X_\infty^*).$$

Note, however, that no uniformity in k is assumed for this convergence; in particular, if k varies in n (e.g., if $k = O(\log n)$), there is now no guarantee that one still has convergence.

Remark 2.5.17. When the underlying objects $X_{n,1}, \ldots, X_{n,k}$ and $X_1, \ldots, X_k$ are self-adjoint, then there is no distinction between convergence in moments and convergence in $*$-moments. However, for non-self-adjoint variables, the latter type of convergence is far stronger, and the former type is usually too weak to be of much use, even in the commutative setting. For instance, let X be a classical random variable drawn uniformly at random from the unit circle $\{z \in \mathbf{C} : |z| = 1\}$. Then the constant sequence $X_n = X$ has all the same moments as the zero random variable 0, and thus converges in the sense of moments to zero, but does not converge in the $*$-moment sense to zero.

It is also clear that if we require that $\mathcal{A}_\infty$ be generated by $X_{\infty,1}, \ldots, X_{\infty,k}$ in the $*$-algebraic sense (i.e., every element of $\mathcal{A}_\infty$ is a polynomial combination of $X_{\infty,1}, \ldots, X_{\infty,k}$ and their adjoints), then a limit in the sense of $*$-moments, if it exists, is unique up to matching joint $*$-moments.

For a sequence X_n of a single, uniformly bounded, self-adjoint element, convergence in moments is equivalent to convergence in distribution:

Exercise 2.5.11. Let $X_n \in \mathcal{A}_n$ be a sequence of self-adjoint elements in non-commutative probability spaces $(\mathcal{A}_n, \tau_n)$ with $\rho(X_n)$ uniformly bounded, and let $X_\infty \in \mathcal{A}_\infty$ be another bounded self-adjoint element in a non-commutative probability space $(\mathcal{A}_\infty, \tau_\infty)$. Show that X_n converges in moments to X_∞ if and only if the spectral measure μ_{X_n} converges in the vague topology to μ_{X_∞}.

Thus, for instance, one can rephrase the Wigner semicircular law (in the convergence in expectation formulation) as the assertion that a sequence $M_n \in L^{\infty-} \otimes M_n(\mathbf{C})$ of Wigner random matrices with (say) sub-Gaussian entries of mean zero and variance one, when viewed as elements of the non-commutative probability space $(L^{\infty-} \otimes M_n(\mathbf{C}), \mathbf{E}\frac{1}{n}\operatorname{tr})$, will converge to any bounded self-adjoint element u of a non-commutative probability space with spectral measure given by the semicircular distribution $\mu_{\mathrm{sc}} := \frac{1}{2\pi}(4 - x^2)_+^{1/2} \, dx$. Such elements are known as *semicircular elements*. Here are some easy examples of semicircular elements:

(i) A classical real random variable u drawn using the probability measure μ_{sc}.

(ii) The identity function $x \mapsto x$ in the Lebesgue space $L^\infty(d\mu_{\mathrm{sc}})$, endowed with the trace $\tau(f) := \int_{\mathbf{R}} f \, d\mu_{\mathrm{sc}}$.

(iii) The function $\theta \mapsto 2\cos\theta$ in the Lebesgue space $L^\infty([0,\pi], \frac{2}{\pi}\sin^2\theta \, d\theta)$.

Here is a more interesting example of a semicircular element:

Exercise 2.5.12. Let $(\mathcal{A}, \tau)$ be the non-commutative space consisting of bounded operators $B(\ell^2(\mathbf{N}))$ on the natural numbers with trace $\tau(X) := \langle e_0, X e_0 \rangle_{\ell^2(\mathbf{N})}$, where $e_0, e_1, \ldots$ is the standard basis of $\ell^2(\mathbf{N})$. Let $U : e_n \mapsto e_{n+1}$ be the right shift on $\ell^2(\mathbf{N})$. Show that $U + U^*$ is a semicircular operator. (*Hint:* One way to proceed here is to use Fourier analysis to identify $\ell^2(\mathbf{N})$ with the space of odd functions $\theta \mapsto f(\theta)$ on $\mathbf{R}/2\pi\mathbf{Z}$, with U being the operator that maps $\sin(n\theta)$ to $\sin((n+1)\theta)$; show that $U + U^*$ is then the operation of multiplication by $2\cos\theta$.) One can also interpret U as a *creation operator* in a *Fock space*, but we will not do so here.

Exercise 2.5.13. With the notation of the previous exercise, show that $\tau((U + U^*)^k)$ is zero for odd k, and is equal to the Catalan number $C_{k/2}$ from Section 2.3 when k is even. Note that this provides a (very) slightly different proof of the semicircular law from that given from the moment method in Section 2.4.

Because we are working in such an abstract setting with so few axioms, limits exist in abundance:

Exercise 2.5.14. For each n, let $X_{n,1}, \ldots, X_{n,k}$ be bounded self-adjoint elements of a tracial non-commutative space $(\mathcal{A}_n, \tau_n)$. Suppose that the spectral radii $\rho(X_{n,1}), \ldots, \rho(X_{n,k})$ are uniformly bounded in n. Show that there exists a subsequence n_j and bounded self-adjoint elements $X_1, \ldots, X_k$ of a tracial non-commutative space $(\mathcal{A}, \tau)$ such that $X_{n_j,1}, \ldots, X_{n_j,k}$ converge in moments to $X_1, \ldots, X_k$ as $j \to \infty$. (*Hint:* Use the Bolzano-Weierstrass theorem and the Arzelá-Ascoli diagonalisation trick to obtain a subsequence in which each of the joint moments of $X_{n_j,1}, \ldots, X_{n_j,k}$ converge as $j \to \infty$. Use these moments to build a non-commutative probability space.)

2.5.3. Free independence. We now come to the fundamental concept in free probability theory, namely that of *free independence*.

Definition 2.5.18 (Free independence). A collection $X_1, \ldots, X_k$ of random variables in a non-commutative probability space $(\mathcal{A}, \tau)$ is *freely independent* (or *free* for short) if one has

$$\tau((P_1(X_{i_1}) - \tau(P_1(X_{i_1}))) \ldots (P_m(X_{i_m}) - \tau(P_m(X_{i_m})))) = 0$$

whenever $P_1, \ldots, P_m$ are polynomials and $i_1, \ldots, i_m \in \{1, \ldots, k\}$ are indices with no two adjacent i_j equal.

A sequence $X_{n,1}, \ldots, X_{n,k}$ of random variables in a non-commutative probability space $(\mathcal{A}_n, \tau_n)$ is *asymptotically freely independent* (or *asymptotically free* for short) if one has

$$\tau_n((P_1(X_{n,i_1}) - \tau(P_1(X_{n,i_1}))) \ldots (P_m(X_{n,i_m}) - \tau(P_m(X_{n,i_m})))) \to 0$$

as $n \to \infty$ whenever $P_1, \ldots, P_m$ are polynomials and $i_1, \ldots, i_m \in \{1, \ldots, k\}$ are indices with no two adjacent i_j equal.

Remark 2.5.19. The above example describes freeness of collections of random variables $\mathcal{A}$. One can more generally define freeness of collections of subalgebras of $\mathcal{A}$, which in some sense is the more natural concept from a category-theoretic perspective, but we will not need this concept here. See e.g., [**Bi2003**] for more discussion.

Thus, for instance, if X, Y are freely independent, then $\tau(P(X)Q(Y)R(X)S(Y))$ will vanish for any polynomials P, Q, R, S for which $\tau(P(X))$, $\tau(Q(Y)), \tau(R(X)), \tau(S(Y))$ all vanish. This is in contrast to classical independence of classical (commutative) random variables, which would only assert that $\tau(P(X)Q(Y)) = 0$ whenever $\tau(P(X)), \tau(Q(Y))$ both vanish.

To contrast free independence with classical independence, suppose that $\tau(X) = \tau(Y) = 0$. If X, Y were freely independent, then $\tau(XYXY) = 0$. If instead X, Y were commuting and classically independent, then we would instead have $\tau(XYXY) = \tau(X^2Y^2) = \tau(X^2)\tau(Y^2)$, which would almost certainly be non-zero.

For a trivial example of free independence, X and Y automatically are freely independent if at least one of X, Y is constant (i.e., a multiple of the identity 1). In the commutative setting, this is basically the only way one can have free independence:

Exercise 2.5.15. Suppose that X, Y are freely independent elements of a faithful non-commutative probability space which also commute. Show that at least one of X, Y is equal to a scalar. (*Hint:* First normalise X, Y to have trace zero, and consider $\tau(XYXY)$.)

A less trivial example of free independence comes from the free group, which provides a clue as to the original motivation of this concept:

Exercise 2.5.16. Let $\mathbf{F}_2$ be the free group on two generators g_1, g_2. Let $\mathcal{A} = B(\ell^2(\mathbf{F}_2))$ be the non-commutative probability space of bounded linear operators on the Hilbert space $\ell^2(\mathbf{F}_2)$, with trace $\tau(X) := \langle Xe_0, e_0 \rangle$, where e_0 is the Kronecker delta function at the identity. Let $U_1, U_2 \in \mathcal{A}$ be the shift operators

$$U_1 f(g) := f(g_1 g); \quad U_2 f(g) := f(g_2 g)$$

for $f \in \ell^2(\mathbf{F}_2)$ and $g \in \mathbf{F}_2$. Show that U_1, U_2 are freely independent.

For classically independent commuting random variables X, Y, knowledge of the individual moments $\tau(X^k)$, $\tau(Y^k)$ gave complete information on the joint moments: $\tau(X^k Y^l) = \tau(X^k)\tau(Y^l)$. The same fact is true for freely independent random variables, though the situation is more complicated. We begin with a simple case: computing $\tau(XY)$ in terms of the moments of X, Y. From free independence we have

$$\tau((X - \tau(X))(Y - \tau(Y))) = 0.$$

Expanding this using the linear nature of trace, one soon sees that

(2.123) $\tau(XY) = \tau(X)\tau(Y).$

So far, this is just the same as with the classically independent case. Next, we consider a slightly more complicated moment, $\tau(XYX)$. If we split $Y = \tau(Y) + (Y - \tau(Y))$, we can write this as

$$\tau(XYX) = \tau(Y)\tau(X^2) + \tau(X(Y - \tau(Y))X).$$

In the classically independent case, we can conclude the latter term would vanish. We cannot immediately say that in the freely independent case, because only one of the factors has mean zero. But from (2.123) we know that $\tau(X(Y - \tau(Y))) = \tau((Y - \tau(Y))X) = 0$. Because of this, we can expand

$$\tau(X(Y - \tau(Y))X) = \tau((X - \tau(X))(Y - \tau(Y))(X - \tau(X)))$$

and now free independence does ensure that this term vanishes, and so

(2.124) $\tau(XYX) = \tau(Y)\tau(X^2).$

So again we have not yet deviated from the classically independent case. But now let us look at $\tau(XYXY)$. We split the second X into $\tau(X)$ and $X - \tau(X)$. Using (2.123) to control the former term, we have

$$\tau(XYXY) = \tau(X)^2\tau(Y^2) + \tau(XY(X - \tau(X))Y).$$

From (2.124) we have $\tau(Y(X - \tau(X))Y) = 0$, so we have

$$\tau(XYXY) = \tau(X)^2\tau(Y^2) + \tau((X - \tau(X))Y(X - \tau(X))Y).$$

Now we split Y into $\tau(Y)$ and $Y - \tau(Y)$. Free independence eliminates all terms except

$$\tau(XYXY) = \tau(X)^2\tau(Y^2) + \tau((X - \tau(X))\tau(Y)(X - \tau(X))\tau(Y))$$

which simplifies to

$$\tau(XYXY) = \tau(X)^2\tau(Y^2) + \tau(X^2)\tau(Y)^2 - \tau(X)^2\tau(Y)^2$$

which differs from the classical independence prediction of $\tau(X^2)\tau(Y^2)$.

This process can be continued:

Exercise 2.5.17. Let $X_1, \ldots, X_k$ be freely independent. Show that any joint moment of $X_1, \ldots, X_k$ can be expressed as a polynomial combination of the individual moments $\tau(X_i^j)$ of the X_i. (*Hint:* Induct on the complexity of the moment.)

The product measure construction allows us to generate classically independent random variables at will (after extending the underlying sample space): see Exercise 1.1.20. There is an analogous construction, called the *amalgamated free product*, that allows one to generate families of freely independent random variables, each of which has a specified distribution. Let us give an illustrative special case of this construction:

Lemma 2.5.20 (Free products). *For each $1 \le i \le k$, let $(\mathcal{A}_i, \tau_i)$ be a non-commutative probability space. Then there exists a non-commutative probability space $(\mathcal{A}, \tau)$ which contain embedded copies of each of the $(\mathcal{A}_i, \tau_i)$, such that whenever $X_i \in \mathcal{A}_i$ for $i = 1, \ldots, k$, then $X_1, \ldots, X_k$ are freely independent.*

Proof (Sketch). Recall that each $\mathcal{A}_i$ can be given an inner product $\langle, \rangle_{L^2(\mathcal{A}_i)}$. One can then orthogonally decompose each space $\mathcal{A}_i$ into the constants $\mathbf{C}$, plus the trace zero elements $\mathcal{A}_i^0 := \{X \in \mathcal{A}_i : \tau(X) = 0\}$.

We now form the *Fock space* $\mathcal{F}$ to be the inner product space formed by the direct sum of tensor products

(2.125) $$\mathcal{A}_{i_1}^0 \otimes \ldots \otimes \mathcal{A}_{i_m}^0$$

where $m \ge 0$, and $i_1, \ldots, i_m \in \{1, \ldots, k\}$ are such that no adjacent pair i_j, i_{j+1} of the $i_1, \ldots, i_m$ are equal. Each element $X_i \in \mathcal{A}_i$ then acts on this Fock space by defining

$$X_i(Y_{i_1} \otimes \ldots \otimes Y_{i_m}) := X_i \otimes Y_{i_1} \otimes \ldots \otimes Y_{i_m}$$

when $i \ne i_1$, and

$$X_i(Y_{i_1} \otimes \ldots \otimes Y_{i_m}) := \tau(X_iY_{i_1})Y_{i_2} \otimes \ldots \otimes Y_{i_m} + (X_iY_{i_1} - \tau(X_iY_{i_1})) \otimes Y_{i_2} \otimes \ldots \otimes Y_{i_m}$$

when $i = i_1$. One can thus map $\mathcal{A}_i$ into the space $\mathcal{A} := \mathrm{Hom}(\mathcal{F}, \mathcal{F})$ of linear maps from $\mathcal{F}$ to itself. The latter can be given the structure of a non-commutative space by defining the trace $\tau(X)$ of an element $X \in \mathcal{A}$ by the formula $\tau(X) := \langle X e_\emptyset, e_\emptyset \rangle_\mathcal{F}$, where $e_\emptyset$ is the *vacuum state* of $\mathcal{F}$, being the unit of the $m = 0$ tensor product. One can verify (Exercise!) that $\mathcal{A}_i$ embeds into $\mathcal{A}$ and that elements from different $\mathcal{A}_i$ are freely independent. $\qquad \square$

Exercise 2.5.18. Complete the proof of Lemma 2.5.20. (*Hint:* You may find it helpful to first do Exercise 2.5.16, as the construction here is an abstraction of the one in that exercise.)

Finally, we illustrate the fundamental connection between free probability and random matrices first observed by Voiculescu [**Vo1991**], namely that (classically) independent families of random matrices are *asymptotically free*. The intuition here is that while a large random matrix M will certainly correlate with itself (so that, for instance, $\mathrm{tr}\, M^*M$ will be large), once one interposes an independent random matrix N of trace zero, the correlation is largely destroyed (thus, for instance, $\mathrm{tr}\, M^*NM$ will usually be quite small).

We give a typical instance of this phenomenon here:

Proposition 2.5.21 (Asymptotic freeness of Wigner matrices). *Let $M_{n,1}$, ..., $M_{n,k}$ be a collection of independent $n \times n$ Wigner matrices, where the coefficients all have uniformly bounded m^{th} moments for each m. Then the random variables $\frac{1}{\sqrt{n}} M_{n,1}, \ldots, \frac{1}{\sqrt{n}} M_{n,k} \in (L^{\infty-} \otimes M_n(\mathbf{C}), \mathbf{E}\frac{1}{n}\mathrm{tr})$ are asymptotically free.*

Proof (Sketch). Let us abbreviate $\frac{1}{\sqrt{n}} M_{n,j}$ as X_j (suppressing the n dependence). It suffices to show that the traces

$$\tau(\prod_{j=1}^m (X_{i_j}^{a_j} - \tau(X_{i_j}^{a_j}))) = o(1)$$

for each fixed choice of natural numbers $a_1, \ldots, a_m$, where no two adjacent i_j, i_{j+1} are equal.

Recall from Section 2.3 that $\tau(X_j^{a_j})$ is (up to errors of $o(1)$) equal to a normalised count of paths of length a_j in which each edge is traversed exactly twice, with the edges forming a tree. After normalisation, this count is equal to 0 when a_j is odd, and equal to the Catalan number $C_{a_j/2}$ when a_j is even.

One can perform a similar computation to compute $\tau(\prod_{j=1}^m X_{i_j}^{a_j})$. Up to errors of $o(1)$, this is a normalised count of *coloured* paths of length $a_1 + \cdots + a_m$, where the first a_1 edges are coloured with colour i_1, the next a_2 with colour i_2, etc. Furthermore, each edge is traversed exactly twice (with the two traversals of each edge being assigned the same colour), and the edges form a tree. As a consequence, there must exist a j for which the block of a_j edges of colour i_j form their own sub-tree, which contributes a factor of $C_{a_j/2}$ or 0 to the final trace. Because of this, when one instead computes the normalised expression $\tau(\prod_{j=1}^m (X_{i_j}^{a_j} - \tau(X_{i_j}^{a_j})))$, all contributions that are not $o(1)$ cancel themselves out, and the claim follows. $\square$

Exercise 2.5.19. Expand the above sketch into a full proof of the above theorem.

Remark 2.5.22. This is by no means the only way in which random matrices can become asymptotically free. For instance, if instead one considers random matrices of the form $M_{n,i} = U_i^* A_i U_i$, where A_i are deterministic Hermitian matrices with uniformly bounded eigenvalues, and the U_i are iid unitary matrices drawn using Haar measure on the unitary group $U(n)$, one can also show that the $M_{n,i}$ are asymptotically free; again, see [**Vo1991**] for details.

2.5.4. Free convolution. When one is summing two classically independent (real-valued) random variables X and Y, the distribution μ_{X+Y} of the sum $X + Y$ is the convolution $\mu_X * \mu_Y$ of the distributions μ_X and μ_Y. This convolution can be computed by means of the characteristic function

$$F_X(t) := \tau(e^{itX}) = \int_{\mathbf{R}} e^{itx}\, d\mu_X(x)$$

by means of the simple formula

$$\tau(e^{it(X+Y)}) = \tau(e^{itX})\tau(e^{itY}).$$

As we saw in Section 2.2, this can be used, in particular, to establish a short proof of the central limit theorem.

There is an analogous theory when summing two freely independent (self-adjoint) non-commutative random variables X and Y; the distribution μ_{X+Y} turns out to be a certain combination $\mu_X \boxplus \mu_Y$, known as the *free convolution* of μ_X and μ_Y. To compute this free convolution, one does not use the characteristic function; instead, the correct tool is the *Stieltjes transform*

$$s_X(z) := \tau((X - z)^{-1}) = \int_{\mathbf{R}} \frac{1}{x - z}\, d\mu_X(x)$$

which has already been discussed earlier.

Here's how to use this transform to compute free convolutions. If one wishes, one can assume that X is bounded so that all series involved converge for z large enough, though actually the entire argument here can be performed at a purely algebraic level, using formal power series, and so the boundedness hypothesis here is not actually necessary.

The trick (which we already saw in Section 2.4) is not to view $s = s_X(z)$ as a function of z, but rather to view $z = z_X(s)$ as a function of s. Given that one asymptotically has $s \sim -1/z$ for z, we expect to be able to perform this inversion for z large and s close to zero; and in any event one can easily invert (2.119) on the level of formal power series.

With this inversion, we thus have

(2.126)
$$s = \tau((X - z_X(s))^{-1})$$

and thus

$$(X - z_X(s))^{-1} = s(1 - E_X)$$

for some $E_X = E_X(s)$ of trace zero. Now we do some (formal) algebraic sleight of hand. We rearrange the above identity as

$$X = z_X(s) + s^{-1}(1 - E_X)^{-1}.$$

Similarly, we have

$$Y = z_Y(s) + s^{-1}(1 - E_Y)^{-1},$$

and so

$$X + Y = z_X(s) + z_Y(s) + s^{-1}[(1 - E_X)^{-1} + (1 - E_Y)^{-1}].$$

We can combine the second two terms via the identity

$$(1 - E_X)^{-1} + (1 - E_Y)^{-1} = (1 - E_X)^{-1}(1 - E_Y + 1 - E_X)(1 - E_Y)^{-1}.$$

Meanwhile,

$$1 = (1 - E_X)^{-1}(1 - E_Y - E_X + E_X E_Y)(1 - E_Y)^{-1},$$

and so

$$X + Y = z_X(s) + z_Y(s) + s^{-1} + s^{-1}[(1 - E_X)^{-1}(1 - E_X E_Y)(1 - E_Y)^{-1}].$$

We can rearrange this a little bit as

$$(X + Y - z_X(s) - z_Y(s) - s^{-1})^{-1} = s[(1 - E_Y)(1 - E_X E_Y)^{-1}(1 - E_X)].$$

We expand out as the (formal) Neumann series:

$$(1 - E_Y)(1 - E_X E_Y)^{-1}(1 - E_X)$$
$$= (1 - E_Y)(1 + E_X E_Y + E_X E_Y E_X E_Y + \ldots)(1 - E_X).$$

This expands out to equal 1 plus a whole string of alternating products of E_X and E_Y.

Now we use the hypothesis that X and Y are free. This easily implies that E_X and E_Y are also free. But they also have trace zero, thus by the definition of free independence, all alternating products of E_X and E_Y have zero trace[47]. We conclude that

$$\tau((1 - E_Y)(1 - E_X E_Y)^{-1}(1 - E_X)) = 1,$$

---

[47]In the case when there are an odd number of terms in the product, one can obtain this zero trace property using the cyclic property of trace and induction.

and so
$$\tau((X + Y - z_X(s) - z_Y(s) - s^{-1})^{-1}) = s.$$
Comparing this against (2.126) for $X + Y$ we conclude that
$$z_{X+Y}(s) = z_X(s) + z_Y(s) + s^{-1}.$$
Thus, if we define the *R-transform* R_X of X to be (formally) given by[48] the formula
$$R_X(s) := z_X(-s) - s^{-1},$$
then we have the addition formula
$$R_{X+Y} = R_X + R_Y.$$

Since one can recover the Stieltjes transform s_X (and hence the R-transform R_X) from the spectral measure μ_X and vice versa, this formula (in principle, at least) lets one compute the spectral measure μ_{X+Y} of $X + Y$ from the spectral measures μ_X, μ_Y, thus allowing one to define free convolution.

For comparison, we have the (formal) addition formula
$$\log F_{X+Y} = \log F_X + \log F_Y$$
for classically independent real random variables X, Y. The following exercises carry this analogy a bit further.

Exercise 2.5.20. Let X be a classical real random variable. Working formally, show that
$$\log F_X(t) = \sum_{k=1}^{\infty} \frac{\kappa_k(X)}{k!}(it)^k$$
where the *cumulants* $\kappa_k(X)$ can be reconstructed from the moments $\tau(X^k)$ by the recursive formula
$$\tau(X^k) = \kappa_k(X) + \sum_{j=1}^{k-1} \kappa_j(X) \sum_{a_1 + \cdots + a_j = k-j} \tau(X^{a_1 + \cdots + a_j})$$
for $k \geq 1$. (*Hint:* Start with the identity $\frac{d}{dt} F_X(t) = (\frac{d}{dt} \log F_X(t)) F_X(t)$.) Thus, for instance, $\kappa_1(X) = \tau(X)$ is the expectation, $\kappa_2(X) = \tau(X^2) - \tau(X)^2$ is the variance, and the third cumulant is given by the formula
$$\kappa_3(X) = \tau(X^3) + 3\tau(X^2)\tau(X) - 4\tau(X)^3.$$
Establish the additional formula
$$\tau(X^k) = \sum_{\pi} \prod_{A \in \pi} C_{|A|}(X)$$
where π ranges over all partitions of $\{1, \ldots, k\}$ into non-empty cells A.

---

[48]The sign conventions may seem odd here, but they are chosen to conform to the notational conventions in the literature, which are based around the moment series $\sum_{n=1}^{\infty} \tau(X^n)z^n$ rather than the (slightly different) Stieltjes transform (2.119).

Exercise 2.5.21. Let X be a non-commutative random variable. Working formally, show that

$$R_X(s) = \sum_{k=1}^{\infty} C_k(X)s^{k-1}$$

where the *free cumulants* $C_k(X)$ can be reconstructed from the moments $\tau(X^k)$ by the recursive formula

$$\tau(X^k) = C_k(X) + \sum_{j=1}^{k-1} C_j(X) \sum_{a_1+\cdots+a_j=k-j} \tau(X^{a_1})\ldots\tau(X^{a_j})$$

for $k \geq 1$. (*Hint:* Start with the identity $s_X(z)R_X(-s_X(z)) = 1 + zs_X(z)$.) Thus, for instance, $C_1(X) = \tau(X)$ is the expectation, $C_2(X) = \tau(X^2) - \tau(X)^2$ is the variance, and the third free cumulant is given by the formula

$$C_3(X) = \tau(X^3) - 3\tau(X^2)\tau(X) + 2\tau(X)^3.$$

Establish the additional formula

$$\tau(X^k) = \sum_{\pi} \prod_{A \in \pi} \kappa_{|A|}(X)$$

where π ranges over all partitions of $\{1, \ldots, k\}$ into non-empty cells A which are *non-crossing*, which means that if $a < b < c < d$ lie in $\{1, \ldots, k\}$, then it cannot be the case that a, c lie in one cell A while b, d lie in a distinct cell A'.

Remark 2.5.23. These computations illustrate a more general principle in free probability, in that the combinatorics of free probability tend to be the "non-crossing" analogue of the combinatorics of classical probability; compare with Remark 2.3.18.

Remark 2.5.24. The R-transform allows for efficient computation of the spectral behaviour of sums $X+Y$ of free random variables. There is an analogous transform, the *S-transform*, for computing the spectral behaviour (or more precisely, the joint moments) of products XY of free random variables; see for instance [**Sp**].

The R-transform clarifies the privileged role of the semicircular elements:

Exercise 2.5.22. Let u be a semicircular element. Show that $R_{\sqrt{t}u}(s) = ts$ for any $t > 0$. In particular, the free convolution of $\sqrt{t}u$ and $\sqrt{t'}u$ is $\sqrt{t+t'}u$.

Exercise 2.5.23. From the above exercise, we see that the effect of adding a free copy of $\sqrt{t}u$ to a non-commutative random variable X is to shift the R-transform by ts. Explain how this is compatible with the Dyson Brownian motion computations in Section 2.4.

It also gives a free analogue of the central limit theorem:

Exercise 2.5.24 (Free central limit theorem). Let X be a self-adjoint random variable with mean zero and variance one (i.e., $\tau(X) = 0$ and $\tau(X^2) = 1$), and let $X_1, X_2, X_3, \ldots$ be free copies of X. Let $S_n := (X_1 + \cdots + X_n)/\sqrt{n}$. Show that the coefficients of the formal power series $R_{S_n}(s)$ converge to that of the identity function s. Conclude that S_n converges in the sense of moments to a semicircular element u.

The free central limit theorem implies the Wigner semicircular law, at least for the GUE ensemble and in the sense of expectation. Indeed, if M_n is an $n \times n$ GUE matrix, then the matrices $\frac{1}{\sqrt{n}} M_n$ are a.s. uniformly bounded (by the Bai-Yin theorem, see Section 2.3), and so (after passing to a subsequence, if necessary), they converge in the sense of moments to some limit u.

On the other hand, if M_n' is an independent copy of M_n, then $M_n + M_n' \equiv \sqrt{2} M_n$ from the properties of Gaussians. Taking limits, we conclude that $u + u' \equiv \sqrt{2} u$, where (by Proposition 2.5.21) u' is a free copy of u. Comparing this with the free central limit theorem (or just the additivity property of R-transforms) we see that u must have the semicircular distribution. Thus the semicircular distribution is the only possible limit point of the $\frac{1}{\sqrt{n}} M_n$, and the Wigner semicircular law then holds (in expectation, and for GUE). Using concentration of measure, we can upgrade the convergence in expectation to a.s. convergence; using the Lindeberg replacement trick one can replace GUE with arbitrary Wigner matrices with (say) bounded coefficients, and then by using the truncation trick one can remove the boundedness hypothesis. (These latter few steps were also discussed in Section 2.4.)

2.6. Gaussian ensembles

Our study of random matrices, to date, has focused on somewhat general ensembles, such as iid random matrices or Wigner random matrices, in which the distribution of the individual entries of the matrices was essentially arbitrary (as long as certain moments, such as the mean and variance, were normalised). In this section, we now focus on two much more special, and much more symmetric, ensembles:

(i) The *Gaussian Unitary Ensemble* (GUE), which is an ensemble of random $n \times n$ Hermitian matrices M_n in which the upper-triangular entries are iid with distribution $N(0,1)_{\mathbf{C}}$, and the diagonal entries are iid with distribution $N(0,1)_{\mathbf{R}}$, and independent of the upper-triangular ones.

(ii) The *Gaussian random matrix ensemble*, which is an ensemble of random $n \times n$ (non-Hermitian) matrices M_n whose entries are iid with distribution $N(0,1)_{\mathbf{C}}$.

The symmetric nature of these ensembles will allow us to compute the spectral distribution by exact algebraic means, revealing a surprising connection with orthogonal polynomials and with determinantal processes. This will, for instance, recover the semicircular law for GUE, but will also reveal *fine* spacing information, such as the distribution of the gap between *adjacent* eigenvalues, which is largely out of reach of tools such as the Stieltjes transform method and the moment method (although the moment method, with some effort, is able to control the extreme edges of the spectrum).

Similarly, we will see for the first time the *circular law* for eigenvalues of non-Hermitian matrices.

There are a number of other highly symmetric ensembles which can also be treated by the same methods, most notably the Gaussian Orthogonal Ensemble (GOE) and the Gaussian Symplectic Ensemble (GSE). However, for simplicity we shall focus just on the above two ensembles. For a systematic treatment of these ensembles, see [**De1999**].

2.6.1. The spectrum of GUE. In Section 3.1 we will use Dyson Brownian motion to establish the Ginibre formula [**Gi1965**]

$$(2.127) \qquad \rho_n(\lambda) = \frac{1}{(2\pi)^{n/2}} e^{-|\lambda|^2/2} |\Delta_n(\lambda)|^2$$

for the density function of the eigenvalues $(\lambda_1, \ldots, \lambda_n) \in \mathbf{R}^n_{\geq}$ of a GUE matrix M_n, where

$$\Delta_n(\lambda) = \prod_{1 \leq i < j \leq n} (\lambda_i - \lambda_j)$$

is the *Vandermonde determinant*. In this section, we give an alternate proof of this result (omitting the exact value of the normalising constant $\frac{1}{(2\pi)^{n/2}}$) that exploits unitary invariance and the change of variables formula (the latter of which we shall do from first principles). The one thing to be careful about is that one has to somehow quotient out by the invariances of the problem before being able to apply the change of variables formula.

One approach here would be to artificially "fix a gauge" and work on some slice of the parameter space which is "transverse" to all the symmetries. With such an approach, one can use the classical change of variables formula. While this can certainly be done, we shall adopt a more "gauge-invariant" approach and carry the various invariances with us throughout the computation[49].

We turn to the details. Let V_n be the space of Hermitian $n \times n$ matrices, then the distribution μ_{M_n} of a GUE matrix M_n is an absolutely continuous

---

[49]For a comparison of the two approaches, see [**Ta2009b**, §1.4].

probability measure on V_n, which can be written using the definition of GUE as

$$\mu_{M_n} = C_n \Big(\prod_{1 \le i < j \le n} e^{-|\xi_{ij}|^2} \Big) \Big(\prod_{1 \le i \le n} e^{-|\xi_{ii}|^2/2} \Big) \, dM_n$$

where dM_n is Lebesgue measure on V_n, ξ_{ij} are the coordinates of M_n, and C_n is a normalisation constant (the exact value of which depends on how one normalises Lebesgue measure on V_n). We can express this more compactly as

$$\mu_{M_n} = C_n e^{-\operatorname{tr}(M_n^2)/2} \, dM_n.$$

Expressed this way, it is clear that the GUE ensemble is invariant under conjugations $M_n \mapsto U M_n U^{-1}$ by any unitary matrix.

Let D be the diagonal matrix whose entries $\lambda_1 \ge \cdots \ge \lambda_n$ are the eigenvalues of M_n in descending order. Then we have $M_n = U D U^{-1}$ for some unitary matrix $U \in U(n)$. The matrix U is not uniquely determined; if R is a diagonal unitary matrix, then R commutes with D, and so one can freely replace U with UR. On the other hand, if the eigenvalues of M are simple, then the diagonal matrices are the *only* matrices that commute with D, and so this freedom to right-multiply U by diagonal unitaries is the only failure of uniqueness here. And in any case, from the unitary invariance of GUE, we see that even after conditioning on D, we may assume, without loss of generality, that U is drawn from the invariant Haar measure on $U(n)$. In particular, U and D can be taken to be independent.

Fix a diagonal matrix $D_0 = \operatorname{diag}(\lambda_1^0, \ldots, \lambda_n^0)$ for some $\lambda_1^0 > \cdots > \lambda_n^0$, let $\varepsilon > 0$ be extremely small, and let us compute the probability

(2.128) $$\mathbf{P}(\|M_n - D_0\|_F \le \varepsilon)$$

that M_n lies within ε of D_0 in the Frobenius norm(2.64). On the one hand, the probability density of M_n is proportional to

$$e^{-\operatorname{tr}(D_0^2)/2} = e^{-|\lambda^0|^2/2}$$

near D_0 (where we write $\lambda^0 := (\lambda_1^0, \ldots, \lambda_n^0)$) and the volume of a ball of radius ε in the n^2-dimensional space V_n is proportional to ε^{n^2}, so (2.128) is equal to

(2.129) $$(C_n' + o(1))\varepsilon^{n^2} e^{-\operatorname{tr}(D_0^2)/2}$$

for some constant $C_n' > 0$ depending only on n, where $o(1)$ goes to zero as $\varepsilon \to 0$ (keeping n and D_0 fixed). On the other hand, if $\|M_n - D_0\|_F \le \varepsilon$, then by the Weyl inequality (1.55) (or Weilandt-Hoffman inequality (1.65)) we have $D = D_0 + O(\varepsilon)$ (we allow implied constants here to depend on n and on D_0). This implies $U D U^{-1} = D + O(\varepsilon)$, thus $U D - D U = O(\varepsilon)$. As a consequence we see that the off-diagonal elements of U are of size $O(\varepsilon)$. We can thus use the inverse function theorem in this local region of parameter space and make the ansatz[50]

$$D = D_0 + \varepsilon E; \quad U = \exp(\varepsilon S)R$$

where E is a bounded diagonal matrix, R is a diagonal unitary matrix, and S is a bounded skew-adjoint matrix with zero diagonal. Note that the map $(R, S) \mapsto \exp(\varepsilon S)R$ has a non-degenerate Jacobian, so the inverse function theorem applies to uniquely specify R, S (and thus E) from U, D in this local region of parameter space.

Conversely, if D, U take the above form, then we can Taylor expand and conclude that

$$M_n = UDU^* = D_0 + \varepsilon E + \varepsilon(SD_0 - D_0 S) + O(\varepsilon^2),$$

and so

$$\|M_n - D_0\|_F = \varepsilon\|E + (SD_0 - D_0 S)\|_F + O(\varepsilon^2).$$

We can thus bound (2.128) from above and below by expressions of the form

(2.130) $$\mathbf{P}(\|E + (SD_0 - D_0 S)\|_F \leq 1 + O(\varepsilon)).$$

As U is distributed using Haar measure on $U(n)$, S is (locally) distributed using $\varepsilon^{n^2 - n}$ times a constant multiple of Lebesgue measure on the space W of skew-adjoint matrices with zero diagonal, which has dimension $n^2 - n$. Meanwhile, E is distributed using $(\rho_n(\lambda^0) + o(1))\varepsilon^n$ times Lebesgue measure on the space of diagonal elements. Thus we can rewrite (2.130) as

$$C_n'' \varepsilon^{n^2} (\rho_n(\lambda^0) + o(1)) \int \int_{\|E + (SD_0 - D_0 S)\|_F \leq 1 + O(\varepsilon)} dE dS$$

where dE and dS denote Lebesgue measure and $C_n'' > 0$ depends only on n.

Observe that the map $S \mapsto SD_0 - D_0 S$ dilates the (complex-valued) ij entry of S by $\lambda_j^0 - \lambda_i^0$, and so the Jacobian of this map is $\prod_{1 \leq i < j \leq n} |\lambda_j^0 - \lambda_i^0|^2 = |\Delta_n(\lambda^0)|^2$. Applying the change of variables, we can express the above as

$$C_n'' \varepsilon^{n^2} \frac{\rho_n(\lambda^0) + o(1)}{|\Delta_n(\lambda^0)|^2} \int \int_{\|E + S\|_F \leq 1 + O(\varepsilon)} dE dS.$$

The integral here is of the form $C_n''' + O(\varepsilon)$ for some other constant $C_n''' > 0$. Comparing this formula with (2.129) we see that

$$\rho_n(\lambda^0) + o(1) = C_n'''' e^{-|\lambda^0|^2/2} |\Delta_n(\lambda^0)|^2 + o(1)$$

for yet another constant $C_n'''' > 0$. Sending $\varepsilon \to 0$ we recover an exact formula

$$\rho_n(\lambda) = C_n''''' e^{-|\lambda|^2/2} |\Delta_n(\lambda)|^2$$

---

[50]Note here the emergence of the freedom to right-multiply U by diagonal unitaries.

when λ is simple. Since almost all Hermitian matrices have simple spectrum (see Exercise 1.3.10), this gives the full spectral distribution of GUE, except for the issue of the unspecified constant.

Remark 2.6.1. In principle, this method should also recover the explicit normalising constant $\frac{1}{(2\pi)^{n/2}}$ in (2.127), but to do this it appears one needs to understand the volume of the fundamental domain of $U(n)$ with respect to the logarithm map, or equivalently to understand the volume of the unit ball of Hermitian matrices in the operator norm. I do not know of a simple way to compute this quantity (though it can be inferred from (2.127) and the above analysis). One can also recover the normalising constant through the machinery of determinantal processes; see below.

Remark 2.6.2. The above computation can be generalised to other $U(n)$-conjugation-invariant ensembles M_n whose probability distribution is of the form
$$\mu_{M_n} = C_n e^{-\operatorname{tr} V(M_n)} \, dM_n$$
for some potential function $V : \mathbf{R} \to \mathbf{R}$ (where we use the spectral theorem to define $V(M_n)$), yielding a density function for the spectrum of the form
$$\rho_n(\lambda) = C'_n e^{-\sum_{j=1}^n V(\lambda_j)} |\Delta_n(\lambda)|^2.$$
Given suitable regularity conditions on V, one can then generalise many of the arguments in this section to such ensembles. See [**De1999**] for details.

2.6.2. The spectrum of Gaussian matrices. The above method also works for Gaussian matrices G, as was first observed by Dyson (though the final formula was first obtained by Ginibre, using a different method). Here, the density function is given by

$$(2.131) \qquad\qquad C_n e^{-\operatorname{tr}(GG^*)} dG = C_n e^{-\|G\|_F^2} dG$$

where $C_n > 0$ is a constant and dG is Lebesgue measure on the space $M_n(\mathbf{C})$ of all complex $n \times n$ matrices. This is invariant under both left and right multiplication by unitary matrices, so in particular, is invariant under unitary conjugations as before.

This matrix G has n complex (generalised) eigenvalues $\sigma(G) = \{\lambda_1, \ldots, \lambda_n\}$, which are usually distinct:

Exercise 2.6.1. Let $n \geq 2$. Show that the space of matrices in $M_n(\mathbf{C})$ with a repeated eigenvalue has codimension 2.

Unlike the Hermitian situation, though, there is no natural way to order these n complex eigenvalues. We will thus consider all $n!$ possible permutations at once, and define the spectral density function $\rho_n(\lambda_1, \ldots, \lambda_n)$ of G

by duality and the formula

$$\int_{\mathbf{C}^n} F(\lambda)\rho_n(\lambda)\, d\lambda := \mathbf{E} \sum_{\{\lambda_1,\dots,\lambda_n\}=\sigma(G)} F(\lambda_1,\dots,\lambda_n)$$

for all test functions F. By the Riesz representation theorem, this uniquely defines ρ_n (as a distribution, at least), although the total mass of ρ_n is $n!$ rather than 1 due to the ambiguity in the spectrum.

Now we compute ρ_n (up to constants). In the Hermitian case, the key was to use the factorisation $M_n = UDU^{-1}$. This particular factorisation is of course unavailable in the non-Hermitian case. However, if the non-Hermitian matrix G has simple spectrum, it can always be factored instead as $G = UTU^{-1}$, where U is unitary and T is upper triangular. Indeed, if one applies the *Gram-Schmidt process* to the eigenvectors of G and uses the resulting orthonormal basis to form U, one easily verifies the desired factorisation. Note that the eigenvalues of G are the same as those of T, which in turn are just the diagonal entries of T.

Exercise 2.6.2. Show that this factorisation is also available when there are repeated eigenvalues. (*Hint:* Use the *Jordan normal form.*)

To use this factorisation, we first have to understand how unique it is, at least in the generic case when there are no repeated eigenvalues. As noted above, if $G = UTU^{-1}$, then the diagonal entries of T form the same set as the eigenvalues of G. We have the freedom to conjugate T by a *permutation matrix* P to obtain $P^{-1}TP$, and right-multiply U by P to counterbalance this conjugation; this permutes the diagonal entries of T around in any one of $n!$ combinations.

Now suppose we fix the diagonal $\lambda_1,\dots,\lambda_n$ of T, which amounts to picking an ordering of the n eigenvalues of G. The eigenvalues of T are $\lambda_1,\dots,\lambda_n$, and furthermore for each $1 \leq j \leq n$, the eigenvector of T associated to λ_j lies in the span of the last $n-j+1$ basis vectors $e_j,\dots,e_n$ of $\mathbf{C}^n$, with a non-zero e_j coefficient (as can be seen by Gaussian elimination or Cramer's rule). As $G = UTU^{-1}$ with U unitary, we conclude that for each $1 \leq j \leq n$, the jth column of U lies in the span of the eigenvectors associated to $\lambda_j,\dots,\lambda_n$. As these columns are orthonormal, they must thus arise from applying the Gram-Schmidt process to these eigenvectors (as discussed earlier). This argument also shows that once the diagonal entries $\lambda_1,\dots,\lambda_n$ of T are fixed, each column of U is determined up to rotation by a unit phase. In other words, the only remaining freedom is to replace U by UR for some unit diagonal matrix R, and then to replace T by $R^{-1}TR$ to counterbalance this change of U.

To summarise, the factorisation $G = UTU^{-1}$ is unique up to right-multiplying U by permutation matrices and diagonal unitary matrices (which together generate the *Weyl group* of the unitary group $U(n)$), and then conjugating T by the same matrix. Given a matrix G, we may apply these symmetries randomly, ending up with a random factorisation UTU^{-1} such that the distribution of T is invariant under conjugation by permutation matrices and diagonal unitary matrices. Also, since G is itself invariant under unitary conjugations, we may also assume that U is distributed uniformly according to the Haar measure of $U(n)$, and independently of T.

To summarise, the Gaussian matrix ensemble G can almost surely be factorised as UTU^{-1}, where $T = (t_{ij})_{1 \le i \le j \le n}$ is an upper-triangular matrix distributed according to some distribution

$$\psi((t_{ij})_{1 \le i \le j \le n}) \prod_{1 \le i \le j \le n} dt_{ij}$$

which is invariant with respect to conjugating T by permutation matrices or diagonal unitary matrices, and U is uniformly distributed according to the Haar measure of $U(n)$, independently of T.

Now let $T_0 = (t_{ij}^0)_{1 \le i \le j \le n}$ be an upper triangular matrix with complex entries whose entries $t_{11}^0, \dots, t_{nn}^0 \in \mathbf{C}$ are distinct. As in the previous section, we consider the probability

(2.132) $\mathbf{P}(\|G - T_0\|_F \le \varepsilon)$.

Since the space $M_n(\mathbf{C})$ of complex $n \times n$ matrices has $2n^2$ real dimensions, we see from (2.131) that this expression is equal to

(2.133) $(C_n' + o(1))e^{-\|T_0\|_F^2} \varepsilon^{2n^2}$

for some constant $C_n' > 0$.

Now we compute (2.132) using the factorisation $G = UTU^{-1}$. Suppose that $\|G - T_0\|_F \le \varepsilon$, so $G = T_0 + O(\varepsilon)$. As the eigenvalues of T_0 are $t_{11}^0, \dots, t_{nn}^0$, which are assumed to be distinct, we see (from the inverse function theorem) that for ε small enough, G has eigenvalues $t_{11}^0 + O(\varepsilon), \dots, t_{nn}^0 + O(\varepsilon)$. Thus the diagonal entries of T are some permutation of $t_{11}^0 + O(\varepsilon), \dots, t_{nn}^0 + O(\varepsilon)$. As we are assuming the distribution of T to be invariant under conjugation by permutation matrices, all permutations here are equally likely, so with probability[51] $1/n!$, we may assume that the diagonal entries of T are given by $t_{11}^0 + O(\varepsilon), \dots, t_{nn}^0 + O(\varepsilon)$ in that order.

Let $u_1^0, \dots, u_n^0$ be the eigenvectors of T_0 associated to $t_{11}^0, \dots, t_{nn}^0$, then the Gram-Schmidt process applied to $u_1, \dots, u_n$ (starting at u_n^0 and working backwards to u_1^0) gives the standard basis $e_1, \dots, e_n$ (in reverse order). By the inverse function theorem, we thus see that we have eigenvectors $u_1 = u_1^0 + O(\varepsilon), \dots, u_n = u_n^0 + O(\varepsilon)$ of G, which when the Gram-Schmidt process is applied, gives a perturbation $e_1 + O(\varepsilon), \dots, e_n + O(\varepsilon)$ in reverse order.

---

[51] The factor of $1/n!$ will eventually be absorbed into one of the unspecified constants.

This gives a factorisation $G = UTU^{-1}$ in which $U = I + O(\varepsilon)$, and hence $T = T_0 + O(\varepsilon)$. This is, however, not the most general factorisation available, even after fixing the diagonal entries of T, due to the freedom to right-multiply U by diagonal unitary matrices R. We thus see that the correct ansatz here is to have

$$U = R + O(\varepsilon); \quad T = R^{-1}T_0 R + O(\varepsilon)$$

for some diagonal unitary matrix R.

In analogy with the GUE case, we can use the inverse function theorem to make the more precise ansatz

$$U = \exp(\varepsilon S)R; \quad T = R^{-1}(T_0 + \varepsilon E)R$$

where S is skew-Hermitian with zero diagonal and size $O(1)$, R is diagonal unitary, and E is an upper triangular matrix of size $O(1)$. From the invariance $U \mapsto UR; T \mapsto R^{-1}TR$ we see that R is distributed uniformly across all diagonal unitaries. Meanwhile, from the unitary conjugation invariance, S is distributed according to a constant multiple of ε^{n^2-n} times Lebesgue measure dS on the $n^2 - n$-dimensional space of skew Hermitian matrices with zero diagonal; and from the definition of ψ, E is distributed according to a constant multiple of the measure

$$(1 + o(1))\varepsilon^{n^2+n}\psi(T_0)\ dE,$$

where dE is Lebesgue measure on the $n^2 + n$-dimensional space of upper-triangular matrices. Furthermore, the invariances ensure that the random variables S, R, E are distributed independently. Finally, we have

$$G = UTU^{-1} = \exp(\varepsilon S)(T_0 + \varepsilon E)\exp(-\varepsilon S).$$

Thus we may rewrite (2.132) as

$$(2.134) \quad (C_n''\psi(T_0) + o(1))\varepsilon^{2n^2}\int\int_{\|\exp(\varepsilon S)(T_0+\varepsilon E)\exp(-\varepsilon S)-T_0\|_F \leq \varepsilon} dSdE$$

for some $C_n'' > 0$ (the R integration being absorbable into this constant C_n''). We can Taylor expand

$$\exp(\varepsilon S)(T_0 + \varepsilon E)\exp(-\varepsilon S) = T_0 + \varepsilon(E + ST_0 - T_0 S) + O(\varepsilon^2)$$

and so we can bound (2.134) above and below by expressions of the form

$$(C_n''\psi(T_0) + o(1))\varepsilon^{2n^2}\int\int_{\|E+ST_0-T_0 S\|_F \leq 1+O(\varepsilon)} dSdE.$$

The Lebesgue measure dE is invariant under translations by upper triangular matrices, so we may rewrite the above expression as

$$(2.135) \quad (C_n''\psi(T_0) + o(1))\varepsilon^{2n^2}\int\int_{\|E+\pi(ST_0-T_0 S)\|_F \leq 1+O(\varepsilon)} dSdE,$$

where $\pi(ST_0 - T_0S)$ is the strictly lower triangular component of $ST_0 - T_0S$.

The next step is to make the (linear) change of variables $V := \pi(ST_0 - T_0S)$. We check dimensions: S ranges in the space of skew-adjoint Hermitian matrices with zero diagonal, which has dimension $(n^2 - n)/2$, as does the space of strictly lower-triangular matrices, which is where V ranges. So we can in principle make this change of variables, but we first have to compute the Jacobian of the transformation (and check that it is non-zero). For this, we switch to coordinates. Write $S = (s_{ij})_{1 \le i,j \le n}$ and $V = (v_{ij})_{1 \le j < i \le n}$. In coordinates, the equation $V = \pi(ST_0 - T_0S)$ becomes

$$v_{ij} = \sum_{k=1}^{j} s_{ik}t_{kj}^0 - \sum_{k=i}^{n} t_{ik}^0 s_{kj},$$

or equivalently,

$$v_{ij} = (t_{jj}^0 - t_{ii}^0)s_{ij} + \sum_{k=1}^{j-1} t_{kj}^0 s_{ik} - \sum_{k=i+1}^{n} t_{ik}^0 s_{kj}.$$

Thus, for instance,

$$v_{n1} = (t_{11}^0 - t_{nn}^0)s_{n1},$$
$$v_{n2} = (t_{22}^0 - t_{nn}^0)s_{n2} + t_{12}^0 s_{n1},$$
$$v_{(n-1)1} = (t_{11}^0 - t_{(n-1)(n-1)}^0)s_{(n-1)1} - t_{(n-1)n}^0 s_{n1},$$
$$v_{n3} = (t_{33}^0 - t_{nn}^0)s_{n3} + t_{13}^0 s_{n1} + t_{23}^0 s_{n2},$$
$$v_{(n-1)2} = (t_{22}^0 - t_{(n-1)(n-1)}^0)s_{(n-1)2} + t_{12}^0 s_{(n-1)1} - t_{(n-1)n}^0 s_{n2},$$
$$v_{(n-2)1} = (t_{11}^0 - t_{(n-2)(n-2)}^0)s_{(n-2)1} - t_{(n-2)(n-1)}^0 s_{(n-1)1} - t_{(n-2)n}^0 s_{n1},$$

etc. We then observe that the transformation matrix from $s_{n1}, s_{n2}, s_{(n-1)1}, \ldots$ to $v_{n1}, v_{n2}, v_{(n-1)1}, \ldots$ is triangular, with diagonal entries given by $t_{jj}^0 - t_{ii}^0$ for $1 \le j < i \le n$. The Jacobian of the (complex-linear) map $S \mapsto V$ is thus given by

$$| \prod_{1 \le j < i \le n} t_{jj}^0 - t_{ii}^0|^2 = |\Delta(t_{11}^0, \ldots, t_{nn}^0)|^2$$

which is non-zero by the hypothesis that the $t_0^{11}, \ldots, t_0^{nn}$ are distinct. We may thus rewrite (2.135) as

$$\frac{C_n''\psi(T_0) + o(1)}{|\Delta(t_{11}^0, \ldots, t_{nn}^0)|^2} \varepsilon^{2n^2} \int \int_{\|E+V\|_F \le 1 + O(\varepsilon)} dS \, dV$$

where dV is Lebesgue measure on strictly lower-triangular matrices. The integral here is equal to $C_n''' + O(\varepsilon)$ for some constant C_n'''. Comparing this with (2.132), cancelling the factor of ε^{2n^2}, and sending $\varepsilon \to 0$, we obtain the formula

$$\psi((t_{ij}^0)_{1 \le i \le j \le n}) = C_n''''|\Delta(t_{11}^0, \ldots, t_{nn}^0)|^2 e^{-\|T_0\|_F^2}$$

for some constant $C_n'''' > 0$. We can expand

$$e^{-\|T_0\|_F^2} = \prod_{1 \leq i \leq j \leq n} e^{-|t_{ij}^0|^2}.$$

If we integrate out the off-diagonal variables t_{ij}^0 for $1 \leq i < j \leq n$, we see that the density function for the diagonal entries $(\lambda_1, \ldots, \lambda_n)$ of T is proportional to

$$|\Delta(\lambda_1, \ldots, \lambda_n)|^2 e^{-\sum_{j=1}^n |\lambda_j|^2}.$$

Since these entries are a random permutation of the eigenvalues of G, we conclude the *Ginibre formula*,

$$(2.136) \qquad \rho_n(\lambda_1, \ldots, \lambda_n) = c_n |\Delta(\lambda_1, \ldots, \lambda_n)|^2 e^{-\sum_{j=1}^n |\lambda_j|^2}$$

for the joint density of the eigenvalues of a Gaussian random matrix, where $c_n > 0$ is a constant.

Remark 2.6.3. Given that (2.127) can be derived using Dyson Brownian motion, it is natural to ask whether (2.136) can be derived by a similar method. It seems that in order to do this, one needs to consider a Dyson-like process not just on the eigenvalues $\lambda_1, \ldots, \lambda_n$, but on the entire triangular matrix T (or more precisely, on the moduli space formed by quotienting out the action of conjugation by unitary diagonal matrices). Unfortunately, the computations seem to get somewhat complicated, and we do not present them here.

2.6.3. Mean field approximation. We can use the formula (2.127) for the joint distribution to heuristically derive the semicircular law, as follows.

It is intuitively plausible that the spectrum $(\lambda_1, \ldots, \lambda_n)$ should concentrate in regions in which $\rho_n(\lambda_1, \ldots, \lambda_n)$ is as large as possible. So it is now natural to ask how to optimise this function. Note that the expression in (2.127) is non-negative, and vanishes whenever two of the λ_i collide, or when one or more of the λ_i go off to infinity, so a maximum should exist away from these degenerate situations.

We may take logarithms and write

$$(2.137) \qquad -\log \rho_n(\lambda_1, \ldots, \lambda_n) = \sum_{j=1}^n \frac{1}{2} |\lambda_j|^2 + \sum \sum_{i \neq j} \log \frac{1}{|\lambda_i - \lambda_j|} + C$$

where $C = C_n$ is a constant whose exact value is not of importance to us. From a mathematical physics perspective, one can interpret (2.137) as a Hamiltonian for n particles at positions $\lambda_1, \ldots, \lambda_n$, subject to a confining harmonic potential (these are the $\frac{1}{2} |\lambda_j|^2$ terms) and a repulsive logarithmic potential between particles (these are the $\frac{1}{|\lambda_i - \lambda_j|}$ terms).

Our objective is now to find a distribution of $\lambda_1, \ldots, \lambda_n$ that minimises this expression.

We know from previous sections that the λ_i should have magnitude $O(\sqrt{n})$. Let us then heuristically make a *mean field approximation*, in that we approximate the discrete spectral measure $\frac{1}{n} \sum_{j=1}^{n} \delta_{\lambda_j/\sqrt{n}}$ by a continuous[52] probability measure $\rho(x) \, dx$. Then we can heuristically approximate (2.137) as

$$n^2 \left(\int_{\mathbf{R}} \frac{1}{2} x^2 \rho(x) \, dx + \int_{\mathbf{R}} \int_{\mathbf{R}} \log \frac{1}{|x-y|} \rho(x) \rho(y) \, dx dy \right) + C_n',$$

so we expect the distribution ρ to minimise the functional

$$(2.138) \qquad \int_{\mathbf{R}} \frac{1}{2} x^2 \rho(x) \, dx + \int_{\mathbf{R}} \int_{\mathbf{R}} \log \frac{1}{|x-y|} \rho(x) \rho(y) \, dx dy.$$

One can compute the Euler-Lagrange equations of this functional:

Exercise 2.6.3. Working formally, and assuming that ρ is a probability measure that minimises (2.138), argue that

$$\frac{1}{2} x^2 + 2 \int_{\mathbf{R}} \log \frac{1}{|x-y|} \rho(y) \, dy = C$$

for some constant C and all x in the support of ρ. For all x outside of the support, establish the inequality

$$\frac{1}{2} x^2 + 2 \int_{\mathbf{R}} \log \frac{1}{|x-y|} \rho(y) \, dy \geq C.$$

There are various ways we can solve this equation for ρ; we sketch here a complex-analytic method. Differentiating in x, we formally obtain

$$x - 2 \, \mathrm{p.\,v.} \int_{\mathbf{R}} \frac{1}{x-y} \rho(y) \, dy = 0$$

on the support of ρ, where p. v. denotes a principal value integral. But recall that if we let

$$s(z) := \int_{\mathbf{R}} \frac{1}{y-z} \rho(y) \, dy$$

be the Stieltjes transform of the probability measure $\rho(x) \, dx$, then we have

$$\mathrm{Im}(s(x+i0^+)) = \pi \rho(x)$$

and

$$\mathrm{Re}(s(x+i0^+)) = - \, \mathrm{p.\,v.} \int_{\mathbf{R}} \frac{1}{x-y} \rho(y) \, dy.$$

We conclude that

$$(x + 2 \, \mathrm{Re}(s(x+i0^+))) \mathrm{Im}(s(x+i0^+))) = 0$$

---

[52]Secretly, we know from the semicircular law that we should be able to take $\rho = \frac{1}{2\pi}(4 - x^2)_+^{1/2}$, but pretend that we do not know this fact yet.

for all x, which we rearrange as

$$\text{Im}(s^2(x + i0^+) + xs(x + i0^+)) = 0.$$

This makes the function $f(z) = s^2(z) + zs(z)$ entire (it is analytic in the upper half-plane, obeys the symmetry $f(\bar{z}) = \overline{f(z)}$, and has no jump across the real line). On the other hand, as $s(z) = \frac{-1+o(1)}{z}$ as $z \to \infty$, f goes to -1 at infinity. Applying *Liouville's theorem*, we conclude that f is constant, thus we have the familiar equation

$$s^2 + zs = -1$$

which can then be solved to obtain the semicircular law as in Section 2.4.

Remark 2.6.4. Recall from Section 3.1 that Dyson Brownian motion can be used to derive the formula (2.127). One can then interpret the Dyson Brownian motion proof of the semicircular law for GUE in Section 2.4 as a rigorous formalisation of the above mean field approximation heuristic argument.

One can perform a similar heuristic analysis for the spectral measure μ_G of a random Gaussian matrix, giving a description of the limiting density:

Exercise 2.6.4. Using heuristic arguments similar to those above, argue that μ_G should be close to a continuous probability distribution $\rho(z)\ dz$ obeying the equation

$$|z|^2 + \int_{\mathbf{C}} \log \frac{1}{|z - w|} \rho(w)\ dw = C$$

on the support of ρ, for some constant C, with the inequality

$$(2.139) \qquad |z|^2 + \int_{\mathbf{C}} \log \frac{1}{|z - w|} \rho(w)\ dw \geq C.$$

Using the *Newton potential* $\frac{1}{2\pi} \log |z|$ for the fundamental solution of the two-dimensional Laplacian $-\partial_x^2 - \partial_y^2$, conclude (non-rigorously) that ρ is equal to $\frac{1}{\pi}$ on its support.

Also argue that ρ should be rotationally symmetric. Use (2.139) and Green's formula to argue why the support of ρ should be simply connected, and then conclude (again non-rigorously) the *circular law*

$$(2.140) \qquad \mu_G \approx \frac{1}{\pi} 1_{|z| \leq 1}\ dz.$$

We will see more rigorous derivations of the circular law later in this text.

2.6.4. Determinantal form of the GUE spectral distribution. In a previous section, we showed (up to constants) that the density function $\rho_n(\lambda_1, \ldots, \lambda_n)$ for the eigenvalues $\lambda_1 \geq \cdots \geq \lambda_n$ of GUE was given by the formula (2.127).

As is well known, the Vandermonde determinant $\Delta(\lambda_1, \ldots, \lambda_n)$ that appears in (2.127) can be expressed up to sign as a determinant of an $n \times n$ matrix, namely the matrix $(\lambda_i^{j-1})_{1 \leq i,j \leq n}$. Indeed, this determinant is clearly a polynomial of degree $n(n-1)/2$ in $\lambda_1, \ldots, \lambda_n$ which vanishes whenever two of the λ_i agree, and the claim then follows from the factor theorem (and inspecting a single coefficient of the Vandermonde determinant, e.g., the $\prod_{j=1}^{n} \lambda_j^{j-1}$ coefficient, to get the sign).

We can square the above fact (or more precisely, multiply the above matrix by its adjoint) and conclude that $|\Delta(\lambda_1, \ldots, \lambda_n)|^2$ is the determinant of the matrix

$$(\sum_{k=0}^{n-1} \lambda_i^k \lambda_j^k)_{1 \leq i,j \leq n}.$$

More generally, if $P_0(x), \ldots, P_{n-1}(x)$ are any sequence of polynomials, in which $P_i(x)$ has degree i, then we see from row operations that the determinant of

$$(P_{j-1}(\lambda_i))_{1 \leq i,j \leq n}$$

is a non-zero constant multiple of $\Delta(\lambda_1, \ldots, \lambda_n)$ (with the constant depending on the leading coefficients of the P_i), and so the determinant of

$$(\sum_{k=0}^{n-1} P_k(\lambda_i) P_k(\lambda_j))_{1 \leq i,j \leq n}$$

is a non-zero constant multiple of $|\Delta(\lambda_1, \ldots, \lambda_n)|^2$. Comparing this with (2.127), we obtain the formula

$$\rho_n(\lambda) = C \det(\sum_{k=0}^{n-1} P_k(\lambda_i) e^{-\lambda_i^2/4} P_k(\lambda_j) e^{-\lambda_j^2/4})_{1 \leq i,j \leq n}$$

for some non-zero constant C.

This formula is valid for any choice of polynomials P_i of degree i. But the formula is particularly useful when we set P_i equal to the (normalised) *Hermite polynomials*, defined[53] by applying the Gram-Schmidt process in $L^2(\mathbf{R})$ to the polynomials $x^i e^{-x^2/4}$ for $i = 0, \ldots, n-1$ to yield $P_i(x) e^{-x^2/4}$. In that case, the expression

$$(2.141) \qquad K_n(x, y) := \sum_{k=0}^{n-1} P_k(x) e^{-x^2/4} P_k(y) e^{-y^2/4}$$

---

[53]Equivalently, the P_i are the *orthogonal polynomials* associated to the measure $e^{-x^2/2} \, dx$.

becomes the integral kernel of the orthogonal projection π_{V_n} operator in $L^2(\mathbf{R})$ to the span of the $x^i e^{-x^2/4}$, thus

$$\pi_{V_n} f(x) = \int_{\mathbf{R}} K_n(x,y) f(y) \, dy$$

for all $f \in L^2(\mathbf{R})$, and so $\rho_n(\lambda)$ is now a constant multiple of

$$\det(K_n(\lambda_i, \lambda_j))_{1 \le i,j \le n}.$$

The reason for working with orthogonal polynomials is that we have the trace identity

(2.142) $$\int_{\mathbf{R}} K_n(x,x) \, dx = \mathrm{tr}(\pi_{V_n}) = n$$

and the reproducing formula

(2.143) $$K_n(x,y) = \int_{\mathbf{R}} K_n(x,z) K_n(z,y) \, dz$$

which reflects the identity $\pi_{V_n} = \pi_{V_n}^2$. These two formulae have an important consequence:

Lemma 2.6.5 (Determinantal integration formula). *Let $K_n : \mathbf{R} \times \mathbf{R} \to \mathbf{R}$ be any symmetric rapidly decreasing function obeying* (2.142), (2.143). *Then for any $k \ge 0$, one has*

(2.144) $$\int_{\mathbf{R}} \det(K_n(\lambda_i,\lambda_j))_{1 \le i,j \le k+1} \, d\lambda_{k+1} = (n-k) \det(K_n(\lambda_i,\lambda_j))_{1 \le i,j \le k}.$$

Remark 2.6.6. This remarkable identity is part of the beautiful algebraic theory of *determinantal processes*, which is discussed further in [**Ta2010b**, §2.6].

Proof. We induct on k. When $k = 0$ this is just (2.142). Now assume that $k \ge 1$ and that the claim has already been proven for $k-1$. We apply *cofactor expansion* to the bottom row of the determinant $\det(K_n(\lambda_i,\lambda_j))_{1 \le i,j \le k+1}$. This gives a principal term

(2.145) $$\det(K_n(\lambda_i,\lambda_j))_{1 \le i,j \le k} K_n(\lambda_{k+1},\lambda_{k+1})$$

plus a sum of k additional terms, the lth term of which is of the form

(2.146) $$(-1)^{k+1-l} K_n(\lambda_l,\lambda_{k+1}) \det(K_n(\lambda_i,\lambda_j))_{1 \le i \le k; 1 \le j \le k+1; j \ne l}.$$

Using (2.142), the principal term (2.145) gives a contribution of

$$n \det(K_n(\lambda_i,\lambda_j))_{1 \le i,j \le k}$$

to (2.144). For each non-principal term (2.146), we use the multilinearity of the determinant to absorb the $K_n(\lambda_l,\lambda_{k+1})$ term into the $j = k+1$ column of the matrix. Using (2.143), we thus see that the contribution of (2.146) to (2.144) can be simplified as

$$(-1)^{k+1-l} \det((K_n(\lambda_i, \lambda_j))_{1 \leq i \leq k; 1 \leq j \leq k; j \neq l}, (K_n(\lambda_i, \lambda_l))_{1 \leq i \leq k})$$

which after row exchange, simplifies to $-\det(K_n(\lambda_i, \lambda_j))_{1 \leq i, j \leq k}$. The claim follows.   □

In particular, if we iterate the above lemma using the Fubini-Tonelli theorem, we see that

$$\int_{\mathbf{R}^n} \det(K_n(\lambda_i, \lambda_j))_{1 \leq i, j \leq n} \, d\lambda_1 \ldots d\lambda_n = n!.$$

On the other hand, if we extend the probability density function $\rho_n(\lambda_1, \ldots, \lambda_n)$ symmetrically from the Weyl chamber $\mathbf{R}^n_{\geq}$ to all of $\mathbf{R}^n$, its integral is also $n!$. Since $\det(K_n(\lambda_i, \lambda_j))_{1 \leq i, j \leq n}$ is clearly symmetric in the $\lambda_1, \ldots, \lambda_n$, we can thus compare constants and conclude the *Gaudin-Mehta formula* [**MeGa1960**]

$$\rho_n(\lambda_1, \ldots, \lambda_n) = \det(K_n(\lambda_i, \lambda_j))_{1 \leq i, j \leq n}.$$

More generally, if we define $\rho_k : \mathbf{R}^k \to \mathbf{R}^+$ to be the function

(2.147) $$\rho_k(\lambda_1, \ldots, \lambda_k) = \det(K_n(\lambda_i, \lambda_j))_{1 \leq i, j \leq k},$$

then the above formula shows that ρ_k is the *k-point correlation function* for the spectrum, in the sense that

(2.148) $$\int_{\mathbf{R}^k} \rho_k(\lambda_1, \ldots, \lambda_k) F(\lambda_1, \ldots, \lambda_k) \, d\lambda_1 \ldots d\lambda_k$$

$$= \mathbf{E} \sum_{1 \leq i_1, \ldots, i_k \leq n, \text{distinct}} F(\lambda_{i_1}(M_n), \ldots, \lambda_{i_k}(M_n))$$

for any test function $F : \mathbf{R}^k \to \mathbf{C}$ supported in the region $\{(x_1, \ldots, x_k) : x_1 \leq \cdots \leq x_k\}$.

In particular, if we set $k = 1$, we obtain the explicit formula

$$\mathbf{E}\mu_{M_n} = \frac{1}{n} K_n(x, x) \, dx$$

for the expected empirical spectral measure of M_n. Equivalently, after renormalising by $\sqrt{n}$, we have

(2.149) $$\mathbf{E}\mu_{M_n/\sqrt{n}} = \frac{1}{n^{1/2}} K_n(\sqrt{n}x, \sqrt{n}x) \, dx.$$

It is thus of interest to understand the kernel K_n better.

To do this, we begin by recalling that the functions $P_i(x)e^{-x^2/4}$ were obtained from $x^i e^{-x^2/4}$ by the Gram-Schmidt process. In particular, each $P_i(x)e^{-x^2/4}$ is orthogonal to $x^j e^{-x^2/4}$ for all $0 \leq j < i$. This implies that $xP_i(x)e^{-x^2/4}$ is orthogonal to $x^j e^{-x^2/4}$ for $0 \leq j < i - 1$. On the other hand, $xP_i(x)$ is a polynomial of degree $i+1$, so $xP_i(x)e^{-x^2/4}$ must lie in the span of $x^j e^{-x^2/4}$ for $0 \le j \le i+1$. Combining the two facts, we see that xP_i must be a linear combination of P_{i-1}, P_i, P_{i+1}, with the P_{i+1} coefficient being non-trivial. We rewrite this fact in the form

(2.150) $$P_{i+1}(x) = (a_i x + b_i)P_i(x) - c_i P_{i-1}(x)$$

for some real numbers a_i, b_i, c_i (with $c_0 = 0$). Taking inner products with P_{i+1} and P_{i-1} we see that

(2.151) $$\int_{\mathbf{R}} x P_i(x) P_{i+1}(x) e^{-x^2/2} \, dx = \frac{1}{a_i}$$

and

$$\int_{\mathbf{R}} x P_i(x) P_{i-1}(x) e^{-x^2/2} \, dx = \frac{c_i}{a_i},$$

and so

(2.152) $$c_i := \frac{a_i}{a_{i-1}}$$

(with the convention $a_{-1} = \infty$).

We will continue the computation of a_i, b_i, c_i later. For now, we pick two distinct real numbers x, y and consider the Wronskian-type expression

$$P_{i+1}(x)P_i(y) - P_i(x)P_{i+1}(y).$$

Using (2.150), (2.152), we can write this as

$$a_i(x - y)P_i(x)P_i(y) + \frac{a_i}{a_{i-1}}(P_{i-1}(x)P_i(y) - P_i(x)P_{i-1}(y)),$$

or in other words,

$$P_i(x)P_i(y) = \frac{P_{i+1}(x)P_i(y) - P_i(x)P_{i+1}(y)}{a_i(x - y)}$$
$$- \frac{P_i(x)P_{i-1}(y) - P_{i-1}(x)P_i(y)}{a_{i-1}(x - y)}.$$

We telescope this and obtain the *Christoffel-Darboux formula* for the kernel (2.141):

(2.153) $$K_n(x, y) = \frac{P_n(x)P_{n-1}(y) - P_{n-1}(x)P_n(y)}{a_{n-1}(x - y)} e^{-(x^2+y^2)/4}.$$

Sending $y \to x$ using *L'Hôpital's rule*, we obtain, in particular, that

(2.154) $$K_n(x, x) = \frac{1}{a_{n-1}}(P'_n(x)P_{n-1}(x) - P'_{n-1}(x)P_n(x))e^{-x^2/2}.$$

Inserting this into (2.149), we see that if we want to understand the expected spectral measure of GUE, we should understand the asymptotic behaviour of P_n and the associated constants a_n. For this, we need to exploit the specific properties of the Gaussian weight $e^{-x^2/2}$. In particular, we have the identity

(2.155) $$xe^{-x^2/2} = -\frac{d}{dx}e^{-x^2/2},$$

so upon integrating (2.151) by parts, we have

$$\int_{\mathbf{R}} (P_i'(x)P_{i+1}(x) + P_i(x)P_{i+1}'(x))e^{-x^2/2}\,dx = \frac{1}{a_i}.$$

As P_i' has degree at most $i-1$, the first term vanishes by the orthonormal nature of the $P_i(x)e^{-x^2/4}$, thus

(2.156) $$\int_{\mathbf{R}} P_i(x)P_{i+1}'(x)e^{-x^2/2}\,dx = \frac{1}{a_i}.$$

To compute this, let us denote the leading coefficient of P_i as k_i. Then P_{i+1}' is equal to $\frac{(i+1)k_{i+1}}{k_i}P_i$ plus lower-order terms, and so we have

$$\frac{(i+1)k_{i+1}}{k_i} = \frac{1}{a_i}.$$

On the other hand, by inspecting the x^{i+1} coefficient of (2.150) we have

$$k_{i+1} = a_i k_i.$$

Combining the two formulae (and making the sign convention that the k_i are always positive), we see that

$$a_i = \frac{1}{\sqrt{i+1}}$$

and

$$k_{i+1} = \frac{k_i}{\sqrt{i+1}}.$$

Meanwhile, a direct computation shows that $P_0(x) = k_0 = \frac{1}{(2\pi)^{1/4}}$, and thus by induction

$$k_i := \frac{1}{(2\pi)^{1/4}\sqrt{i!}}.$$

A similar method lets us compute the b_i. Indeed, taking inner products of (2.150) with $P_i(x)e^{-x^2/2}$ and using orthonormality we have

$$b_i = -a_i \int_{\mathbf{R}} xP_i(x)^2 e^{-x^2/2}\,dx,$$

which upon integrating by parts using (2.155) gives

$$b_i = -2a_i \int_{\mathbf{R}} P_i(x)P_i'(x)e^{-x^2/2}\,dx.$$

As P_i' is of degree strictly less than i, the integral vanishes by orthonormality, thus $b_i = 0$. The identity (2.150) thus becomes *Hermite recurrence relation*

$$(2.157) \qquad P_{i+1}(x) = \frac{1}{\sqrt{i+1}} x P_i(x) - \frac{\sqrt{i}}{\sqrt{i+1}} P_{i-1}(x).$$

Another recurrence relation arises by considering the integral

$$\int_{\mathbf{R}} P_j(x) P_{i+1}'(x) e^{-x^2/2} \, dx.$$

On the one hand, as P_{i+1}' has degree at most i, this integral vanishes if $j > i$ by orthonormality. On the other hand, integrating by parts using (2.155), we can write the integral as

$$\int_{\mathbf{R}} (xP_j - P_j')(x) P_{i+1}(x) e^{-x^2/2} \, dx.$$

If $j < i$, then $xP_j - P_j'$ has degree less than $i+1$, so the integral again vanishes. Thus the integral is non-vanishing only when $j = i$. Using (2.156), we conclude that

$$(2.158) \qquad P_{i+1}' = \frac{1}{a_i} P_i = \sqrt{i+1} P_i.$$

We can combine (2.158) with (2.157) to obtain the formula

$$\frac{d}{dx} (e^{-x^2/2} P_i(x)) = -\sqrt{i+1} e^{-x^2/2} P_{i+1}(x),$$

which together with the initial condition $P_0 = \frac{1}{(2\pi)^{1/4}}$ gives the explicit representation

$$(2.159) \qquad P_n(x) := \frac{(-1)^n}{(2\pi)^{1/4}\sqrt{n!}} e^{x^2/2} \frac{d^n}{dx^n} e^{-x^2/2}$$

for the Hermite polynomials. Thus, for instance, at $x = 0$ one sees from Taylor expansion that

$$(2.160) \qquad P_n(0) = \frac{(-1)^{n/2}\sqrt{n!}}{(2\pi)^{1/4}2^{n/2}(n/2)!}; \quad P_n'(0) = 0$$

when n is even, and

$$(2.161) \qquad P_n(0) = 0; \quad P_n'(0) = \frac{(-1)^{(n+1)/2}(n+1)\sqrt{n!}}{(2\pi)^{1/4}2^{(n+1)/2}((n+1)/2)!}$$

when n is odd.

In principle, the formula (2.159), together with (2.154), gives us an explicit description of the kernel $K_n(x,x)$ (and thus of $\mathbf{E}\mu_{M_n/\sqrt{n}}$, by (2.149)). However, to understand the asymptotic behaviour as $n \to \infty$, we would have to understand the asymptotic behaviour of $\frac{d^n}{dx^n} e^{-x^2/2}$ as $n \to \infty$, which is not immediately discernable by inspection. However, one can obtain such asymptotics by a variety of means. We give two such methods here: a method based on ODE analysis, and a complex-analytic method, based on the *method of steepest descent*.

We begin with the ODE method. Combining (2.157) with (2.158) we see that each polynomial P_m obeys the *Hermite differential equation*

$$P_m''(x) - xP_m'(x) + mP_m(x) = 0.$$

If we look instead at the Hermite functions $\phi_m(x) := P_m(x)e^{-x^2/4}$, we obtain the differential equation

$$L\phi_m(x) = (m + \frac{1}{2})\phi_m$$

where L is the *harmonic oscillator operator*

$$L\phi := -\phi'' + \frac{x^2}{4}\phi.$$

Note that the self-adjointness of L here is consistent with the orthogonal nature of the ϕ_m.

Exercise 2.6.5. Use (2.141), (2.154), (2.159), (2.157), (2.158) to establish the identities

$$K_n(x, x) = \sum_{j=0}^{n-1} \phi_j(x)^2$$

$$= \phi_n'(x)^2 + (n - \frac{x^2}{4})\phi_n(x)^2$$

and thus by (2.149),

$$\mathbf{E}\mu_{M_n/\sqrt{n}} = \frac{1}{\sqrt{n}}\sum_{j=0}^{n-1} \phi_j(\sqrt{n}x)^2\ dx$$

$$= [\frac{1}{\sqrt{n}}\phi_n'(\sqrt{n}x)^2 + \sqrt{n}(1 - \frac{x^2}{4})\phi_n(\sqrt{n}x)^2]\ dx.$$

It is thus natural to look at the rescaled functions

$$\tilde{\phi}_m(x) := \sqrt{n}\phi_m(\sqrt{n}x)$$

which are orthonormal in $L^2(\mathbf{R})$ and solve the equation

$$L_{1/\sqrt{n}}\tilde{\phi}_m(x) = \frac{m + 1/2}{n}\tilde{\phi}_m$$

where L_h is the *semiclassical harmonic oscillator operator*

$$L_h\phi := -h^2\phi'' + \frac{x^2}{4}\phi,$$

thus

$$\mathbf{E}\mu_{M_n/\sqrt{n}} = \frac{1}{n}\sum_{j=0}^{n-1}\tilde{\phi}_j(x)^2\ dx$$

(2.162)
$$= [\frac{1}{n}\tilde{\phi}'_n(x)^2 + (1 - \frac{x^2}{4})\tilde{\phi}_n(x)^2]\ dx.$$

The projection π_{V_n} is then the spectral projection operator of $L_{1/\sqrt{n}}$ to $[0,1]$. According to *semi-classical analysis*, with h being interpreted as analogous to Planck's constant, the operator L_h has symbol $p^2 + \frac{x^2}{4}$, where $p := -ih\frac{d}{dx}$ is the momentum operator, so the projection π_{V_n} is a projection to the region $\{(x,p) : p^2 + \frac{x^2}{4} \leq 1\}$ of phase space, or equivalently to the region $\{(x,p) : |p| < (4 - x^2)_+^{1/2}\}$. In the semi-classical limit $h \to 0$, we thus expect the diagonal $K_n(x,x)$ of the normalised projection $h^2\pi_{V_n}$ to be proportional to the projection of this region to the x variable, i.e., proportional to $(4 - x^2)_+^{1/2}$. We are thus led to the semicircular law via semi-classical analysis.

It is possible to make the above argument rigorous, but this would require developing the theory of *microlocal analysis*, which would be overkill given that we are just dealing with an ODE rather than a PDE here (and an extremely classical ODE at that); but see Section 3.3. We instead use a more basic semiclassical approximation, the *WKB approximation*, which we will make rigorous using the classical *method of variation of parameters* (one could also proceed using the closely related *Prüfer transformation*, which we will not detail here). We study the eigenfunction equation

$$L_h\phi = \lambda\phi$$

where we think of $h > 0$ as being small, and λ as being close to 1. We rewrite this as

(2.163)
$$\phi'' = -\frac{1}{h^2}k(x)^2\phi$$

where $k(x) := \sqrt{\lambda - x^2/4}$, where we will only work in the "classical" region $x^2/4 < \lambda$ (so $k(x) > 0$) for now.

Recall that the general solution to the constant coefficient ODE $\phi'' = -\frac{1}{h^2}k^2\phi$ is given by $\phi(x) = Ae^{ikx/h} + Be^{-ikx/h}$. Inspired by this, we make the ansatz

$$\phi(x) = A(x)e^{i\Psi(x)/h} + B(x)e^{-i\Psi(x)/h}$$

where $\Psi(x) := \int_0^x k(y)\ dy$ is the antiderivative of k. Differentiating this, we have

$$\phi'(x) = \frac{ik(x)}{h}(A(x)e^{i\Psi(x)/h} - B(x)e^{-i\Psi(x)/h})$$
$$+ A'(x)e^{i\Psi(x)/h} + B'(x)e^{-i\Psi(x)/h}.$$

Because we are representing a single function ϕ by two functions A, B, we have the freedom to place an additional constraint on A, B. Following the usual variation of parameters strategy, we will use this freedom to eliminate the last two terms in the expansion of ϕ, thus

$$(2.164) \qquad A'(x)e^{i\Psi(x)/h} + B'(x)e^{-i\Psi(x)/h} = 0.$$

We can now differentiate again and obtain

$$\phi''(x) = -\frac{k(x)^2}{h^2}\phi(x) + \frac{ik'(x)}{h}(A(x)e^{i\Psi(x)/h} - B(x)e^{-i\Psi(x)/h})$$

$$+ \frac{ik(x)}{h}(A'(x)e^{i\Psi(x)/h} - B'(x)e^{-i\Psi(x)/h}).$$

Comparing this with (2.163) we see that

$$A'(x)e^{i\Psi(x)/h} - B'(x)e^{-i\Psi(x)/h} = -\frac{k'(x)}{k(x)}(A(x)e^{i\Psi(x)/h} - B(x)e^{-i\Psi(x)/h}).$$

Combining this with (2.164), we obtain equations of motion for A and B:

$$A'(x) = -\frac{k'(x)}{2k(x)}A(x) + \frac{k'(x)}{2k(x)}B(x)e^{-2i\Psi(x)/h},$$

$$B'(x) = -\frac{k'(x)}{2k(x)}B(x) + \frac{k'(x)}{2k(x)}A(x)e^{2i\Psi(x)/h}.$$

We can simplify this using the integrating factor substitution

$$A(x) = k(x)^{-1/2}a(x); \quad B(x) = k(x)^{-1/2}b(x)$$

to obtain

$$(2.165) \qquad a'(x) = \frac{k'(x)}{2k(x)}b(x)e^{-2i\Psi(x)/h},$$

$$(2.166) \qquad b'(x) = \frac{k'(x)}{2k(x)}a(x)e^{2i\Psi(x)/h}.$$

The point of doing all these transformations is that the role of the h-parameter no longer manifests itself through amplitude factors, and instead only is present in a phase factor. In particular, we have

$$a', b' = O(|a| + |b|)$$

on any compact interval I in the interior of the classical region $x^2/4 < \lambda$ (where we allow implied constants to depend on I), which by *Gronwall's inequality* gives the bounds

$$a'(x), b'(x), a(x), b(x) = O(|a(0)| + |b(0)|)$$

on this interval I. We can then insert these bounds into (2.165), (2.166) again and integrate by parts (taking advantage of the non-stationary nature of Ψ) to obtain the improved bounds[54]
(2.167)
$$a(x) = a(0) + O(h(|a(0)| + |b(0)|)); \quad b(x) = b(0) + O(h(|a(0)| + |b(0)|))$$

on this interval. This is already enough to get the asymptotics that we need:

Exercise 2.6.6. Use (2.162) to show that on any compact interval I in $(-2, 2)$, the density of $\mathbf{E}\mu_{M_n/\sqrt{n}}$ is given by

$$(|a|^2(x) + |b|^2(x))(\sqrt{1 - x^2/4} + o(1)) + O(|a(x)||b(x)|)$$

where a, b are as above with $\lambda = 1 + \frac{1}{2n}$ and $h = \frac{1}{n}$. Combining this with (2.167), (2.160), (2.161), and Stirling's formula, conclude that $\mathbf{E}\mu_{M_n/\sqrt{n}}$ converges in the vague topology to the semicircular law $\frac{1}{2\pi}(4 - x^2)_+^{1/2} dx$. (Note that once one gets convergence inside $(-2, 2)$, the convergence outside of $[-2, 2]$ can be obtained for free since $\mu_{M_n/\sqrt{n}}$ and $\frac{1}{2\pi}(4 - x^2)_+^{1/2} dx$ are both probability measures.)

We now sketch out the approach using the *method of steepest descent*. The starting point is the Fourier inversion formula

$$e^{-x^2/2} = \frac{1}{\sqrt{2\pi}} \int_{\mathbf{R}} e^{itx} e^{-t^2/2} \, dt$$

which upon repeated differentiation gives

$$\frac{d^n}{dx^n} e^{-x^2/2} = \frac{i^n}{\sqrt{2\pi}} \int_{\mathbf{R}} t^n e^{itx} e^{-t^2/2} \, dt$$

and thus by (2.159),

$$P_n(x) = \frac{(-i)^n}{(2\pi)^{3/4}\sqrt{n!}} \int_{\mathbf{R}} t^n e^{-(t-ix)^2/2} \, dt$$

and thus

$$\tilde{\phi}_n(x) = \frac{(-i)^n}{(2\pi)^{3/4}\sqrt{n!}} n^{(n+1)/2} \int_{\mathbf{R}} e^{n\phi(t)} \, dt$$

where

$$\phi(t) := \log t - (t - ix)^2/2 - x^2/4$$

where we use a suitable branch of the complex logarithm to handle the case of negative t.

The idea of the principle of steepest descent is to shift the contour of integration to where the real part of $\phi(z)$ is as small as possible. For this, it

---

[54]More precise asymptotic expansions can be obtained by iterating this procedure, but we will not need them here.

turns out that the stationary points of $\phi(z)$ play a crucial role. A brief calculation using the quadratic formula shows that there are two such stationary points, at

$$z = \frac{ix \pm \sqrt{4 - x^2}}{2}.$$

When $|x| < 2$, ϕ is purely imaginary at these stationary points, while for $|x| > 2$ the real part of ϕ is negative at both points. One then draws a contour through these two stationary points in such a way that near each such point, the imaginary part of $\phi(z)$ is kept fixed, which keeps oscillation to a minimum and allows the real part to decay as steeply as possible (which explains the name of the method). After a certain tedious amount of computation, one obtains the same type of asymptotics for $\tilde{\phi}_n$ that were obtained by the ODE method when $|x| < 2$ (and exponentially decaying estimates for $|x| > 2$).

Exercise 2.6.7. Let $f : \mathbf{C} \to \mathbf{C}$, $g : \mathbf{C} \to \mathbf{C}$ be functions which are analytic near a complex number z_0, with $f'(z_0) = 0$ and $f''(z_0) \neq 0$. Let $\varepsilon > 0$ be a small number, and let γ be the line segment $\{z_0 + tv : -\varepsilon < t < \varepsilon\}$, where v is a complex phase such that $f''(z_0)v^2$ is a negative real. Show that for ε sufficiently small, one has

$$\int_\gamma e^{\lambda f(z)} g(z) \, dz = (1 + o(1)) \frac{\sqrt{2\pi} v}{\sqrt{f''(z_0)\lambda}} e^{\lambda f(z_0)} g(z_0)$$

as $\lambda \to +\infty$. This is the basic estimate behind the method of steepest descent; readers who are also familiar with the *method of stationary phase* may see a close parallel.

Remark 2.6.7. The method of steepest descent requires an explicit representation of the orthogonal polynomials as contour integrals, and as such is largely restricted to the classical orthogonal polynomials (such as the Hermite polynomials). However, there is a non-linear generalisation of the method of steepest descent developed by Deift and Zhou, in which one solves a matrix Riemann-Hilbert problem rather than a contour integral; see [**De1999**] for details. Using these sorts of tools, one can generalise much of the above theory to the spectral distribution of $U(n)$-conjugation-invariant discussed in Remark 2.6.2, with the theory of Hermite polynomials being replaced by the more general theory of orthogonal polynomials; this is discussed in [**De1999**] or [**DeGi2007**].

The computations performed above for the diagonal kernel $K_n(x, x)$ can be summarised by the asymptotic

$$K_n(\sqrt{n}x, \sqrt{n}x) = \sqrt{n}(\rho_{\mathrm{sc}}(x) + o(1))$$

whenever $x \in \mathbf{R}$ is fixed and $n \to \infty$, and $\rho_{sc}(x) := \frac{1}{2\pi}(4 - x^2)_+^{1/2}$ is the semicircular law distribution. It is reasonably straightforward to generalise these asymptotics to the off-diagonal case as well, obtaining the more general result

$$(2.168)$$
$$K_n(\sqrt{n}x + \frac{y_1}{\sqrt{n}\rho_{sc}(x)}, \sqrt{n}x + \frac{y_2}{\sqrt{n}\rho_{sc}(x)}) = \sqrt{n}(\rho_{sc}(x)K(y_1, y_2) + o(1))$$

for fixed $x \in (-2, 2)$ and $y_1, y_2 \in \mathbf{R}$, where K is the *Dyson sine kernel*

$$K(y_1, y_2) := \frac{\sin(\pi(y_1 - y_2))}{\pi(y_1 - y_2)}.$$

In the language of semi-classical analysis, what is going on here is that the rescaling in the left-hand side of (2.168) is transforming the phase space region $\{(x, p) : p^2 + \frac{x^2}{4} \leq 1\}$ to the region $\{(x, p) : |p| \leq 1\}$ in the limit $n \to \infty$, and the projection to the latter region is given by the Dyson sine kernel. A formal proof of (2.168) can be given by using either the ODE method or the steepest descent method to obtain asymptotics for Hermite polynomials, and thence (via the Christoffel-Darboux formula) to asymptotics for K_n; we do not give the details here; but see for instance [**AnGuZi2010**].

From (2.168) and (2.147), (2.148) we obtain the asymptotic formula

$$\mathbf{E} \sum_{1 \leq i_1 < \ldots < i_k \leq n} F(\sqrt{n}\rho_{sc}(x)(\lambda_{i_1}(M_n) - \sqrt{n}x), \ldots,$$
$$\sqrt{n}\rho_{sc}(x)(\lambda_{i_k}(M_n) - \sqrt{n}x))$$
$$\to \int_{\mathbf{R}^k} F(y_1, \ldots, y_k) \det(K(y_i, y_j))_{1 \leq i, j \leq k} \, dy_1 \ldots dy_k$$

for the local statistics of eigenvalues. By means of further algebraic manipulations (using the general theory of determinantal processes), this allows one to control such quantities as the distribution of eigenvalue gaps near $\sqrt{n}x$, normalised at the scale $\frac{1}{\sqrt{n}\rho_{sc}(x)}$, which is the average size of these gaps as predicted by the semicircular law. For instance, for any $s_0 > 0$, one can show (basically by the above formulae combined with the inclusion-exclusion principle) that the proportion of eigenvalues λ_i with normalised gap $\sqrt{n}\frac{\lambda_{i+1}-\lambda_i}{\rho_{sc}(t_{i/n})}$ less than s_0 converges as $n \to \infty$ to $\int_0^{s_0} \frac{d^2}{ds^2} \det(1 - K)_{L^2[0,s]} \, ds$, where $t_c \in [-2, 2]$ is defined by the formula $\int_{-2}^{t_c} \rho_{sc}(x) \, dx = c$, and K is the integral operator with kernel $K(x, y)$ (this operator can be verified to be *trace class*, so the determinant can be defined in a *Fredholm sense*). See for instance[55] [**Me2004**].

---

[55] A finitary version of this inclusion-exclusion argument can also be found at [**Ta2010b**, §2.6].

Remark 2.6.8. One can also analyse the distribution of the eigenvalues at the edge of the spectrum, i.e., close to $\pm 2\sqrt{n}$. This ultimately hinges on understanding the behaviour of the projection π_{V_n} near the corners $(0, \pm 2)$ of the phase space region $\Omega = \{(p, x) : p^2 + \frac{x^2}{4} \leq 1\}$, or of the Hermite polynomials $P_n(x)$ for x close to $\pm 2\sqrt{n}$. For instance, by using steepest descent methods, one can show that

$$n^{1/12}\phi_n(2\sqrt{n} + \frac{x}{n^{1/6}}) \to \text{Ai}(x)$$

as $n \to \infty$ for any fixed x, y, where Ai is the *Airy function*

$$\text{Ai}(x) := \frac{1}{\pi} \int_0^\infty \cos(\frac{t^3}{3} + tx) \, dt.$$

This asymptotic and the Christoffel-Darboux formula then give the asymptotic

$$(2.169) \qquad n^{1/6} K_n(2\sqrt{n} + \frac{x}{n^{1/6}}, 2\sqrt{n} + \frac{y}{n^{1/6}}) \to K_{\text{Ai}}(x, y)$$

for any fixed x, y, where K_{Ai} is the *Airy kernel*

$$K_{\text{Ai}}(x, y) := \frac{\text{Ai}(x)\,\text{Ai}'(y) - \text{Ai}'(x)\,\text{Ai}(y)}{x - y}.$$

This then gives an asymptotic description of the largest eigenvalues of a GUE matrix, which cluster in the region $2\sqrt{n} + O(n^{-1/6})$. For instance, one can use the above asymptotics to show that the largest eigenvalue λ_1 of a GUE matrix obeys the *Tracy-Widom law*

$$\mathbf{P}\left(\frac{\lambda_1 - 2\sqrt{n}}{n^{-1/6}} < t\right) \to \det(1 - A)_{L^2([t, +\infty))}$$

for any fixed t, where A is the integral operator with kernel K_{Ai}. See [**AnGuZi2010**] and Section 3.3 for further discussion.

2.6.5. Tridiagonalisation. We now discuss another approach to studying the spectral statistics of GUE matrices, based on applying the *tridiagonal matrix algorithm* to convert a GUE matrix into a (real symmetric) *tridiagonal matrix*—a matrix which has non-zero entries only on the diagonal, and on the entries adjacent to the diagonal. The key observation (due to Trotter) is the following:

Proposition 2.6.9 (Tridiagonal form of GUE [**Tr1984**]). *Let M'_n be the random tridiagonal real symmetric matrix*

$$M'_n = \begin{pmatrix} a_1 & b_1 & 0 & \dots & 0 & 0 \\ b_1 & a_2 & b_2 & \dots & 0 & 0 \\ 0 & b_2 & a_3 & \dots & 0 & 0 \\ \vdots & \vdots & \vdots & \ddots & \vdots & \vdots \\ 0 & 0 & 0 & \dots & a_{n-1} & b_{n-1} \\ 0 & 0 & 0 & \dots & b_{n-1} & a_n \end{pmatrix}$$

where the $a_1, \dots, a_n, b_1, \dots, b_{n-1}$ are jointly independent real random variables, with $a_1, \dots, a_n \equiv N(0,1)_{\mathbf{R}}$ being standard real Gaussians, and each b_i having a χ-distribution,

$$b_i = (\sum_{j=1}^{i} |z_{i,j}|^2)^{1/2}$$

where $z_{i,j} \equiv N(0,1)_{\mathbf{C}}$ are iid complex Gaussians. Let M_n be drawn from GUE. Then the joint eigenvalue distribution of M_n is identical to the joint eigenvalue distribution of M'_n.

Proof. Let M_n be drawn from GUE. We can write

$$M_n = \begin{pmatrix} M_{n-1} & X_n \\ X_n^* & a_n \end{pmatrix}$$

where M_{n-1} is drawn from the $n - 1 \times n - 1$ GUE, $a_n \equiv N(0,1)_{\mathbf{R}}$, and $X_n \in \mathbf{C}^{n-1}$ is a random Gaussian vector with all entries iid with distribution $N(0,1)_{\mathbf{C}}$. Furthermore, M_{n-1}, X_n, a_n are jointly independent.

We now apply the tridiagonal matrix algorithm. Let $b_{n-1} := |X_n|$, then b_n has the χ-distribution indicated in the proposition. We then conjugate M_n by a unitary matrix U that preserves the final basis vector e_n, and maps X to $b_{n-1} e_{n-1}$. Then we have

$$U M_n U^* = \begin{pmatrix} \tilde{M}_{n-1} & b_{n-1} e_{n-1} \\ b_{n-1} e_{n-1}^* & a_n \end{pmatrix}$$

where $\tilde{M}_{n-1}$ is conjugate to M_{n-1}. Now we make the crucial observation: because M_{n-1} is distributed according to GUE (which is a unitarily invariant ensemble), and U is a unitary matrix independent of M_{n-1}, $\tilde{M}_{n-1}$ is also distributed according to GUE, and remains independent of both b_{n-1} and a_n.

We continue this process, expanding $U M_n U^*$ as

$$\begin{pmatrix} M_{n-2} & X_{n-1} & 0 \\ X_{n-1}^* & a_{n-1} & b_{n-1} \\ 0 & b_{n-1} & a_n \end{pmatrix}.$$

Applying a further unitary conjugation that fixes e_{n-1}, e_n but maps X_{n-1} to $b_{n-2}e_{n-2}$, we may replace X_{n-1} by $b_{n-2}e_{n-2}$ while transforming M_{n-2} to another GUE matrix $\tilde{M}_{n-2}$ independent of $a_n, b_{n-1}, a_{n-1}, b_{n-2}$. Iterating this process, we eventually obtain a coupling of M_n to M'_n by unitary conjugations, and the claim follows. $\square$

Because of this proposition, any fact about the spectrum of a GUE matrix M_n (such as the semicircular law) is equivalent to the corresponding fact about the spectrum of the tridiagonal matrix M'_n. Because of its much sparser form, this matrix can be significantly simpler to analyse than the original GUE matrix (though at the cost of not having many of the symmetries enjoyed by GUE). We illustrate this with Trotter's proof of the semicircular law for GUE. Let A_n denote the deterministic real symmetric tridiagonal matrix

$$
A_n := \begin{pmatrix}
0 & \sqrt{1} & 0 & \cdots & 0 & 0 \\
\sqrt{1} & 0 & \sqrt{2} & \cdots & 0 & 0 \\
0 & \sqrt{2} & 0 & \cdots & 0 & 0 \\
\vdots & \vdots & \vdots & \ddots & \vdots & \vdots \\
0 & 0 & 0 & \cdots & 0 & \sqrt{n-1} \\
0 & 0 & 0 & \cdots & \sqrt{n-1} & 0
\end{pmatrix}.
$$

Exercise 2.6.8. Show that the eigenvalues of A_n are given by the zeroes $x_1, \ldots, x_n$ of the nth Hermite polynomial P_n, with an orthogonal basis of eigenvectors given by

$$
\begin{pmatrix}
P_0(x_i) \\
\cdots \\
P_{n-1}(x_i)
\end{pmatrix}
$$

for $i = 1, \ldots, n$. Using the asymptotics for P_n, conclude that $\mu_{\frac{1}{\sqrt{n}}A_n}$ converges in the vague topology to the semicircular law as $n \to \infty$.

Exercise 2.6.9. Let M'_n be the tridiagonal matrix from Proposition 2.6.9. Show that

$$
\mathbf{E}\|M'_n - A_n\|_F^2 \ll n
$$

where $\|\ \|_F$ is the Frobenius norm. Using the preceding exercise and the Weilandt-Hoffman inequality (1.68), deduce that $\mu_{\frac{1}{\sqrt{n}}M'_n}$ (and hence $\mu_{\frac{1}{\sqrt{n}}M_n}$) converges in the vague topology to the semicircular law as $n \to \infty$.

2.6.6. Determinantal form of the Gaussian matrix distribution. One can perform an analogous analysis of the joint distribution function (2.136) of Gaussian random matrices. Indeed, given any family $P_0(z), \ldots, P_{n-1}(z)$ of polynomials, with each P_i of degree i, much the same arguments as before show that (2.136) is equal to a constant multiple of

$$\det\left(\sum_{k=0}^{n-1} P_k(\lambda_i)e^{-|\lambda_i|^2/2}\overline{P_k(\lambda_j)}e^{-|\lambda_j|^2/2}\right)_{1\leq i,j\leq n}.$$

One can then select $P_k(z)e^{-|z|^2/2}$ to be orthonormal in $L^2(\mathbf{C})$. Actually, in this case, the polynomials are very simple, being given explicitly by the formula

$$P_k(z) := \frac{1}{\sqrt{\pi k!}}z^k.$$

Exercise 2.6.10. Verify that the $P_k(z)e^{-|z|^2/2}$ are indeed orthonormal, and then conclude that (2.136) is equal to $\det(K_n(\lambda_i, \lambda_j))_{1\leq i,j\leq n}$, where

$$K_n(z, w) := \frac{1}{\pi}e^{-(|z|^2+|w|^2)/2}\sum_{k=0}^{n-1}\frac{(z\overline{w})^k}{k!}.$$

Conclude further that the m-point correlation functions $\rho_m(z_1,\ldots,z_m)$ are given as

$$\rho_m(z_1,\ldots,z_m) = \det(K_n(z_i, z_j))_{1\leq i,j\leq m}.$$

Exercise 2.6.11. Show that as $n\to\infty$, one has

$$K_n(\sqrt{n}z, \sqrt{n}z) = \frac{1}{\pi}1_{|z|\leq 1} + o(1)$$

for almost every $z\in\mathbf{C}$, and deduce that the expected spectral measure $\mathbf{E}\mu_{G/\sqrt{n}}$ converges vaguely to the circular measure $\mu_c := \frac{1}{\pi}1_{|z|\leq 1}\,dz$; this is a special case of the *circular law*.

Remark 2.6.10. One can use the above formulae as the starting point for many other computations on the spectrum of random Gaussian matrices; to give just one example, one can show that the expected number of eigenvalues which are real is of the order of $\sqrt{n}$ (see [**Ed1996**] for more precise results of this nature). It remains a challenge to extend these results to more general ensembles than the Gaussian ensemble.

2.7. The least singular value

Now we turn attention to another important spectral statistic, the *least singular value* $\sigma_n(M)$ of an $n\times n$ matrix M (or, more generally, the least nontrivial singular value $\sigma_p(M)$ of an $n\times p$ matrix with $p\leq n$). This quantity controls the invertibility of M. Indeed, M is invertible precisely when $\sigma_n(M)$ is non-zero, and the operator norm $\|M^{-1}\|_{\mathrm{op}}$ of M^{-1} is given by $1/\sigma_n(M)$. This quantity is also related to the *condition number* $\sigma_1(M)/\sigma_n(M) = \|M\|_{\mathrm{op}}\|M^{-1}\|_{\mathrm{op}}$ of M, which is of importance in numerical linear algebra.

As we shall see in Section 2.8, the least singular value of M (and more generally, of the shifts $\frac{1}{\sqrt{n}}M - zI$ for complex z) will be of importance in rigorously
establishing the *circular law* for iid random matrices M, as it plays a key
role in computing the *Stieltjes transform* $\frac{1}{n}\operatorname{tr}(\frac{1}{\sqrt{n}}M - zI)^{-1}$ of such matrices, which as we have already seen is a powerful tool in understanding the
spectra of random matrices.

The least singular value

$$\sigma_n(M) = \inf_{\|x\|=1} \|Mx\|,$$

which sits at the "hard edge" of the spectrum, bears a superficial similarity
to the operator norm

$$\|M\|_{\mathrm{op}} = \sigma_1(M) = \sup_{\|x\|=1} \|Mx\|$$

at the "soft edge" of the spectrum, that was discussed back in Section 2.3, so
one may at first think that the methods that were effective in controlling the
latter, namely the epsilon-net argument and the moment method, would also
work to control the former. The epsilon-net method does indeed have some
effectiveness when dealing with rectangular matrices (in which the spectrum
stays well away from zero), but the situation becomes more delicate for
square matrices; it can control some "low entropy" portions of the infimum
that arise from "structured" or "compressible" choices of x, but are not
able to control the "generic" or "incompressible" choices of x, for which
new arguments will be needed. As for the moment method, this can give
the coarse order of magnitude (for instance, for rectangular matrices with
$p = yn$ for $0 < y < 1$, it gives an upper bound of $(1 - \sqrt{y} + o(1))n$ for the
singular value with high probability, thanks to the *Marcenko-Pastur law*),
but again this method begins to break down for square matrices, although
one can make some partial headway by considering *negative* moments such
as $\operatorname{tr} M^{-2}$, though these are more difficult to compute than positive moments
$\operatorname{tr} M^k$.

So one needs to supplement these existing methods with additional tools.
It turns out that the key issue is to understand the distance between one of
the n rows $X_1, \ldots, X_n \in \mathbf{C}^n$ of the matrix M, and the hyperplane spanned
by the other $n-1$ rows. The reason for this is as follows. First suppose that
$\sigma_n(M) = 0$, so that M is non-invertible, and there is a linear dependence
between the rows $X_1, \ldots, X_n$. Thus, one of the X_i will lie in the hyperplane
spanned by the other rows, and so one of the distances mentioned above will
vanish; in fact, one expects many of the n distances to vanish. Conversely,
whenever one of these distances vanishes, one has a linear dependence, and
so $\sigma_n(M) = 0$.

More generally, if the least singular value $\sigma_n(M)$ is small, one generically expects many of these n distances to be small also, and conversely. Thus, control of the least singular value is morally equivalent to control of the distance between a row X_i and the hyperplane spanned by the other rows. This latter quantity is basically the dot product of X_i with a unit normal n_i of this hyperplane.

When working with random matrices with jointly independent coefficients, we have the crucial property that the unit normal n_i (which depends on all the rows other than X_i) is *independent* of X_i, so even after conditioning n_i to be fixed, the entries of X_i remain independent. As such, the dot product $X_i \cdot n_i$ is a familiar scalar random walk, and can be controlled by a number of tools, most notably Littlewood-Offord theorems and the Berry-Esséen central limit theorem. As it turns out, this type of control works well except in some rare cases in which the normal n_i is "compressible" or otherwise highly structured; but epsilon-net arguments can be used to dispose of these cases[56].

These methods rely quite strongly on the joint independence on all the entries; it remains a challenge to extend them to more general settings. Even for Wigner matrices, the methods run into difficulty because of the non-independence of some of the entries (although it turns out one can understand the least singular value in such cases by rather different methods).

To simplify the exposition, we shall focus primarily on just one specific ensemble of random matrices, the *Bernoulli ensemble* $M = (\xi_{ij})_{1 \le i,j \le n}$ of random sign matrices, where $\xi_{ij} = \pm 1$ are independent Bernoulli signs. However, the results can extend to more general classes of random matrices, with the main requirement being that the coefficients are jointly independent.

2.7.1. The epsilon-net argument. We begin by using the epsilon net argument to establish a lower bound in the rectangular case, first established in [**LiPaRuTo2005**]:

Theorem 2.7.1 (Lower bound). *Let $M = (\xi_{ij})_{1 \le i \le p; 1 \le j \le n}$ be an $n \times p$ Bernoulli matrix, where $1 \le p \le (1 - \delta)n$ for some $\delta > 0$ (independent of n). Then with exponentially high probability (i.e., $1 - O(e^{-cn})$ for some $c > 0$), one has $\sigma_p(M) \ge c\sqrt{n}$, where $c > 0$ depends only on δ.*

This should be compared with the upper bound established in Section 2.3, which asserts that

$$(2.170) \qquad \|M\|_{\mathrm{op}} = \sigma_1(M) \le C\sqrt{n}$$

---

[56]This general strategy was first developed for the technically simpler singularity problem in [**Ko1967**], and then extended to the least singular value problem in [**Ru2008**].

holds with overwhelming probability for some absolute constant C (indeed, one can take any $C > 2$ here).

We use the epsilon net argument introduced in Section 2.3, but with a smaller value of $\varepsilon > 0$ than used for the largest singular value. We write

$$\sigma_p(M) = \inf_{x \in \mathbf{C}^p : \|x\| = 1} \|Mx\|.$$

Taking Σ to be a maximal ε-net of the unit sphere in $\mathbf{C}^p$, with $\varepsilon > 0$ to be chosen later, we have that

$$\sigma_p(M) \geq \inf_{x \in \Sigma} \|Mx\| - \varepsilon \|M\|_{\mathrm{op}}$$

and thus by (2.170), we have with overwhelming probability that

$$\sigma_p(M) \geq \inf_{x \in \Sigma} \|Mx\| - C\varepsilon\sqrt{n},$$

and so it suffices to show that

$$\mathbf{P}(\inf_{x \in \Sigma} \|Mx\| \leq 2C\varepsilon\sqrt{n})$$

is exponentially small in n. From the union bound, we can upper bound this by

$$\sum_{x \in \Sigma} \mathbf{P}(\|Mx\| \leq 2C\varepsilon\sqrt{n}).$$

From the volume packing argument we have

(2.171) $$|\Sigma| \leq O(1/\varepsilon)^p \leq O(1/\varepsilon)^{(1-\delta)n}.$$

So we need to upper bound, for each $x \in \Sigma$, the probability

$$\mathbf{P}(\|Mx\| \leq 2C\varepsilon\sqrt{n}).$$

If we let $Y_1, \ldots, Y_n \in \mathbf{C}^p$ be the rows of M, we can write this as

$$\mathbf{P}(\sum_{j=1}^{n} |Y_j \cdot x|^2 \leq 4C^2\varepsilon^2 n).$$

By Markov's inequality (1.14), the only way that this event can hold is if we have

$$|Y_j \cdot x|^2 \leq 8C^2\varepsilon^2$$

for at least $n/2$ values of j. We do not know in advance what the set of j is for which this event holds; but the number of possible values of such sets of j is at most 2^n. Applying the union bound (and paying the entropy cost of 2^n) and using symmetry, we may thus bound the above probability by[57]

$$\leq 2^n \mathbf{P}(|Y_j \cdot x|^2 \leq 8C^2\varepsilon^2 \text{ for } 1 \leq j \leq n/2).$$

---

[57]We will take n to be even for sake of notation, although it makes little essential difference.

Now observe that the random variables $Y_j \cdot x$ are independent, and so we can bound this expression by

$$\leq 2^n \mathbf{P}(|Y \cdot x| \leq \sqrt{8}C\varepsilon)^{n/2}$$

where $Y = (\xi_1, \ldots, \xi_n)$ is a random vector of iid Bernoulli signs.

We write $x = (x_1, \ldots, x_n)$, so that $Y \cdot x$ is a random walk

$$Y \cdot x = \xi_1 x_1 + \cdots + \xi_n x_n.$$

To understand this walk, we apply (a slight variant) of the Berry-Esséen theorem from Section 2.2:

Exercise 2.7.1. Show[58] that

$$\sup_t \mathbf{P}(|Y \cdot x - t| \leq r) \ll \frac{r}{\|x\|^2} + \frac{1}{\|x\|^3} \sum_{j=1}^n |x_j|^3$$

for any $r > 0$ and any non-zero x. (*Hint:* First normalise $\|x\| = 1$, then adapt the proof of the Berry-Esséen theorem.)

Conclude, in particular, that if

$$\sum_{j:|x_j| \leq \varepsilon^{100}} |x_j|^2 \geq \varepsilon^{10},$$

(say) then

$$\sup_t \mathbf{P}(|Y \cdot x - t| \leq \sqrt{8}C\varepsilon) \ll \varepsilon.$$

(*Hint:* Condition out all the x_j with $|x_j| > 1/2$.)

Let us temporarily call x *incompressible* if

$$\sum_{j:|x_j| \leq \varepsilon^{100}} |x_j|^2 < \varepsilon^{10}$$

and *compressible* otherwise. If we only look at the incompressible elements of Σ, we can now bound

$$\mathbf{P}(\|Mx\| \leq 2C\varepsilon\sqrt{n}) \ll O(\varepsilon)^n,$$

and comparing this against the entropy cost (2.171) we obtain an acceptable contribution for ε small enough (here we are crucially using the rectangular condition $p \leq (1 - \delta)n$).

It remains to deal with the compressible vectors. Observe that such vectors lie within ε of a *sparse* unit vector which is only supported in at most ε^{-200} positions. The ε-entropy of these sparse vectors (i.e., the number of balls of radius ε needed to cover this space) can easily be computed to be

---

[58]Actually, for the purposes of this section, it would suffice to establish a weaker form of the Berry-Esséen theorem with $\sum_{j=1}^3 |x_j|^3/\|x\|^3$ replaced by $(\sum_{j=1}^3 |x_j|^3/\|x\|^3)^c$ for any fixed $c > 0$.

of polynomial size $O(n^{O_\varepsilon(1)})$ in n. Meanwhile, we have the following crude bound:

Exercise 2.7.2. For any unit vector x, show that

$$\mathbf{P}(|Y \cdot x| \leq \kappa) \leq 1 - \kappa$$

for $\kappa > 0$ small enough. (*Hint:* Use the *Paley-Zygmund inequality*, Exercise 1.1.9. Bounds on higher moments on $|Y \cdot x|$ can be obtained, for instance, using Hoeffding's inequality, or by direct computation.) Use this to show that

$$\mathbf{P}(\|Mx\| \leq 2C\varepsilon\sqrt{n}) \ll \exp(-cn)$$

for all such x and ε sufficiently small, with $c > 0$ independent of ε and n.

Thus the compressible vectors give a net contribution of $O(n^{O_\varepsilon(1)}) \times \exp(-cn)$, which is acceptable. This concludes the proof of Theorem 2.7.1.

2.7.2. Singularity probability. Now we turn to square Bernoulli matrices $M = (\xi_{ij})_{1 \leq i,j \leq n}$. Before we investigate the size of the least singular value, we first tackle the easier problem of bounding the *singularity probability*

$$\mathbf{P}(\sigma_n(M) = 0),$$

i.e., the probability that M is not invertible. The problem of computing this probability exactly is still not completely settled. Since M is singular whenever the first two rows (say) are identical, we obtain a lower bound

$$\mathbf{P}(\sigma_n(M) = 0) \geq \frac{1}{2^n},$$

and it is conjectured that this bound is essentially tight in the sense that

$$\mathbf{P}(\sigma_n(M) = 0) = (\frac{1}{2} + o(1))^n,$$

but this remains open; the best bound currently is [**BoVuWo2010**], and gives

$$\mathbf{P}(\sigma_n(M) = 0) \leq (\frac{1}{\sqrt{2}} + o(1))^n.$$

We will not prove this bound here, but content ourselves with a weaker bound, essentially due to Komlós [**Ko1967**]:

Proposition 2.7.2. *We have* $\mathbf{P}(\sigma_n(M) = 0) \ll 1/n^{1/2}$.

To show this, we need the following combinatorial fact, due to Erdős [**Er1945**]:

Proposition 2.7.3 (Erdős Littlewood-Offord theorem). *Let* $x = (x_1, \ldots, x_n)$ *be a vector with at least k non-zero entries, and let* $Y = (\xi_1, \ldots, \xi_n)$ *be a random vector of iid Bernoulli signs. Then* $\mathbf{P}(Y \cdot x = 0) \ll k^{-1/2}$.

Proof. By taking real and imaginary parts we may assume that x is real. By eliminating zero coefficients of x we may assume that $k = n$; reflecting we may then assume that all the x_i are positive. Observe that the set of $Y = (\xi_1, \ldots, \xi_n) \in \{-1, 1\}^n$ with $Y \cdot x = 0$ forms an *antichain*[59] in $\{-1, 1\}^n$ with the product partial ordering. The claim now easily follows from *Sperner's theorem* and Stirling's formula (Section 1.2). □

Note that we also have the obvious bound

$$(2.172) \qquad \mathbf{P}(Y \cdot x = 0) \le 1/2$$

for any non-zero x.

Now we prove the proposition. In analogy with the arguments of Section 2.7, we write

$$\mathbf{P}(\sigma_n(M) = 0) = \mathbf{P}(Mx = 0 \text{ for some non-zero } x \in \mathbf{C}^n)$$

(actually we can take $x \in \mathbf{R}^n$ since M is real). We divide into compressible and incompressible vectors as before, but our definition of compressibility and incompressibility is slightly different now. Also, one has to do a certain amount of technical maneuvering in order to preserve the crucial independence between rows and columns.

Namely, we pick an $\varepsilon > 0$ and call x *compressible* if it is supported on at most εn coordinates, and *incompressible* otherwise.

Let us first consider the contribution of the event that $Mx = 0$ for some non-zero compressible x. Pick an x with this property which is as sparse as possible, say k sparse for some $1 \le k < \varepsilon n$. Let us temporarily fix k. By paying an entropy cost of $\lfloor \varepsilon n \rfloor \binom{n}{k}$, we may assume that it is the first k entries that are non-zero for some $1 \le k \le \varepsilon n$. This implies that the first k columns $Y_1, \ldots, Y_k$ of M have a linear dependence given by x; by minimality, $Y_1, \ldots, Y_{k-1}$ are linearly independent. Thus, x is uniquely determined (up to scalar multiples) by $Y_1, \ldots, Y_k$. Furthermore, as the $n \times k$ matrix formed by $Y_1, \ldots, Y_k$ has rank $k - 1$, there is some $k \times k$ minor which already determines x up to constants; by paying another entropy cost of $\binom{n}{k}$, we may assume that it is the top left minor which does this. In particular, we can now use the first k rows $X_1, \ldots, X_k$ to determine x up to constants. But the remaining $n - k$ rows are independent of $X_1, \ldots, X_k$ and still need to be orthogonal to x; by Proposition 2.7.3, this happens with probability

---

[59] An *antichain* in a partially ordered set X is a subset S of X such that no two elements in S are comparable in the order. The product partial ordering on $\{-1, 1\}^n$ is defined by requiring $(x_1, \ldots, x_n) \le (y_1, \ldots, y_n)$ iff $x_i \le y_i$ for all i. *Sperner's theorem* asserts that all anti-chains in $\{-1, 1\}^n$ have cardinality at most $\binom{n}{\lfloor n/2 \rfloor}$.

at most $O(\sqrt{k})^{-(n-k)}$, giving a total cost of

$$\sum_{1 \le k \le \varepsilon n} \binom{n}{k}^2 O(\sqrt{k})^{-(n-k)},$$

which by Stirling's formula (Section 1.2) is acceptable (in fact this gives an exponentially small contribution).

The same argument gives that the event that $y^* M = 0$ for some non-zero compressible y also has exponentially small probability. The only remaining event to control is the event that $Mx = 0$ for some incompressible x, but that $Mz \ne 0$ and $y^* M \ne 0$ for all non-zero compressible z, y. Call this event E.

Since $Mx = 0$ for some incompressible x, we see that for at least εn values of $k \in \{1, \ldots, n\}$, the row X_k lies in the vector space V_k spanned by the remaining $n-1$ rows of M. Let E_k denote the event that E holds, and that X_k lies in V_k; then we see from double counting that

$$\mathbf{P}(E) \le \frac{1}{\varepsilon n} \sum_{k=1}^{n} \mathbf{P}(E_k).$$

By symmetry, we thus have

$$\mathbf{P}(E) \le \frac{1}{\varepsilon} \mathbf{P}(E_n).$$

To compute $\mathbf{P}(E_n)$, we freeze $X_1, \ldots, X_{n-1}$ and consider a normal vector x to V_{n-1}; note that we can select x depending only on $X_1, \ldots, X_{n-1}$. We may assume that an incompressible normal vector exists, since otherwise the event E_n would be empty. We make the crucial observation that X_n is still independent of x. By Proposition 2.7.3, we thus see that the conditional probability that $X_n \cdot x = 0$, for fixed $X_1, \ldots, X_{n-1}$, is $O_\varepsilon(n^{-1/2})$. We thus see that $\mathbf{P}(E) \ll_\varepsilon 1/n^{1/2}$, and the claim follows.

Remark 2.7.4. Further progress has been made on this problem by a finer analysis of the concentration probability $\mathbf{P}(Y \cdot x = 0)$, and in particular, in classifying those x for which this concentration probability is large (this is known as the *inverse Littlewood-Offord problem*). Important breakthroughs in this direction were made by Halász [**Ha1977**] (introducing Fourier-analytic tools) and by Kahn, Komlós, and Szemerédi [**KaKoSz1995**] (introducing an efficient "swapping" argument). In [**TaVu2007**] tools from additive combinatorics (such as Freiman's theorem) were introduced to obtain further improvements, leading eventually to the results from [**BoVuWo2010**] mentioned earlier.

2.7.3. Lower bound for the least singular value. Now we return to the least singular value $\sigma_n(M)$ of an iid Bernoulli matrix, and establish a lower bound. Given that there are n singular values between 0 and $\sigma_1(M)$, which is typically of size $O(\sqrt{n})$, one expects the least singular value to be of size about $1/\sqrt{n}$ on the average. Another argument supporting this heuristic comes from the following identity:

Exercise 2.7.3 (Negative second moment identity). Let M be an invertible $n \times n$ matrix, let $X_1, \ldots, X_n$ be the rows of M, and let $R_1, \ldots, R_n$ be the columns of M^{-1}. For each $1 \le i \le n$, let V_i be the hyperplane spanned by all the rows $X_1, \ldots, X_n$ other than X_i. Show that $\|R_i\| = \mathrm{dist}(X_i, V_i)^{-1}$ and $\sum_{i=1}^n \sigma_i(M)^{-2} = \sum_{i=1}^n \mathrm{dist}(X_i, V_i)^2$.

From Talagrand's inequality (Theorem 2.1.13), we expect each $\mathrm{dist}(X_i, V_i)$ to be of size $O(1)$ on the average, which suggests that $\sum_{i=1}^n \sigma_i(M)^{-2} = O(n)$; this is consistent with the heuristic that the eigenvalues $\sigma_i(M)$ should be roughly evenly spaced in the interval $[0, 2\sqrt{n}]$ (so that $\sigma_{n-i}(M)$ should be about $(i+1)/\sqrt{n}$).

Now we give a rigorous lower bound:

Theorem 2.7.5 (Lower tail estimate for the least singular value). *For any* $\lambda > 0$, *one has*

$$\mathbf{P}(\sigma_n(M) \le \lambda/\sqrt{n}) \ll o_{\lambda \to 0}(1) + o_{n \to \infty; \lambda}(1)$$

where $o_{\lambda \to 0}(1)$ *goes to zero as* $\lambda \to 0$ *uniformly in* n, *and* $o_{n \to \infty; \lambda}(1)$ *goes to zero as* $n \to \infty$ *for each fixed* λ.

This is a weaker form of a result of Rudelson and Vershynin [**RuVe2008**] (which obtains a bound of the form $O(\lambda) + O(c^n)$ for some $c < 1$), which builds upon the earlier works [**Ru2008**], [**TaVu2009**], which obtained variants of the above result.

The scale $1/\sqrt{n}$ that we are working at here is too fine to use epsilon net arguments (unless one has a *lot* of control on the entropy, which can be obtained in some cases thanks to powerful inverse Littlewood-Offord theorems, but is difficult to obtain in general.) We can prove this theorem along similar lines to the arguments in the previous section; we sketch the method as follows. We can take λ to be small. We write the probability to be estimated as

$$\mathbf{P}(\|Mx\| \le \lambda/\sqrt{n} \text{ for some unit vector } x \in \mathbf{C}^n).$$

We can assume that $\|M\|_{\mathrm{op}} \le C\sqrt{n}$ for some absolute constant C, as the event that this fails has exponentially small probability.

We pick an $\varepsilon > 0$ (not depending on λ) to be chosen later. We call a unit vector $x \in \mathbf{C}^n$ *compressible* if x lies within a distance ε of a εn-sparse vector. Let us first dispose of the case in which $\|Mx\| \leq \lambda\sqrt{n}$ for some compressible x. By paying an entropy cost of $\binom{n}{\lfloor \varepsilon n \rfloor}$, we may assume that x is within ε of a vector y supported in the first $\lfloor \varepsilon n \rfloor$ coordinates. Using the operator norm bound on M and the triangle inequality, we conclude that

$$\|My\| \leq (\lambda + C\varepsilon)\sqrt{n}.$$

Since y has norm comparable to 1, this implies that the least singular value of the first $\lfloor \varepsilon n \rfloor$ columns of M is $O((\lambda + \varepsilon)\sqrt{n})$. But by Theorem 2.7.1, this occurs with probability $O(\exp(-cn))$ (if λ, ε are small enough). So the total probability of the compressible event is at most $\binom{n}{\lfloor \varepsilon n \rfloor}O(\exp(-cn))$, which is acceptable if ε is small enough.

Thus we may assume now that $\|Mx\| > \lambda/\sqrt{n}$ for all compressible unit vectors x; we may similarly assume that $\|y^*M\| > \lambda/\sqrt{n}$ for all compressible unit vectors y. Indeed, we may also assume that $\|y^*M_i\| > \lambda/\sqrt{n}$ for every i, where M_i is M with the ith column removed.

The remaining case is if $\|Mx\| \leq \lambda/\sqrt{n}$ for some incompressible x. Let us call this event E. Write $x = (x_1, \ldots, x_n)$, and let $Y_1, \ldots, Y_n$ be the column of M, thus

$$\|x_1 Y_1 + \cdots + x_n Y_n\| \leq \lambda/\sqrt{n}.$$

Letting W_i be the subspace spanned by all the $Y_1, \ldots, Y_n$ except for Y_i, we conclude upon projecting to the orthogonal complement of W_i that

$$|x_i|\, \text{dist}(Y_i, W_i) \leq \lambda/\sqrt{n}$$

for all i (compare with Exercise 2.7.3). On the other hand, since x is incompressible, we see that $|x_i| \geq \varepsilon/\sqrt{n}$ for at least εn values of i, and thus

$$(2.173) \qquad\qquad \text{dist}(Y_i, W_i) \leq \lambda/\varepsilon$$

for at least εn values of i. If we let E_i be the event that E and (2.173) both hold, we thus have from double-counting that

$$\mathbf{P}(E) \leq \frac{1}{\varepsilon n}\sum_{i=1}^{n}\mathbf{P}(E_i)$$

and thus by symmetry

$$\mathbf{P}(E) \leq \frac{1}{\varepsilon}\mathbf{P}(E_n)$$

(say). However, if E_n holds, then setting y to be a unit normal vector to W_i (which is necessarily incompressible, by the hypothesis on M_i), we have

$$|Y_i \cdot y| \leq \lambda/\varepsilon.$$

Again, the crucial point is that Y_i and y are independent. The incompressibility of y, combined with a Berry-Esséen type theorem, then gives

Exercise 2.7.4. Show that

$$\mathbf{P}(|Y_i \cdot y| \leq \lambda/\varepsilon) \ll \varepsilon^2$$

(say) if λ is sufficiently small depending on ε, and n is sufficiently large depending on ε.

This gives a bound of $O(\varepsilon)$ for $\mathbf{P}(E)$ if λ is small enough depending on ε, and n is large enough; this gives the claim.

Remark 2.7.6. A variant of these arguments, based on inverse Littlewood-Offord theorems rather than the Berry-Esséen theorem, gives the variant estimate

$$(2.174) \qquad \sigma_n(\frac{1}{\sqrt{n}}M_n - zI) \geq n^{-A}$$

with high probability for some $A > 0$, and any z of polynomial size in n. There are several results of this type, with overlapping ranges of generality (and various values of A) [**GoTi2007, PaZh2010, TaVu2008**], and the exponent A is known to degrade if one has too few moment assumptions on the underlying random matrix M. This type of result (with an unspecified A) is important for the circular law, discussed in the next section.

2.7.4. Upper bound for the least singular value. One can complement the lower tail estimate with an upper tail estimate:

Theorem 2.7.7 (Upper tail estimate for the least singular value). *For any* $\lambda > 0$, *one has*

$$(2.175) \qquad \mathbf{P}(\sigma_n(M) \geq \lambda/\sqrt{n}) \ll o_{\lambda \to \infty}(1) + o_{n \to \infty; \lambda}(1).$$

We prove this using an argument of Rudelson and Vershynin [**RuVe2009**]. Suppose that $\sigma_n(M) > \lambda/\sqrt{n}$, then

$$(2.176) \qquad \|y^*M^{-1}\| \leq \sqrt{n}\|y\|/\lambda$$

for all y.

Next, let $X_1, \ldots, X_n$ be the rows of M, and let $R_1, \ldots, R_n$ be the columns of M^{-1}, thus $R_1, \ldots, R_n$ is a *dual basis* for $X_1, \ldots, X_n$. From (2.176) we have

$$\sum_{i=1}^{n} |y \cdot R_i|^2 \leq n\|y\|^2/\lambda^2.$$

We apply this with y equal to $X_n - \pi_n(X_n)$, where π_n is the orthogonal projection to the space V_{n-1} spanned by $X_1, \ldots, X_{n-1}$. On the one hand, we have

$$\|y\|^2 = \text{dist}(X_n, V_{n-1})^2$$

and, on the other hand, we have for any $1 \leq i < n$ that

$$y \cdot R_i = -\pi_n(X_n) \cdot R_i = -X_n \cdot \pi_n(R_i),$$

and so

(2.177) $$\sum_{i=1}^{n-1} |X_n \cdot \pi_n(R_i)|^2 \leq n \operatorname{dist}(X_n, V_{n-1})^2/\lambda^2.$$

If (2.177) holds, then $|X_n \cdot \pi_n(R_i)|^2 = O(\operatorname{dist}(X_n, V_{n-1})^2/\lambda^2)$ for at least half of the i, so the probability in (2.175) can be bounded by

$$\ll \frac{1}{n} \sum_{i=1}^{n-1} \mathbf{P}(|X_n \cdot \pi_n(R_i)|^2 = O(\operatorname{dist}(X_n, V_{n-1})^2/\lambda^2))$$

which by symmetry can be bounded by

$$\ll \mathbf{P}(|X_n \cdot \pi_n(R_1)|^2 = O(\operatorname{dist}(X_n, V_{n-1})^2/\lambda^2)).$$

Let $\varepsilon > 0$ be a small quantity to be chosen later. From Talagrand's inequality (Theorem 2.1.13) we know that $\operatorname{dist}(X_n, V_{n-1}) = O_\varepsilon(1)$ with probability $1 - O(\varepsilon)$, so we obtain a bound of

$$\ll \mathbf{P}(X_n \cdot \pi_n(R_1) = O_\varepsilon(1/\lambda)) + O(\varepsilon).$$

Now a key point is that the vectors $\pi_n(R_1), \ldots, \pi_n(R_{n-1})$ depend only on $X_1, \ldots, X_{n-1}$ and not on X_n; indeed, they are the dual basis for $X_1, \ldots, X_{n-1}$ in V_{n-1}. Thus, after conditioning $X_1, \ldots, X_{n-1}$ and thus $\pi_n(R_1)$ to be fixed, X_n is still a Bernoulli random vector. Applying a Berry-Esséen inequality, we obtain a bound of $O(\varepsilon)$ for the conditional probability that $X_n \cdot \pi_n(R_1) = O_\varepsilon(1/\lambda)$ for λ sufficiently small depending on ε, unless $\pi_n(R_1)$ is compressible (in the sense that, say, it is within ε of an εn-sparse vector). But this latter possibility can be controlled (with exponentially small probability) by the same type of arguments as before; we omit the details.

2.7.5. Asymptotic for the least singular value. The distribution of singular values of a Gaussian random matrix can be computed explicitly by techniques similar to those employed in Section 2.6. In particular, if M is a real Gaussian matrix (with all entries iid with distribution $N(0,1)_{\mathbf{R}}$), it was shown in [**Ed1988**] that $\sqrt{n}\sigma_n(M)$ converges in distribution to the distribution $\mu_E := \frac{1+\sqrt{x}}{2\sqrt{x}} e^{-x/2-\sqrt{x}} \, dx$ as $n \to \infty$. It turns out that this result can be extended to other ensembles with the same mean and variance. In particular, we have the following result from [**TaVu2010**]:

Theorem 2.7.8. *If M is an iid Bernoulli matrix, then $\sqrt{n}\sigma_n(M)$ also converges in distribution to μ_E as $n \to \infty$. (In fact there is a polynomial rate of convergence.)*

This should be compared with Theorems 2.7.5, 2.7.7, which show that $\sqrt{n}\sigma_n(M)$ have a tight sequence of distributions in $(0, +\infty)$. The arguments from [**TaVu2010**] thus provide an alternate proof of these two theorems. The same result in fact holds for all iid ensembles obeying a finite moment condition.

The arguments used to prove Theorem 2.7.8 do not establish the limit μ_E directly, but instead use the result of [**Ed1988**] as a black box, focusing instead on establishing the *universality* of the limiting distribution of $\sqrt{n}\sigma_n(M)$, and in particular, that this limiting distribution is the same whether one has a Bernoulli ensemble or a Gaussian ensemble.

The arguments are somewhat technical and we will not present them in full here, but instead give a sketch of the key ideas.

In previous sections we have already seen the close relationship between the least singular value $\sigma_n(M)$, and the distances $\text{dist}(X_i, V_i)$ between a row X_i of M and the hyperplane V_i spanned by the other $n-1$ rows. It is not hard to use the above machinery to show that as $n \to \infty$, $\text{dist}(X_i, V_i)$ converges in distribution to the absolute value $|N(0,1)_\mathbf{R}|$ of a Gaussian regardless of the underlying distribution of the coefficients of M (i.e., it is asymptotically universal). The basic point is that one can write $\text{dist}(X_i, V_i)$ as $|X_i \cdot n_i|$ where n_i is a unit normal of V_i (we will assume here that M is non-singular, which by previous arguments is true asymptotically almost surely). The previous machinery lets us show that n_i is incompressible with high probability, and the claim then follows from the Berry-Esséen theorem.

Unfortunately, despite the presence of suggestive relationships such as Exercise 2.7.3, the asymptotic universality of the distances $\text{dist}(X_i, V_i)$ does not directly imply asymptotic universality of the least singular value. However, it turns out that one can obtain a higher-dimensional version of the universality of the scalar quantities $\text{dist}(X_i, V_i)$, as follows. For any small k (say, $1 \le k \le n^c$ for some small $c > 0$) and any distinct $i_1, \ldots, i_k \in \{1, \ldots, n\}$, a modification of the above argument shows that the covariance matrix

$$(2.178) \qquad (\pi(X_{i_a}) \cdot \pi(X_{i_b}))_{1 \le a,b \le k}$$

of the orthogonal projections $\pi(X_{i_1}), \ldots, \pi(X_{i_k})$ of the k rows $X_{i_1}, \ldots, X_{i_k}$ to the complement $V_{i_1,\ldots,i_k}^\perp$ of the space $V_{i_1,\ldots,i_k}$ spanned by the other $n - k$ rows of M, is also universal, converging in distribution to the covariance[60] matrix $(G_a \cdot G_b)_{1 \le a,b \le k}$ of k iid Gaussians $G_a \equiv N(0,1)_\mathbf{R}$ (note that the convergence of $\text{dist}(X_i, V_i)$ to $|N(0,1)_\mathbf{R}|$ is the $k = 1$ case of this claim). The key point is that one can show that the complement $V_{i_1,\ldots,i_k}^\perp$ is usually "incompressible" in a certain technical sense, which implies that the

---

[60]These covariance matrix distributions are also known as *Wishart distributions*.

projections $\pi(X_{i_a})$ behave like iid Gaussians on that projection thanks to a multidimensional Berry-Esséen theorem.

On the other hand, the covariance matrix (2.178) is closely related to the inverse matrix M^{-1}:

Exercise 2.7.5. Show that (2.178) is also equal to A^*A, where A is the $n \times k$ matrix formed from the $i_1, \ldots, i_k$ columns of M^{-1}.

In particular, this shows that the singular values of k randomly selected columns of M^{-1} have a universal distribution.

Recall that our goal is to show that $\sqrt{n}\sigma_n(M)$ has an asymptotically universal distribution, which is equivalent to asking that $\frac{1}{\sqrt{n}}\|M^{-1}\|_{\text{op}}$ has an asymptotically universal distribution. The goal is then to extract the operator norm of M^{-1} from looking at a random $n \times k$ minor B of this matrix. This comes from the following application of the second moment method:

Exercise 2.7.6. Let A be an $n \times n$ matrix with columns $R_1, \ldots, R_n$, and let B be the $n \times k$ matrix formed by taking k of the columns $R_1, \ldots, R_n$ at random. Show that

$$\mathbf{E}\|A^*A - \frac{n}{k}B^*B\|_F^2 \leq \frac{n}{k}\sum_{k=1}^{n}\|R_k\|^4,$$

where $\|\|_F$ is the Frobenius norm(2.64).

Recall from Exercise 2.7.3 that $\|R_k\| = 1/\text{dist}(X_k, V_k)$, so we expect each $\|R_k\|$ to have magnitude about $O(1)$. This, together with the Wielandt-Hoeffman inequality (1.68) means that we expect $\sigma_1((M^{-1})^*(M^{-1})) = \sigma_n(M)^{-2}$ to differ by $O(n^2/k)$ from $\frac{n}{k}\sigma_1(B^*B) = \frac{n}{k}\sigma_1(B)^2$. In principle, this gives us asymptotic universality on $\sqrt{n}\sigma_n(M)$ from the already established universality of B.

There is one technical obstacle remaining, however: while we know that each $\text{dist}(X_k, V_k)$ is distributed like a Gaussian, so that each individual R_k is going to be of size $O(1)$ with reasonably good probability, in order for the above exercise to be useful, one needs to bound *all* of the R_k *simultaneously* with high probability. A naive application of the union bound leads to terrible results here. Fortunately, there is a strong correlation between the R_k: they tend to be large together or small together, or equivalently that the distances $\text{dist}(X_k, V_k)$ tend to be small together or large together. Here is one indication of this:

Lemma 2.7.9. *For any* $1 \leq k < i \leq n$, *one has*

$$\text{dist}(X_i, V_i) \geq \frac{\|\pi_i(X_i)\|}{1 + \sum_{j=1}^{k}\frac{\|\pi_i(X_j)\|}{\|\pi_i(X_i)\|\,\text{dist}(X_j,V_j)}},$$

where π_i is the orthogonal projection onto the space spanned by $X_1, \ldots, X_k, X_i$.

Proof. We may relabel so that $i = k + 1$; then projecting everything by π_i we may assume that $n = k + 1$. Our goal is now to show that

$$\text{dist}(X_n, V_{n-1}) \geq \frac{\|X_n\|}{1 + \sum_{j=1}^{n-1} \frac{\|X_j\|}{\|X_n\| \, \text{dist}(X_j, V_j)}}.$$

Recall that $R_1, \ldots, R_n$ is a dual basis to $X_1, \ldots, X_n$. This implies, in particular, that

$$x = \sum_{j=1}^{n} (x \cdot X_j) R_j$$

for any vector x; applying this to X_n we obtain

$$X_n = \|X_n\|^2 R_n + \sum_{j=1}^{n-1} (X_j \cdot X_n) R_j$$

and hence by the triangle inequality

$$\|X_n\|^2 \|R_n\| \leq \|X_n\| + \sum_{j=1}^{n-1} \|X_j\| \|X_n\| \|R_j\|.$$

Using the fact that $\|R_j\| = 1/\text{dist}(X_j, R_j)$, the claim follows.   □

In practice, once k gets moderately large (e.g., $k = n^c$ for some small $c > 0$), one can control the expressions $\|\pi_i(X_j)\|$ appearing here by Talagrand's inequality (Theorem 2.1.13), and so this inequality tells us that once $\text{dist}(X_j, V_j)$ is bounded away from zero for $j = 1, \ldots, k$, it is bounded away from zero for all other k also. This turns out to be enough to get uniform control on the R_j to make Exercise 2.7.6 useful, and ultimately to complete the proof of Theorem 2.7.8.

2.8. The circular law

In this section, we leave the realm of self-adjoint matrix ensembles, such as Wigner random matrices, and consider instead the simplest examples of non-self-adjoint ensembles, namely the iid matrix ensembles.

The basic result in this area is

Theorem 2.8.1 (Circular law). *Let M_n be an $n \times n$ iid matrix, whose entries ξ_{ij}, $1 \leq i, j \leq n$ are iid with a fixed (complex) distribution $\xi_{ij} \equiv \xi$ of mean zero and variance one. Then the spectral measure $\mu_{\frac{1}{\sqrt{n}} M_n}$ converges both in probability and almost surely to the circular law $\mu_{\text{circ}} := \frac{1}{\pi} 1_{|x|^2 + |y|^2 \leq 1} \, dx dy$, where x, y are the real and imaginary coordinates of the complex plane.*

This theorem has a long history; it is analogous to the semicircular law, but the non-Hermitian nature of the matrices makes the spectrum so unstable that key techniques that are used in the semicircular case, such as truncation and the moment method, no longer work; significant new ideas are required. In the case of random Gaussian matrices, this result was established by Mehta [**Me2004**] (in the complex case) and by Edelman [**Ed1996**] (in the real case), as was sketched out in Section 2.6. In 1984, Girko [**Gi1984**] laid out a general strategy for establishing the result for non-Gaussian matrices, which formed the base of all future work on the subject; however, a key ingredient in the argument, namely a bound on the least singular value of shifts $\frac{1}{\sqrt{n}} M_n - zI$, was not fully justified at the time. A rigorous proof of the circular law was then established by Bai [**Ba1997**], assuming additional moment and boundedness conditions on the individual entries. These additional conditions were then slowly removed in a sequence of papers [**GoTi2007, Gi2004, PaZh2010, TaVu2008**], with the last moment condition being removed in [**TaVuKr2010**].

At present, the known methods used to establish the circular law for general ensembles rely very heavily on the joint independence of all the entries. It is a key challenge to see how to weaken this joint independence assumption.

2.8.1. Spectral instability. One of the basic difficulties present in the non-Hermitian case is *spectral instability*: small perturbations in a large matrix can lead to large fluctuations in the spectrum. In order for any sort of analytic technique to be effective, this type of instability must somehow be precluded.

The canonical example of spectral instability comes from perturbing the right shift matrix

$$U_0 := \begin{pmatrix} 0 & 1 & 0 & \ldots & 0 \\ 0 & 0 & 1 & \ldots & 0 \\ \vdots & \vdots & \vdots & \ddots & \vdots \\ 0 & 0 & 0 & \ldots & 0 \end{pmatrix}$$

to the matrix

$$U_\varepsilon := \begin{pmatrix} 0 & 1 & 0 & \ldots & 0 \\ 0 & 0 & 1 & \ldots & 0 \\ \vdots & \vdots & \vdots & \ddots & \vdots \\ \varepsilon & 0 & 0 & \ldots & 0 \end{pmatrix}$$

for some $\varepsilon > 0$.

The matrix U_0 is nilpotent: $U_0^n = 0$. Its characteristic polynomial is $(-\lambda)^n$, and it thus has n repeated eigenvalues at the origin. In contrast, U_ε obeys the equation $U_\varepsilon^n = \varepsilon I$, its characteristic polynomial is $(-\lambda)^n - \varepsilon(-1)^n$, and it thus has n eigenvalues at the nth roots $\varepsilon^{1/n} e^{2\pi i j/n}$, $j = 0, \ldots, n - 1$ of ε. Thus, even for exponentially small values of ε, say $\varepsilon = 2^{-n}$, the eigenvalues for U_ε can be quite far from the eigenvalues of U_0, and can wander all over the unit disk. This is in sharp contrast with the Hermitian case, where eigenvalue inequalities such as the Weyl inequalities (1.64) or Wielandt-Hoffman inequalities (1.68) ensure stability of the spectrum.

One can explain the problem in terms of *pseudospectrum*[61]. The only spectrum of U_0 is at the origin, so the resolvents $(U_0 - zI)^{-1}$ of U_0 are finite for all non-zero z. However, while these resolvents are finite, they can be extremely large. Indeed, from the nilpotent nature of U_0 we have the Neumann series

$$(U_0 - zI)^{-1} = -\frac{1}{z} - \frac{U_0}{z^2} - \cdots - \frac{U_0^{n-1}}{z^n},$$

so for $|z| < 1$ we see that the resolvent has size roughly $|z|^{-n}$, which is exponentially large in the interior of the unit disk. This exponentially large size of resolvent is consistent with the exponential instability of the spectrum:

Exercise 2.8.1. Let M be a square matrix, and let z be a complex number. Show that $\|(M - zI)^{-1}\|_{\text{op}} \geq R$ if and only if there exists a perturbation $M + E$ of M with $\|E\|_{\text{op}} \leq 1/R$ such that $M + E$ has z as an eigenvalue.

This already hints strongly that if one wants to rigorously prove control on the spectrum of M near z, one needs some sort of upper bound on $\|(M - zI)^{-1}\|_{\text{op}}$, or equivalently one needs some sort of lower bound on the least singular value $\sigma_n(M - zI)$ of $M - zI$.

Without such a bound, though, the instability precludes the direct use of the *truncation method*, which was so useful in the Hermitian case. In particular, there is no obvious way to reduce the proof of the circular law to the case of bounded coefficients, in contrast to the semicircular law where this reduction follows easily from the Wielandt-Hoffman inequality (see Section 2.4). Instead, we must continue working with unbounded random variables throughout the argument (unless, of course, one makes an additional decay hypothesis, such as assuming certain moments are finite; this helps explain the presence of such moment conditions in many papers on the circular law).

2.8.2. Incompleteness of the moment method. In the Hermitian case, the moments

$$\frac{1}{n} \text{tr}(\frac{1}{\sqrt{n}} M)^k = \int_{\mathbf{R}} x^k \, d\mu_{\frac{1}{\sqrt{n} M_n}}(x)$$

---

[61] The *pseudospectrum* of an operator T is the set of complex numbers z for which the operator norm $\|(T - zI)^{-1}\|_{\text{op}}$ is either infinite, or larger than a fixed threshold $1/\varepsilon$. See [**Tr1991**] for further discussion.

of a matrix can be used (in principle) to understand the distribution $\mu_{\frac{1}{\sqrt{n}M_n}}$ completely (at least, when the measure $\mu_{\frac{1}{\sqrt{n}M_n}}$ has sufficient decay at infinity). This is ultimately because the space of real polynomials $P(x)$ is dense in various function spaces (the Weierstrass approximation theorem).

In the non-Hermitian case, the spectral measure $\mu_{\frac{1}{\sqrt{n}M_n}}$ is now supported on the complex plane rather than the real line. One still has the formula

$$\frac{1}{n}\operatorname{tr}(\frac{1}{\sqrt{n}}M)^k = \int_{\mathbf{R}} z^k \, d\mu_{\frac{1}{\sqrt{n}M_n}}(z),$$

but it is much less useful now, because the space of complex polynomials $P(z)$ no longer has any good density properties[62]. In particular, the moments no longer uniquely determine the spectral measure.

This can be illustrated with the shift examples given above. It is easy to see that U_0 and U_ε have vanishing moments up to $(n-1)^{\text{th}}$ order, i.e.,

$$\frac{1}{n}\operatorname{tr}(\frac{1}{\sqrt{n}}U_0)^k = \frac{1}{n}\operatorname{tr}(\frac{1}{\sqrt{n}}U_\varepsilon)^k = 0$$

for $k = 1, \ldots, n-1$. Thus we have

$$\int_{\mathbf{R}} z^k \, d\mu_{\frac{1}{\sqrt{n}}U_0}(z) = \int_{\mathbf{R}} z^k \, d\mu_{\frac{1}{\sqrt{n}}U_\varepsilon}(z) = 0$$

for $k = 1, \ldots, n-1$. Despite this enormous number of matching moments, the spectral measures $\mu_{\frac{1}{\sqrt{n}}U_0}$ and $\mu_{\frac{1}{\sqrt{n}}U_\varepsilon}$ are dramatically different; the former is a Dirac mass at the origin, while the latter can be arbitrarily close to the unit circle. Indeed, even if we set *all* moments equal to zero,

$$\int_{\mathbf{R}} z^k \, d\mu = 0$$

for $k = 1, 2, \ldots$, then there are an uncountable number of possible (continuous) probability measures that could still be the (asymptotic) spectral measure μ; for instance, any measure which is rotationally symmetric around the origin would obey these conditions.

If one could somehow control the mixed moments

$$\int_{\mathbf{R}} z^k \bar{z}^l \, d\mu_{\frac{1}{\sqrt{n}}M_n}(z) = \frac{1}{n}\sum_{j=1}^{n}(\frac{1}{\sqrt{n}}\lambda_j(M_n))^k(\frac{1}{\sqrt{n}}\overline{\lambda_j}(M_n))^l$$

of the spectral measure, then this problem would be resolved, and one could use the moment method to reconstruct the spectral measure accurately. However, there does not appear to be any obvious way to compute this quantity; the obvious guess of $\frac{1}{n}\operatorname{tr}(\frac{1}{\sqrt{n}}M_n)^k(\frac{1}{\sqrt{n}}M_n^*)^l$ works when the matrix

---

[62]For instance, the uniform closure of the space of polynomials on the unit disk is not the space of continuous functions, but rather the space of holomorphic functions that are continuous on the closed unit disk.

M_n is *normal*, as M_n and M_n^* then share the same basis of eigenvectors, but generically one does not expect these matrices to be normal.

Remark 2.8.2. The failure of the moment method to control the spectral measure is consistent with the instability of spectral measure with respect to perturbations, because moments are stable with respect to perturbations.

Exercise 2.8.2. Let $k \geq 1$ be an integer, and let M_n be an iid matrix whose entries have a fixed distribution ξ with mean zero, variance 1, and with k^{th} moment finite. Show that $\frac{1}{n} \operatorname{tr}(\frac{1}{\sqrt{n}} M_n)^k$ converges to zero as $n \to \infty$ in expectation, in probability, and in the almost sure sense. Thus we see that $\int_{\mathbf{R}} z^k \, d\mu_{\frac{1}{\sqrt{n}} M_n}(z)$ converges to zero in these three senses also. This is of course consistent with the circular law, but does not come close to establishing that law, for the reasons given above.

The failure of the moment method also shows that methods of free probability (Section 2.5) do not work directly. For instance, observe that for fixed ε, U_0 and U_ε (in the non-commutative probability space $(\operatorname{Mat}_n(\mathbf{C}), \frac{1}{n} \operatorname{tr})$) both converge in the sense of $*$-moments as $n \to \infty$ to that of the right shift operator on $\ell^2(\mathbf{Z})$ (with the trace $\tau(T) = \langle e_0, T e_0 \rangle$, with e_0 being the Kronecker delta at 0); but the spectral measures of U_0 and U_ε are different. Thus the spectral measure cannot be read off directly from the free probability limit.

2.8.3. The logarithmic potential.

With the moment method out of consideration, attention naturally turns to the Stieltjes transform

$$s_n(z) = \frac{1}{n} \operatorname{tr}(\frac{1}{\sqrt{n}} M_n - zI)^{-1} = \int_{\mathbf{C}} \frac{d\mu_{\frac{1}{\sqrt{n}} M_n}(w)}{w - z}.$$

Even though the measure $\mu_{\frac{1}{\sqrt{n}} M_n}$ is now supported on $\mathbf{C}$ rather than $\mathbf{R}$, the Stieltjes transform is still well-defined. The Plemelj formula for reconstructing spectral measure from the Stieltjes transform that was used in previous sections is no longer applicable, but there are other formulae one can use instead, in particular, one has

Exercise 2.8.3. Show that

$$\mu_{\frac{1}{\sqrt{n}} M_n} = \frac{1}{\pi} \partial_{\bar{z}} s_n(z)$$

in the sense of distributions, where

$$\partial_{\bar{z}} := \frac{1}{2}(\frac{\partial}{\partial x} + i\frac{\partial}{\partial y})$$

is the Cauchy-Riemann operator.

One can control the Stieltjes transform quite effectively away from the origin. Indeed, for iid matrices with sub-Gaussian entries, one can show (using the methods from Section 2.3) that the operator norm of $\frac{1}{\sqrt{n}} M_n$ is $1 + o(1)$ almost surely; this, combined with (2.8.2) and Laurent expansion, tells us that $s_n(z)$ almost surely converges to $-1/z$ locally uniformly in the region $\{z : |z| > 1\}$, and that the spectral measure $\mu_{\frac{1}{\sqrt{n}} M_n}$ converges almost surely to zero in this region (which can of course also be deduced directly from the operator norm bound). This is of course consistent with the circular law, but is not sufficient to prove it (for instance, the above information is also consistent with the scenario in which the spectral measure collapses towards the origin). One also needs to control the Stieltjes transform inside the disk $\{z : |z| \le 1\}$ in order to fully control the spectral measure.

For this, existing methods (such as predecessor comparison) are not particularly effective (mainly because of the spectral instability, and also because of the lack of analyticity in the interior of the spectrum). Instead, one proceeds by relating the Stieltjes transform to the *logarithmic potential*

$$f_n(z) := \int_{\mathbf{C}} \log|w - z| d\mu_{\frac{1}{\sqrt{n}} M_n}(w).$$

It is easy to see that $s_n(z)$ is essentially the (distributional) gradient of $f_n(z)$:

$$s_n(z) = (-\frac{\partial}{\partial x} + i\frac{\partial}{\partial y}) f_n(z),$$

and thus f_n is related to the spectral measure by the distributional formula[63]

(2.179) $$\mu_{\frac{1}{\sqrt{n}} M_n} = \frac{1}{2\pi} \Delta f_n$$

where $\Delta := \frac{\partial^2}{\partial x^2} + \frac{\partial^2}{\partial y^2}$ is the Laplacian.

In analogy to previous continuity theorems, we have

Theorem 2.8.3 (Logarithmic potential continuity theorem). *Let M_n be a sequence of random matrices, and suppose that for almost every complex number z, $f_n(z)$ converges almost surely (resp., in probability) to*

$$f(z) := \int_{\mathbf{C}} \log|z - w| d\mu(w)$$

for some probability measure μ. Then $\mu_{\frac{1}{\sqrt{n}} M_n}$ converges almost surely (resp., in probability) to μ in the vague topology.

Proof. We prove the almost sure version of this theorem, and leave the convergence in the probability version as an exercise.

---

[63]This formula just reflects the fact that $\frac{1}{2\pi} \log|z|$ is the *Newtonian potential* in two dimensions.

On any bounded set K in the complex plane, the functions $\log|\cdot - w|$ lie in $L^2(K)$ uniformly in w. From Minkowski's integral inequality, we conclude that the f_n and f are uniformly bounded in $L^2(K)$. On the other hand, almost surely the f_n converge pointwise to f. From the dominated convergence theorem this implies that $\min(|f_n - f|, M)$ converges in $L^1(K)$ to zero for any M; using the uniform bound in $L^2(K)$ to compare $\min(|f_n - f|, M)$ with $|f_n - f|$ and then sending $M \to \infty$, we conclude that f_n converges to f in $L^1(K)$. In particular, f_n converges to f in the sense of distributions; taking distributional Laplacians using (2.179) we obtain the claim. $\qquad\square$

Exercise 2.8.4. Establish the convergence in probability version of Theorem 2.8.3.

Thus, the task of establishing the circular law then reduces to showing, for almost every z, that the logarithmic potential $f_n(z)$ converges (in probability or almost surely) to the right limit $f(z)$.

Observe that the logarithmic potential

$$f_n(z) = \frac{1}{n} \sum_{j=1}^{n} \log \left| \frac{\lambda_j(M_n)}{\sqrt{n}} - z \right|$$

can be rewritten as a log-determinant:

$$f_n(z) = \frac{1}{n} \log \left| \det\left(\frac{1}{\sqrt{n}} M_n - zI \right) \right|.$$

To compute this determinant, we recall that the determinant of a matrix A is not only the product of its eigenvalues, but also has a magnitude equal to the product of its *singular* values:

$$|\det A| = \prod_{j=1}^{n} \sigma_j(A) = \prod_{j=1}^{n} \lambda_j(A^*A)^{1/2}$$

and thus

$$f_n(z) = \frac{1}{2} \int_0^\infty \log x \; d\nu_{n,z}(x)$$

where $d\nu_{n,z}$ is the spectral measure of the matrix $(\frac{1}{\sqrt{n}}M_n - zI)^*(\frac{1}{\sqrt{n}}M_n - zI)$.

The advantage of working with this spectral measure, as opposed to the original spectral measure $\mu_{\frac{1}{\sqrt{n}}M_n}$, is that the matrix $(\frac{1}{\sqrt{n}}M_n - zI)^*(\frac{1}{\sqrt{n}}M_n - zI)$ is *self-adjoint*, and so methods such as the moment method or free probability can now be safely applied to compute the limiting spectral distribution. Indeed, Girko [**Gi1984**] established that for almost every z, $\nu_{n,z}$ converged both in probability and almost surely to an explicit (though slightly complicated) limiting measure ν_z in the vague topology. Formally, this implied that $f_n(z)$ would converge pointwise (almost surely and in probability) to

$$\frac{1}{2} \int_0^\infty \log x \, d\nu_z(x).$$

A lengthy but straightforward computation then showed that this expression was indeed the logarithmic potential $f(z)$ of the circular measure μ_{circ}, so that the circular law would then follow from the logarithmic potential continuity theorem.

Unfortunately, the vague convergence of $\nu_{n,z}$ to ν_z only allows one to deduce the convergence of $\int_0^\infty F(x) \, d\nu_{n,z}$ to $\int_0^\infty F(x) \, d\nu_z$ for F continuous and compactly supported. Unfortunately, $\log x$ has singularities at zero and at infinity, and so the convergence

$$\int_0^\infty \log x \, d\nu_{n,z}(x) \to \int_0^\infty \log x \, d\nu_z(x)$$

can fail if the spectral measure $\nu_{n,z}$ sends too much of its mass to zero or to infinity.

The latter scenario can be easily excluded, either by using operator norm bounds on M_n (when one has enough moment conditions) or even just the Frobenius norm bounds (which require no moment conditions beyond the unit variance). The real difficulty is with preventing mass from going to the origin.

The approach of Bai [**Ba1997**] proceeded in two steps. First, he established a polynomial lower bound

$$\sigma_n(\frac{1}{\sqrt{n}} M_n - zI) \geq n^{-C}$$

asymptotically almost surely for the least singular value of $\frac{1}{\sqrt{n}} M_n - zI$. This has the effect of capping off the $\log x$ integrand to be of size $O(\log n)$. Next, by using Stieltjes transform methods, the convergence of $\nu_{n,z}$ to ν_z in an appropriate metric (e.g., the Levi distance metric) was shown to be polynomially fast, so that the distance decayed like $O(n^{-c})$ for some $c > 0$. The $O(n^{-c})$ gain can safely absorb the $O(\log n)$ loss, and this leads to a proof of the circular law assuming enough boundedness and continuity hypotheses to ensure the least singular value bound and the convergence rate. This basic paradigm was also followed by later works [**GoTi2007, PaZh2010, TaVu2008**], with the main new ingredient being the advances in the understanding of the least singular value (Section 2.7).

Unfortunately, to get the polynomial convergence rate, one needs some moment conditions beyond the zero mean and unit variance rate (e.g., finite $2+\eta^{\text{th}}$ moment for some $\eta > 0$). In [**TaVuKr2010**] the additional tool of the Talagrand concentration inequality (Theorem 2.1.13) was used to eliminate the need for the polynomial convergence. Intuitively, the point is that only a small fraction of the singular values of $\frac{1}{\sqrt{n}}M_n - zI$ are going to be as small as n^{-c}; most will be much larger than this, and so the $O(\log n)$ bound is only going to be needed for a small fraction of the measure. To make this rigorous, it turns out to be convenient to work with a slightly different formula for the determinant magnitude $|\det(A)|$ of a square matrix than the product of the eigenvalues, namely the base-times-height formula

$$|\det(A)| = \prod_{j=1}^{n} \text{dist}(X_j, V_j)$$

where X_j is the jth row and V_j is the span of $X_1, \ldots, X_{j-1}$.

Exercise 2.8.5. Establish the inequality

$$\prod_{j=n+1-m}^{n} \sigma_j(A) \leq \prod_{j=1}^{m} \text{dist}(X_j, V_j) \leq \prod_{j=1}^{m} \sigma_j(A)$$

for any $1 \leq m \leq n$. (*Hint:* The middle product is the product of the singular values of the first m rows of A, and so one should try to use the Cauchy interlacing inequality for singular values, see Section 1.3.3.) Thus we see that $\text{dist}(X_j, V_j)$ is a variant of $\sigma_j(A)$.

The least singular value bounds, translated in this language (with $A := \frac{1}{\sqrt{n}}M_n - zI$), tell us that $\text{dist}(X_j, V_j) \geq n^{-C}$ with high probability; this lets us ignore the most dangerous values of j, namely those j that are equal to $n - O(n^{0.99})$ (say). For low values of j, say $j \leq (1-\delta)n$ for some small δ, one can use the moment method to get a good lower bound for the distances and the singular values, to the extent that the logarithmic singularity of $\log x$ no longer causes difficulty in this regime; the limit of this contribution can then be seen by moment method or Stieltjes transform techniques to be *universal* in the sense that it does not depend on the precise distribution of the components of M_n. In the medium regime $(1-\delta)n < j < n - n^{0.99}$, one can use Talagrand's inequality (Theorem 2.1.13) to show that $\text{dist}(X_j, V_j)$ has magnitude about $\sqrt{n-j}$, giving rise to a net contribution to $f_n(z)$ of the form $\frac{1}{n}\sum_{(1-\delta)n<j<n-n^{0.99}} O(\log \sqrt{n-j})$, which is small. Putting all this together, one can show that $f_n(z)$ converges to a universal limit as $n \to \infty$ (independent of the component distributions); see [**TaVuKr2010**] for details. As a consequence, once the circular law is established for one class of iid matrices, such as the complex Gaussian random matrix ensemble, it automatically holds for all other ensembles also.

2.8.4. Brown measure. We mentioned earlier that due to eigenvalue instability (or equivalently, due to the least singular value of shifts possibly going to zero), the moment method (and thus, by extension, free probability) was not sufficient by itself to compute the asymptotic spectral measure of non-Hermitian matrices in the large n limit. However, this method can be used to give a heuristic *prediction* as to what that measure is, known as the *Brown measure* [**Br1986**]. While Brown measure is not *always* the limiting spectral measure of a sequence of matrices, it turns out in practice that this measure can (with some effort) be shown to be the limiting spectral measure in key cases. As Brown measure can be computed (again, after some effort) in many cases, this gives a general strategy towards computing asymptotic spectral measure for various ensembles.

To define Brown measure, we use the language of free probability (Section 2.5). Let u be a bounded element (not necessarily self-adjoint) of a non-commutative probability space $(\mathcal{A}, \tau)$, which we will assume to be tracial. To derive Brown measure, we mimic the Girko strategy used for the circular law. First, for each complex number z, we let ν_z be the spectral measure of the non-negative self-adjoint element $(u - z)^*(u - z)$.

Exercise 2.8.6. Verify that the spectral measure of a positive element u^*u is automatically supported on the non-negative real axis. (*Hint:* Show that $\tau(P(u^*u)u^*uP(u^*u)) \geq 0$ for any real polynomial P, and use the spectral theorem.)

By the above exercise, ν_z is a compactly supported probability measure on $[0, +\infty)$. We then define the logarithmic potential $f(z)$ by the formula

$$f(z) = \frac{1}{2} \int_0^\infty \log x \; d\nu_z(x).$$

Note that f may equal $-\infty$ at some points.

To understand this determinant, we introduce the regularised determinant

$$f_\varepsilon(z) := \frac{1}{2} \int_0^\infty \log(\varepsilon + x) \; d\nu_z(x)$$

for $\varepsilon > 0$. From the monotone convergence theorem we see that $f_\varepsilon(z)$ decreases pointwise to $f(z)$ as $\varepsilon \to 0$.

We now invoke the Gelfand-Naimark theorem (Exercise 2.5.10) and embed[64] $\mathcal{A}$ into the space of bounded operators on $L^2(\tau)$, so that we may now obtain a functional calculus. Then we can write

$$f_\varepsilon(z) = \frac{1}{2}\tau(\log(\varepsilon + (u - z)^*(u - z))).$$

One can compute the first variation of f_ε:

---

[64]If τ is not faithful, this embedding need not be injective, but this will not be an issue in what follows.

Exercise 2.8.7. Let $\varepsilon > 0$. Show that the function f_ε is continuously differentiable with

$$\partial_x f_\varepsilon(z) = -\operatorname{Re}\tau((\varepsilon + (u-z)^*(u-z))^{-1}(u-z))$$

and

$$\partial_y f_\varepsilon(z) = -\operatorname{Im}\tau((\varepsilon + (u-z)^*(u-z))^{-1}(u-z)).$$

Then, one can compute the second variation at, say, the origin:

Exercise 2.8.8. Let $\varepsilon > 0$. Show that the function f_ε is twice continuously differentiable with

$$\partial_{xx} f_\varepsilon(0) = \operatorname{Re}\tau((\varepsilon + u^*u)^{-1} - (\varepsilon + u^*u)^{-1}(u + u^*)(\varepsilon + u^*u)^{-1}u)$$

and

$$\partial_{yy} f_\varepsilon(0) = \operatorname{Re}\tau((\varepsilon + u^*u)^{-1} - (\varepsilon + u^*u)^{-1}(u^* - u)(\varepsilon + u^*u)^{-1}u).$$

We conclude, in particular, that

$$\Delta f_\varepsilon(0) = 2\operatorname{Re}\tau((\varepsilon + u^*u)^{-1} - (\varepsilon + u^*u)^{-1}u^*(\varepsilon + u^*u)^{-1}u),$$

or equivalently,

$$\Delta f_\varepsilon(0) = 2(\|(\varepsilon + u^*u)^{-1/2}\|_{L^2(\tau)}^2 - \|(\varepsilon + u^*u)^{-1/2}u(\varepsilon + u^*u)^{-1/2}\|_{L^2(\tau)}^2).$$

Exercise 2.8.9. Show that

$$\|(\varepsilon + u^*u)^{-1/2}u(\varepsilon + u^*u)^{-1/2}\|_{L^2(\tau)} \leq \|(\varepsilon + u^*u)^{-1/2}\|_{L^2(\tau)}.$$

(*Hint:* Adapt the proof of Lemma 2.5.13.)

We conclude that Δf_ε is non-negative at zero. Translating u by any complex number we see that Δf_ε is non-negative everywhere, that is to say that f_ε is subharmonic. Taking limits we see that f is subharmonic also; thus if we define the *Brown measure* $\mu = \mu_u$ of u as

$$\mu := \frac{1}{2\pi}\Delta f$$

(cf. (2.179)), then μ is a non-negative measure.

Exercise 2.8.10. Show that for $|z| > \rho(u) := \rho(u^*u)^{1/2}$, f is continuously differentiable with

$$\partial_x f(z) = -\operatorname{Re}\tau((u-z)^{-1})$$

and

$$\partial_y f(z) = \operatorname{Im}\tau((u-z)^{-1})$$

and conclude that f is harmonic in this region; thus Brown measure is supported in the disk $\{z : |z| \leq \rho(u)\}$. Using Green's theorem, conclude also that Brown measure is a probability measure.

Exercise 2.8.11. In a finite-dimensional non-commutative probability space $(\text{Mat}_n(\mathbf{C}), \frac{1}{n}\text{tr})$, show that Brown measure is the same as spectral measure.

Exercise 2.8.12. In a commutative probability space $(L^\infty(\Omega), \mathbf{E})$, show that Brown measure is the same as the probability distribution.

Exercise 2.8.13. If u is the left shift on $\ell^2(\mathbf{Z})$ (with the trace $\tau(T) := \langle Te_0, e_0 \rangle$), show that the Brown measure of u is the uniform measure on the unit circle $\{z \in \mathbf{C} : |z| = 1\}$.

This last exercise illustrates the limitations of Brown measure for understanding asymptotic spectral measure. The shift U_0 and the perturbed shift U_ε introduced in previous sections both converge in the sense of $*$-moments as $n \to \infty$ (holding ε fixed) to the left shift u. For non-zero ε, the spectral measure of U_ε does indeed converge to the Brown measure of u, but for $\varepsilon = 0$ this is not the case. This illustrates a more general principle[65], that Brown measure is the right asymptotic limit for "generic" matrices, but not for exceptional matrices.

The machinery used to establish the circular law in full generality can be used to show that Brown measure is the correct asymptotic spectral limit for other models:

Theorem 2.8.4. *Let M_n be a sequence of random matrices whose entries are joint independent and with all moments uniformly bounded, with variance uniformly bounded from below, and which converges in the sense of $*$-moments to an element u of a non-commutative probability space. Then the spectral measure $\mu_{\frac{1}{\sqrt{n}}M_n}$ converges almost surely and in probability to the Brown measure of u.*

This theorem is essentially [**TaVuKr2010**, Theorem 1.20]. The main ingredients are those mentioned earlier, namely a polynomial lower bound on the least singular value, and the use of Talagrand's inequality (Theorem 2.1.13) to control medium singular values (or medium codimension distances to subspaces). Of the two ingredients, the former is more crucial, and is much more heavily dependent at present on the joint independence hypothesis; it would be of interest to see how to obtain lower bounds on the least singular value in more general settings. Some recent progress in this direction can be found in [**GuKrZe2009**], [**BoCaCh2008**]. See also [**BiLe2001**] for extensive computations of Brown measure for various random matrix models.

---

[65]See [**Sn2002**] for a precise formulation of this heuristic, using Gaussian regularisation.

Chapter 3

Related articles

3.1. Brownian motion and Dyson Brownian motion

One theme in this text will be the central nature played by the *Gaussian random variables* $X \equiv N(\mu, \sigma^2)$. Gaussians have an incredibly rich algebraic structure, and many results about general random variables can be established by first using this structure to verify the result for Gaussians, and then using universality techniques (such as the Lindeberg exchange strategy) to extend the results to more general variables.

One way to exploit this algebraic structure is to continuously deform the variance $t := \sigma^2$ from an initial variance of zero (so that the random variable is deterministic) to some final level T. We would like to use this to give a continuous family $t \mapsto X_t$ of random variables $X_t \equiv N(\mu, t)$ as t (viewed as a "time" parameter) runs from 0 to T.

At present, we have not completely specified what X_t should be, because we have only described the individual distribution $X_t \equiv N(\mu, t)$ of each X_t, and not the joint distribution. However, there is a very natural way to specify a joint distribution of this type, known as *Brownian motion*. In this section we lay the necessary probability theory foundations to set up this motion, and indicate its connection with the heat equation, the central limit theorem, and the Ornstein-Uhlenbeck process. This is the beginning of *stochastic calculus*, which we will not develop fully here.

We will begin with one-dimensional Brownian motion, but it is a simple matter to extend the process to higher dimensions. In particular, we can define Brownian motion on vector spaces of matrices, such as the space of $n \times n$ Hermitian matrices. This process is equivariant with respect to conjugation by unitary matrices, and so we can quotient out by this conjugation and obtain a new process on the quotient space, or in other words, on the *spectrum* of $n \times n$ Hermitian matrices. This process is called *Dyson Brownian motion*, and turns out to have a simple description in terms of ordinary Brownian motion; it will play a key role in this text.

3.1.1. Formal construction of Brownian motion.
We begin with constructing one-dimensional Brownian motion, following the classical method of Lévy. We shall model this motion using the machinery of *Wiener processes*:

Definition 3.1.1 (Wiener process). Let $\mu \in \mathbf{R}$, and let $\Sigma \subset [0, +\infty)$ be a set of times containing 0. A (one-dimensional) *Wiener process* on Σ with initial position μ is a collection $(X_t)_{t \in \Sigma}$ of real random variables X_t for each time $t \in \Sigma$, with the following properties:

(i) $X_0 = \mu$.

(ii) Almost surely, the map $t \mapsto X_t$ is a continuous function on Σ.

(iii) For every $0 \leq t_- < t_+$ in Σ, the increment $X_{t_+} - X_{t_-}$ has the distribution of $N(0, t_+ - t_-)_{\mathbf{R}}$. (In particular, $X_t \equiv N(\mu, t)_{\mathbf{R}}$ for every $t > 0$.)

(iv) For every $t_0 \leq t_1 \leq \cdots \leq t_n$ in Σ, the increments $X_{t_i} - X_{t_{i-1}}$ for $i = 1, \ldots, n$ are jointly independent.

If Σ is discrete, we say that $(X_t)_{t \in \Sigma}$ is a *discrete Wiener process*; if $\Sigma = [0, +\infty)$, then we say that $(X_t)_{t \in \Sigma}$ is a *continuous Wiener process*.

Remark 3.1.2. Collections of random variables $(X_t)_{t \in \Sigma}$, where Σ is a set of times, will be referred to as *stochastic processes*, thus Wiener processes are a (very) special type of stochastic process.

Remark 3.1.3. In the case of discrete Wiener processes, the continuity requirement (ii) is automatic. For continuous Wiener processes, there is a minor technical issue: the event that $t \mapsto X_t$ is continuous need not be a measurable event (one has to take uncountable intersections to define this event). Because of this, we interpret (ii) by saying that there exists a measurable event of probability 1, such that $t \mapsto X_t$ is continuous on all of this event, while also allowing for the possibility that $t \mapsto X_t$ could also sometimes be continuous outside of this event also. One can view the collection $(X_t)_{t \in \Sigma}$ as a single random variable, taking values in the product space $\mathbf{R}^\Sigma$ (with the product σ-algebra, of course).

Remark 3.1.4. One can clearly normalise the initial position μ of a Wiener process to be zero by replacing X_t with $X_t - \mu$ for each t.

We shall abuse notation somewhat and identify continuous Wiener processes with Brownian motion in our informal discussion, although technically the former is merely a model for the latter. To emphasise this link with Brownian motion, we shall often denote continuous Wiener processes as $(B_t)_{t \in [0, +\infty)}$ rather than $(X_t)_{t \in [0, +\infty)}$.

It is not yet obvious that Wiener processes exist, and to what extent they are unique. The situation is easily clarified though for discrete processes:

Proposition 3.1.5 (Discrete Brownian motion). *Let Σ be a discrete subset of $[0, +\infty)$ containing 0, and let $\mu \in \mathbf{R}$. Then (after extending the sample space if necessary) there exists a Wiener process $(X_t)_{t \in \Sigma}$ with base point μ. Furthermore, any other Wiener process $(X'_t)_{t \in \Sigma}$ with base point μ has the same distribution as μ.*

Proof. As Σ is discrete and contains 0, we can write it as $\{t_0, t_1, t_2, \ldots\}$ for some

$$0 = t_0 < t_1 < t_2 < \ldots.$$

Let $(dX_i)_{i=1}^\infty$ be a collection of jointly independent random variables with $dX_i \equiv N(0, t_i - t_{i-1})_{\mathbf{R}}$ (the existence of such a collection, after extending the sample space, is guaranteed by Exercise 1.1.20). If we then set

$$X_{t_i} := \mu + dX_1 + \cdots + dX_i$$

for all $i = 0, 1, 2, \ldots$, then one easily verifies (using Exercise 2.1.9) that $(X_t)_{t \in \Sigma}$ is a Wiener process.

Conversely, if $(X_t')_{t \in \Sigma}$ is a Wiener process, and we define $dX_i' := X_i' - X_{i-1}'$ for $i = 1, 2, \ldots$, then from the definition of a Wiener process we see that the dX_i' have distribution $N(0, t_i - t_{i-1})_{\mathbf{R}}$ and are jointly independent (i.e., any finite subcollection of the dX_i' are jointly independent). This implies for any finite n that the random variables $(dX_i)_{i=1}^n$ and $(dX_i')_{i=1}^n$ have the same distribution, and thus $(X_t)_{t \in \Sigma'}$ and $(X_t')_{t \in \Sigma'}$ have the same distribution for any finite subset Σ' of Σ. From the construction of the product σ-algebra we conclude that $(X_t)_{t \in \Sigma}$ and $(X_t')_{t \in \Sigma}$ have the same distribution, as required. $\qquad\square$

Now we pass from the discrete case to the continuous case.

Proposition 3.1.6 (Continuous Brownian motion). *Let $\mu \in \mathbf{R}$. Then (after extending the sample space if necessary) there exists a Wiener process $(X_t)_{t \in [0, +\infty)}$ with base point μ. Furthermore, any other Wiener process $(X_t')_{t \in [0, +\infty)}$ with base point μ has the same distribution as μ.*

Proof. The uniqueness claim follows by the same argument used to prove the uniqueness component of Proposition 3.1.5, so we just prove existence here. The iterative construction we give here is somewhat analogous to that used to create self-similar fractals, such as the *Koch snowflake*. (Indeed, Brownian motion can be viewed as a probabilistic analogue of a self-similar fractal.)

The idea is to create a sequence of increasingly fine discrete Brownian motions, and then to take a limit. Proposition 3.1.5 allows one to create each individual discrete Brownian motion, but the key is to *couple* these discrete processes together in a consistent manner.

Here's how. We start with a discrete Wiener process $(X_t)_{t \in \mathbf{N}}$ on the natural numbers $\mathbf{N} = \{0, 1, 2 \ldots\}$ with initial position μ, which exists by Proposition 3.1.5. We now extend this process to the denser set of times $\frac{1}{2}\mathbf{N} := \{\frac{1}{2}n : n \in \mathbf{N}\}$ by setting

$$X_{t+\frac{1}{2}} := \frac{X_t + X_{t+1}}{2} + Y_{t,0}$$

for $t = 0, 1, 2, \ldots$, where $(Y_{t,0})_{t \in \mathbf{N}}$ are iid copies of $N(0, 1/4)_{\mathbf{R}}$, which are jointly independent of the $(X_t)_{t \in \mathbf{N}}$. It is a routine matter to use Exercise

2.1.9 to show that this creates a discrete Wiener process $(X_t)_{t \in \frac{1}{2}\mathbf{N}}$ on $\frac{1}{2}\mathbf{N}$ which extends the previous process.

Next, we extend the process further to the denser set of times $\frac{1}{4}\mathbf{N}$ by defining

$$X_{t+\frac{1}{4}} := \frac{X_t + X_{t+1/2}}{2} + Y_{t,1}$$

for $t \in \frac{1}{2}\mathbf{N}$, where $(Y_{t,1})_{t \in \frac{1}{2}\mathbf{N}}$ are iid copies of $N(0, 1/8)_{\mathbf{R}}$, jointly independent of $(X_t)_{t \in \frac{1}{2}\mathbf{N}}$. Again, it is a routine matter to show that this creates a discrete Wiener process $(X_t)_{t \in \frac{1}{4}\mathbf{N}}$ on $\frac{1}{4}\mathbf{N}$.

Iterating this procedure a countable number[1] of times, we obtain a collection of discrete Wiener processes $(X_t)_{t \in \frac{1}{2^k}\mathbf{N}}$ for $k = 0, 1, 2, \ldots$ which are consistent with each other, in the sense that the earlier processes in this collection are restrictions of later ones.

Now we establish a Hölder continuity property. Let θ be any exponent between 0 and 1/2, and let $T > 0$ be finite. Observe that for any $k = 0, 1, \ldots$ and any $j \in \mathbf{N}$, we have $X_{(j+1)/2^k} - X_{j/2^k} \equiv N(0, 1/2^k)_{\mathbf{R}}$ and hence (by the sub-Gaussian nature of the normal distribution)

$$\mathbf{P}(|X_{(j+1)/2^k} - X_{j/2^k}| \geq 2^{-k\theta}) \leq C \exp(-c2^{k(1-2\theta)})$$

for some absolute constants C, c. The right-hand side is summable as j, k run over $\mathbf{N}$ subject to the constraint $j/2^k \leq T$. Thus, by the Borel-Cantelli lemma, for each fixed T, we almost surely have that

$$|X_{(j+1)/2^k} - X_{j/2^k}| \leq 2^{-k\theta}$$

for all but finitely many $j, k \in \mathbf{N}$ with $j/2^k \leq T$. In particular, this implies that for each fixed T, the function $t \mapsto X_t$ is almost surely *Hölder continuous*[2] of exponent θ on the dyadic rationals $j/2^k$ in $[0, T]$, and thus (by the countable union bound) is almost surely locally Hölder continuous of exponent θ on the dyadic rationals in $[0, +\infty)$. In particular, they are almost surely locally uniformly continuous on this domain.

As the dyadic rationals are dense in $[0, +\infty)$, we can thus almost surely[3] extend $t \mapsto X_t$ uniquely to a continuous function on all of $[0, +\infty)$. Note that if t_n is any sequence in $[0, +\infty)$ converging to t, then X_{t_n} converges almost surely to X_t, and thus also converges in probability and in distribution.

---

[1] This requires a countable number of extensions of the underlying sample space, but one can capture all of these extensions into a single extension via the machinery of *inverse limits* of probability spaces; it is also not difficult to manually build a single extension sufficient for performing all the above constructions.

[2] In other words, there exists a constant C_T such that $|X_s - X_t| \leq C_T |s - t|^\theta$ for all $s, t \in [0, T]$.

[3] On the remaining probability zero event, we extend $t \mapsto X_t$ in some arbitrary measurable fashion.

Similarly, for differences such as $X_{t_{+},n} - X_{t_{-},n}$. Using this, we easily verify that $(X_t)_{t\in[0,+\infty)}$ is a continuous Wiener process, as required.   □

Remark 3.1.7. One could also have used the *Kolmogorov extension theorem* (see e.g. [**Ta2011**]) to establish the limit.

Exercise 3.1.1. Let $(X_t)_{t\in[0,+\infty)}$ be a continuous Wiener process. We have already seen that if $0 < \theta < 1/2$, that the map $t \mapsto X_t$ is almost surely Hölder continuous of order θ. Show that if $1/2 \le \theta \le 1$, then the map $t \mapsto X_t$ is almost surely *not* Hölder continuous of order θ.

Show also that the map $t \mapsto X_t$ is almost surely nowhere differentiable. Thus, Brownian motion provides a (probabilistic) example of a continuous function which is nowhere differentiable.

Remark 3.1.8. In the above constructions, the initial position μ of the Wiener process was deterministic. However, one can easily construct Wiener processes in which the initial position X_0 is itself a random variable. Indeed, one can simply set

$$X_t := X_0 + B_t$$

where $(B_t)_{t\in[0,+\infty)}$ is a continuous Wiener process with initial position 0 which is independent of X_0. Then we see that X_t obeys properties (ii), (iii), (iv) of Definition 3.1.1, but the distribution of X_t is no longer $N(\mu, t)_{\mathbf{R}}$, but is instead the convolution of the law of X_0, and the law of $N(0, t)_{\mathbf{R}}$.

3.1.2. Connection with random walks. We saw how to construct Brownian motion as a limit of discrete Wiener processes, which were partial sums of independent Gaussian random variables. The central limit theorem (see Section 2.2) allows one to interpret Brownian motion in terms of limits of partial sums of more general independent random variables, otherwise known as (independent) *random walks*.

Definition 3.1.9 (Random walk). Let ΔX be a real random variable, let $\mu \in \mathbf{R}$ be an initial position, and let $\Delta t > 0$ be a time step. We define a *discrete random walk* with initial position μ, time step Δt and step distribution ΔX (or $\mu_{\Delta X}$) to be a process $(X_t)_{t\in\Delta t\cdot\mathbf{N}}$ defined by

$$X_{n\Delta t} := \mu + \sum_{i=1}^{n} \Delta X_{i\Delta t}$$

where $(\Delta X_{i\Delta t})_{i=1}^{\infty}$ are iid copies of ΔX.

Example 3.1.10. From the proof of Proposition 3.1.5, we see that a discrete Wiener process on $\Delta t \cdot \mathbf{N}$ with initial position μ is nothing more than a discrete random walk with step distribution of $N(0, \Delta t)_{\mathbf{R}}$. Another basic example is *simple random walk*, in which ΔX is equal to $(\Delta t)^{1/2}$ times a signed Bernoulli variable, thus we have $X_{(n+1)\Delta t} = X_{n\Delta t} \pm (\Delta t)^{1/2}$, where the signs $\pm$ are unbiased and are jointly independent in n.

Exercise 3.1.2 (Central limit theorem). Let X be a real random variable with mean zero and variance 1, and let $\mu \in \mathbf{R}$. For each $\Delta t > 0$, let $(X_t^{(\Delta t)})_{t \in [0,+\infty)}$ be a process formed by starting with a random walk $(X_t^{(\Delta t)})_{t \in \Delta t \cdot \mathbf{N}}$ with initial position μ, time step Δt, and step distribution $(\Delta t)^{1/2} X$, and then extending to other times in $[0, +\infty)$, in a piecewise linear fashion, thus

$$X_{(n+\theta)\Delta t}^{(\Delta t)} := (1-\theta) X_{n\Delta t}^{(\Delta t)} + \theta X_{(n+1)\Delta t}^{(\Delta t)}$$

for all $n \in \mathbf{N}$ and $0 < \theta < 1$. Show that as $\Delta t \to 0$, the process $(X_t^{(\Delta t)})_{t \in [0,+\infty)}$ converges in distribution to a continuous Wiener process with initial position μ. (*Hint:* From the Riesz representation theorem (or the Kolmogorov extension theorem), it suffices to establish this convergence for every finite set of times in $[0, +\infty)$. Now use the central limit theorem; treating the piecewise linear modifications to the process as an error term.)

3.1.3. Connection with the heat equation. Let $(B_t)_{t \in [0,+\infty)}$ be a Wiener process with base point μ, and let $F : \mathbf{R} \to \mathbf{R}$ be a smooth function with all derivatives bounded. Then, for each time t, the random variable $F(B_t)$ is bounded and thus has an expectation $\mathbf{E}F(B_t)$. From the almost sure continuity of B_t and the dominated convergence theorem we see that the map $t \mapsto \mathbf{E}F(B_t)$ is continuous. In fact, it is differentiable, and obeys the following differential equation:

Lemma 3.1.11 (Equation of motion). *For all times $t \geq 0$, we have*

$$\frac{d}{dt} \mathbf{E}F(B_t) = \frac{1}{2} \mathbf{E}F_{xx}(B_t)$$

where F_{xx} is the second derivative of F. In particular, $t \mapsto \mathbf{E}F(B_t)$ is continuously differentiable (because the right-hand side is continuous).

Proof. We work from first principles. It suffices to show for fixed $t \geq 0$, that

$$\mathbf{E}F(B_{t+dt}) = \mathbf{E}F(B_t) + \frac{1}{2} dt \mathbf{E}F_{xx}(B_t) + o(dt)$$

as $dt \to 0$. We shall establish this just for non-negative dt; the claim for negative dt (which only needs to be considered for $t > 0$) is similar and is left as an exercise.

Write $dB_t := B_{t+dt} - B_t$. From Taylor expansion and the bounded third derivative of F, we have

$$(3.1) \qquad F(B_{t+dt}) = F(B_t) + F_x(B_t)dB_t + \frac{1}{2}F_{xx}(B_t)|dB_t|^2 + O(|dB_t|^3).$$

We take expectations. Since $dB_t \equiv N(0, dt)_{\mathbf{R}}$, we have $\mathbf{E}|dB_t|^3 = O((dt)^{3/2})$, so in particular,

$$\mathbf{E}F(B_{t+dt}) = \mathbf{E}F(B_t) + \mathbf{E}F_x(B_t)dB_t + \frac{1}{2}\mathbf{E}F_{xx}(B_t)|dB_t|^2 + o(dt).$$

Now observe that dB_t is independent of B_t, and has mean zero and variance dt. The claim follows. $\square$

Exercise 3.1.3. Complete the proof of the lemma by considering negative values of dt. (*Hint:* One has to exercise caution because dB_t is not independent of B_t in this case. However, it will be independent of B_{t+dt}. Also, use the fact that $\mathbf{E}F_x(B_t)$ and $\mathbf{E}F_{xx}(B_t)$ are continuous in t. Alternatively, one can deduce the formula for the left-derivative from that of the right-derivative via a careful application of the fundamental theorem of calculus, paying close attention to the hypotheses of that theorem.)

Remark 3.1.12. In the language of *Ito calculus*, we can write (3.1) as

$$(3.2) \qquad\qquad dF(B_t) = F_x(B_t)dB_t + \frac{1}{2}F_{xx}(B_t)dt.$$

Here, $dF(B_t) := F(B_{t+dt}) - F(B_t)$, and dt should either be thought of as being infinitesimal, or being very small, though in the latter case the equation (3.2) should not be viewed as being exact, but instead only being true up to errors of mean $o(dt)$ and third moment $O(dt^3)$. This is a special case of *Ito's formula*. It should be compared against the chain rule

$$dF(X_t) = F_x(X_t)dX_t$$

when $t \mapsto X_t$ is a *smooth* process. The non-smooth nature of Brownian motion causes the quadratic term in the Taylor expansion to be non-negligible, which explains[4] the additional term in (3.2), although the Hölder continuity of this motion is sufficient to still be able to ignore terms that are of cubic order or higher.

Let $\rho(t, x)\, dx$ be the probability density function of B_t; by inspection of the normal distribution, this is a smooth function for $t > 0$, but is a Dirac mass at μ at time $t = 0$. By definition of density function,

$$\mathbf{E}F(B_t) = \int_{\mathbf{R}} F(x)\rho(t, x)\, dx$$

for any *Schwartz function* F. Applying Lemma 3.1.11 and integrating by parts, we see that

$$(3.3) \qquad\qquad \partial_t \rho = \frac{1}{2}\partial_{xx}\rho$$

---

[4] In this spirit, one can summarise (the differential side of) Ito calculus informally by the heuristic equations $dB_t = O((dt)^{1/2})$ and $|dB_t|^2 = dt$, with the understanding that all terms that are $o(dt)$ are discarded.

in the sense of (tempered) distributions (see e.g. [**Ta2010**, §1.13]). In other words, ρ is a (tempered distributional) solution to the *heat equation* (3.3). Indeed, since ρ is the Dirac mass at μ at time $t = 0$, ρ for later times t is the *fundamental solution* of that equation from initial position μ.

From the theory of PDE one can solve[5] the (distributional) heat equation with this initial data to obtain the unique solution

$$\rho(t, x) = \frac{1}{\sqrt{2\pi t}} e^{-|x-\mu|^2/2t}.$$

Of course, this is also the density function of $N(\mu, t)_{\mathbf{R}}$, which is (unsurprisingly) consistent with the fact that $B_t \equiv N(\mu, t)$. Thus we see why the normal distribution of the central limit theorem involves the same type of functions (i.e. Gaussians) as the fundamental solution of the heat equation. Indeed, one can use this argument to heuristically *derive* the central limit theorem from the fundamental solution of the heat equation (cf. Section 2.2.7), although the derivation is only heuristic because one first needs to know that *some* limiting distribution already exists (in the spirit of Exercise 3.1.2).

Remark 3.1.13. Because we considered a Wiener process with a deterministic initial position μ, the density function ρ was a Dirac mass at time $t = 0$. However, one can run exactly the same arguments for Wiener processes with stochastic initial position (see Remark 3.1.8), and one will still obtain the same heat equation (3.1.8), but now with a more general initial condition.

We have related one-dimensional Brownian motion to the one-dimensional heat equation, but there is no difficulty establishing a similar relationship in higher dimensions. In a vector space $\mathbf{R}^n$, define a (continuous) *Wiener process* $(X_t)_{t \in [0, +\infty)}$ in $\mathbf{R}^n$ with an initial position $\mu = (\mu_1, \ldots, \mu_n) \in \mathbf{R}^n$ to be a process whose components $(X_{t,i})_{t \in [0, +\infty)}$ for $i = 1, \ldots, n$ are independent Wiener processes with initial position μ_i. It is easy to see that such processes exist, are unique in distribution, and obey the same sort of properties as in Definition 3.1.1, but with the one-dimensional Gaussian distribution $N(\mu, \sigma^2)_{\mathbf{R}}$ replaced by the n-dimensional analogue $N(\mu, \sigma^2 I)_{\mathbf{R}^n}$, which is given by the density function

$$\frac{1}{(2\pi\sigma)^{n/2}} e^{-|x-\mu|^2/\sigma^2} \, dx$$

where dx is now Lebesgue measure on $\mathbf{R}^n$.

Exercise 3.1.4. If $(B_t)_{t \in [0, +\infty)}$ is an n-dimensional continuous Wiener process, show that

$$\frac{d}{dt} \mathbf{E} F(B_t) = \frac{1}{2} \mathbf{E}(\Delta F)(B_t)$$

---

[5]See for instance [**Ta2010**, §1.12].

whenever $F : \mathbf{R}^n \to \mathbf{R}$ is smooth with all derivatives bounded, where

$$\Delta F := \sum_{i=1}^{n} \frac{\partial^2}{\partial x_i^2} F$$

is the Laplacian of F. Conclude, in particular, that the density function $\rho(t, x) \, dx$ of B_t obeys the (distributional) heat equation

$$\partial_t \rho = \frac{1}{2} \Delta \rho.$$

A simple but fundamental observation is that n-dimensional Brownian motion is rotation-invariant; more precisely, if $(X_t)_{t \in [0,+\infty)}$ is an n-dimensional Wiener process with initial position 0, and $U \in O(n)$ is any orthogonal transformation on $\mathbf{R}^n$, then $(UX_t)_{t \in [0,+\infty)}$ is another Wiener process with initial position 0, and thus has the same distribution:

$$(3.4) \qquad\qquad (UX_t)_{t \in [0,+\infty)} \equiv (X_t)_{t \in [0,+\infty)}.$$

This is ultimately because the n-dimensional normal distributions $N(0, \sigma^2 I)_{\mathbf{R}^n}$ are manifestly rotation-invariant (see Exercise 2.2.13).

Remark 3.1.14. One can also relate variable-coefficient heat equations to variable-coefficient Brownian motion $(X_t)_{t \in [0,+\infty)}$, in which the variance of an increment dX_t is now only proportional to dt for infinitesimal dt rather than being equal to dt, with the constant of proportionality allowed to depend on the time t and on the position X_t. One can also add drift terms by allowing the increment dX_t to have a non-zero mean (which is also proportional to dt). This can be accomplished through the machinery of *stochastic calculus*, which we will not discuss in detail in this text. In a similar fashion, one can construct Brownian motion (and heat equations) on manifolds or on domains with boundary, though we will not discuss this topic here.

Exercise 3.1.5. Let X_0 be a real random variable of mean zero and variance 1. Define a stochastic process $(X_t)_{t \in [0,+\infty)}$ by the formula

$$X_t := e^{-t}(X_0 + B_{e^{2t}-1})$$

where $(B_t)_{t \in [0,+\infty)}$ is a Wiener process with initial position zero that is independent of X_0. This process is known as an *Ornstein-Uhlenbeck process*.

- Show that each X_t has mean zero and variance 1.
- Show that X_t converges in distribution to $N(0, 1)_{\mathbf{R}}$ as $t \to \infty$.
- If $F : \mathbf{R} \to \mathbf{R}$ is smooth with all derivatives bounded, show that

$$\frac{d}{dt} \mathbf{E} F(X_t) = \mathbf{E} L F(X_t)$$

  where L is the *Ornstein-Uhlenbeck operator*

$$LF := F_{xx} - x F_x.$$

Conclude that the density function $\rho(t, x)$ of X_t obeys (in a distributional sense, at least) the *Ornstein-Uhlenbeck equation*

$$\partial_t \rho = L^* \rho$$

where the adjoint operator L^* is given by

$$L^* \rho := \rho_{xx} + \partial_x(x\rho).$$

- Show that the only probability density function ρ for which $L^*\rho = 0$ is the Gaussian $\frac{1}{\sqrt{2\pi}} e^{-x^2/2} \, dx$; further, show that $\mathrm{Re}\langle \rho, L^*\rho\rangle_{L^2(\mathbf{R})} \leq 0$ for all probability density functions ρ in the Schwartz space with mean zero and variance 1. Discuss how this fact relates to the preceding two parts of this exercise.

Remark 3.1.15. The heat kernel $\frac{1}{(\sqrt{2\pi t})^d} e^{-|x-\mu|^2/2t}$ in d dimensions is absolutely integrable in time away from the initial time $t = 0$ for dimensions $d \geq 3$, but becomes divergent in dimension 1 and (just barely) divergent for $d = 2$. This causes the qualitative behaviour of Brownian motion B_t in $\mathbf{R}^d$ to be rather different in the two regimes. For instance, in dimensions $d \geq 3$ Brownian motion is *transient*; almost surely one has $B_t \to \infty$ as $t \to \infty$. But in dimension $d = 1$ Brownian motion is *recurrent*: for each $x_0 \in \mathbf{R}$, one almost surely has $B_t = x_0$ for infinitely many t. In the critical dimension $d = 2$, Brownian motion turns out to not be recurrent, but is instead *neighbourhood recurrent*; almost surely, B_t revisits every neighbourhood of x_0 at arbitrarily large times, but does not visit x_0 itself for any positive time t. The study of Brownian motion and its relatives is in fact a huge and active area of study in modern probability theory, but will not be discussed in this text.

3.1.4. Dyson Brownian motion. The space V of $n \times n$ Hermitian matrices can be viewed as a real vector space of dimension n^2 using the *Frobenius norm*

$$A \mapsto \mathrm{tr}(A^2)^{1/2} = \left(\sum_{i=1}^n a_{ii}^2 + 2 \sum_{1 \leq i < j \leq n} \mathrm{Re}(a_{ij})^2 + \mathrm{Im}(a_{ij})^2 \right)^{1/2}$$

where a_{ij} are the coefficients of A. One can then identify V explicitly with $\mathbf{R}^{n^2}$ via the identification

$$(a_{ij})_{1 \leq i,j \leq n} \equiv ((a_{ii})_{i=1}^n, (\sqrt{2}\,\mathrm{Re}(a_{ij}), \sqrt{2}\,\mathrm{Im}(a_{ij}))_{1 \leq i < j \leq n}).$$

Now that one has this identification, for each Hermitian matrix $A_0 \in V$ (deterministic or stochastic) we can define a Wiener process $(A_t)_{t \in [0,+\infty)}$ on V with initial position A_0. By construction, we see that $t \mapsto A_t$ is almost surely continuous, and each increment $A_{t_+} - A_{t_-}$ is equal to $(t_+ - t_-)^{1/2}$ times a matrix drawn from the *Gaussian Unitary Ensemble* (GUE), with disjoint increments being jointly independent. In particular, the diagonal entries of $A_{t_+} - A_{t_-}$ have distribution $N(0, t_+ - t_-)_{\mathbf{R}}$, and the off-diagonal entries have distribution $N(0, t_+ - t_-)_{\mathbf{C}}$.

Given any Hermitian matrix A, one can form the *spectrum* $(\lambda_1(A), \ldots, \lambda_n(A))$, which lies in the *Weyl chamber* $\mathbf{R}_{\geq}^n := \{(\lambda_1, \ldots, \lambda_n) \in \mathbf{R}^n : \lambda_1 \geq \cdots \geq \lambda_n\}$. Taking the spectrum of the Wiener process $(A_t)_{t \in [0, +\infty)}$, we obtain a process

$$(\lambda_1(A_t), \ldots, \lambda_n(A_t))_{t \in [0, +\infty)}$$

in the Weyl chamber. We abbreviate $\lambda_i(A_t)$ as λ_i.

For $t > 0$, we see that A_t is absolutely continuously distributed in V. In particular, since almost every Hermitian matrix has simple spectrum, we see that A_t has almost surely simple spectrum for $t > 0$. (The same is true for $t = 0$ if we assume that A_0 also has an absolutely continuous distribution.)

The stochastic dynamics of this evolution can be described by Dyson Brownian motion [**Dy1962**]:

Theorem 3.1.16 (Dyson Brownian motion). *Let $t > 0$, and let $dt > 0$, and let $\lambda_1, \ldots, \lambda_n$ be as above. Then we have*

$$(3.5) \qquad d\lambda_i = dB_i + \sum_{1 \leq j \leq n: j \neq i} \frac{dt}{\lambda_i - \lambda_j} + \ldots$$

for all $1 \leq i \leq n$, where $d\lambda_i := \lambda_i(A_{t+dt}) - \lambda_i(A_t)$, and $dB_1, \ldots, dB_n$ are iid copies of $N(0, dt)_{\mathbf{R}}$ which are jointly independent of $(A_{t'})_{t' \in [0,t]}$, and the error term $\ldots$ has mean $o(dt)$ and third moment $O(dt^3)$ in the limit $dt \to 0$ (holding t and n fixed).

Using the language of Ito calculus, one usually views dt as infinitesimal and drops the $\ldots$ error, thus giving the elegant formula

$$d\lambda_i = dB_i + \sum_{1 \leq j \leq n: j \neq i} \frac{dt}{\lambda_i - \lambda_j}$$

that shows that the eigenvalues λ_i evolve by Brownian motion, combined with a deterministic *repulsion* force that repels nearby eigenvalues from each other with a strength inversely proportional to the separation. One can extend the theorem to the $t = 0$ case by a limiting argument provided that A_0 has an absolutely continuous distribution. Note that the decay rate of the error $\ldots$ can depend on n, so it is not safe to let n go off to infinity while holding dt fixed. However, it is safe to let dt go to zero *first*, and *then* send n off to infinity[6].

---

[6]It is also possible, by being more explicit with the error terms, to work with dt being a specific negative power of n; see [**TaVu2009b**].

Proof. Fix t. We can write $A_{t+dt} = A_t + (dt)^{1/2}G$, where G is independent[7] of A_t and has the GUE distribution. We now condition A_t to be fixed, and establish (3.5) for almost every fixed choice of A_t; the general claim then follows upon undoing the conditioning (and applying the dominated convergence theorem). Due to independence, observe that G continues to have the GUE distribution even after conditioning A_t to be fixed.

Almost surely, A_t has simple spectrum; so we may assume that the fixed choice of A_t has simple spectrum also. The eigenvalues λ_i now vary smoothly near t, so we may Taylor expand

$$\lambda_i(A_{t+dt}) = \lambda_i + (dt)^{1/2}\nabla_G\lambda_i + \frac{1}{2}dt\nabla_G^2\lambda_i + O((dt)^{3/2}\|G\|^3)$$

for sufficiently small dt, where ∇_G is directional differentiation in the G direction, and the implied constants in the $O()$ notation can depend on A_t and n. In particular, we do not care what norm is used to measure G in.

As G has the GUE distribution, the expectation and variance of $\|G\|^3$ is bounded (possibly with constant depending on n), so the error here has mean $o(dt)$ and third moment $O(dt^3)$. We thus have

$$d\lambda_i = (dt)^{1/2}\nabla_G\lambda_i + \frac{1}{2}dt\nabla_G^2\lambda_i + \dots.$$

Next, from the first and second Hadamard variation formulae (1.73), (1.74) we have

$$\nabla_G\lambda_i = u_i^*Gu_i$$

and

$$\nabla_G^2\lambda_i = 2\sum_{i\neq j}\frac{|u_j^*Gu_i|^2}{\lambda_i - \lambda_j}$$

where $u_1, \dots, u_n$ are an orthonormal eigenbasis for A_t, and thus

$$d\lambda_i = (dt)^{1/2}u_i^*Gu_i + dt\sum_{i\neq j}\frac{|u_j^*Gu_i|^2}{\lambda_i - \lambda_j} + \dots.$$

Now we take advantage of the unitary invariance of the Gaussian unitary ensemble (that is, that $UGU^* \equiv G$ for all unitary matrices G; this is easiest to see by noting that the probability density function of G is proportional to $\exp(-\|G\|_F^2/2)$). From this invariance, we can assume without loss of generality that $u_1, \dots, u_n$ is the standard orthonormal basis of $\mathbf{C}^n$, so that we now have

$$d\lambda_i = (dt)^{1/2}\xi_{ii} + dt\sum_{i\neq j}\frac{|\xi_{ij}|^2}{\lambda_i - \lambda_j} + \dots$$

---

[7]Strictly speaking, G depends on dt, but this dependence will not concern us.

where ξ_{ij} are the coefficients of G. But the ξ_{ii} are iid copies of $N(0,1)_{\mathbf{R}}$, and the ξ_{ij} are iid copies of $N(0,1)_{\mathbf{C}}$, and the claim follows (note that $dt \sum_{i \neq j} \frac{|\xi_{ij}|^2 - 1}{\lambda_i - \lambda_j}$ has mean zero and third moment $O(dt^3)$.) $\qquad\square$

Remark 3.1.17. Interestingly, one can interpret Dyson Brownian motion in a different way, namely as the motion of n independent Wiener processes $\lambda_i(t)$ *after* one conditions the λ_i to be non-intersecting for all time; see [**Gr1999**]. It is intuitively reasonable that this conditioning would cause a repulsion effect, though we do not know of a simple heuristic reason why this conditioning should end up giving the specific repulsion force present in (3.5).

In the previous section, we saw how a Wiener process led to a PDE (the heat flow equation) that could be used to derive the probability density function for each component X_t of that process. We can do the same thing here:

Exercise 3.1.6. Let $\lambda_1, \ldots, \lambda_n$ be as above. Let $F : \mathbf{R}^n \to \mathbf{R}$ be a smooth function with bounded derivatives. Show that for any $t \geq 0$, one has

$$\partial_t \mathbf{E} F(\lambda_1, \ldots, \lambda_n) = \mathbf{E} D^* F(\lambda_1, \ldots, \lambda_n)$$

where D^* is the *adjoint Dyson operator*

$$D^* F := \frac{1}{2} \sum_{i=1}^n \partial_{\lambda_i}^2 F + \sum_{1 \leq i,j \leq n: i \neq j} \frac{\partial_{\lambda_i} F}{\lambda_i - \lambda_j}.$$

If we let $\rho : [0, +\infty) \times \mathbf{R}^n_{\geq} \to \mathbf{R}$ denote the density function $\rho(t, \cdot) : \mathbf{R}^n_{\geq} \to \mathbf{R}$ of $(\lambda_1(t), \ldots, \lambda_n(t))$ at time $t \in [0, +\infty)$, deduce the *Dyson partial differential equation*

$$(3.6) \qquad\qquad\qquad \partial_t \rho = D\rho$$

(in the sense of distributions, at least, and on the interior of $\mathbf{R}^n_{\geq}$), where D is the *Dyson operator*

$$(3.7) \qquad D\rho := \frac{1}{2} \sum_{i=1}^n \partial_{\lambda_i}^2 \rho - \sum_{1 \leq i,j \leq n: i \neq j} \partial_{\lambda_i} \left(\frac{\rho}{\lambda_i - \lambda_j} \right).$$

The Dyson partial differential equation (3.6) looks a bit complicated, but it can be simplified (formally, at least) by introducing the *Vandermonde determinant*

$$(3.8) \qquad\qquad \Delta_n(\lambda_1, \ldots, \lambda_n) := \prod_{1 \leq i < j \leq n} (\lambda_i - \lambda_j).$$

Exercise 3.1.7. Show that (3.8) is the determinant of the matrix $(\lambda_i^{j-1})_{1 \leq i,j \leq n}$, and is also the sum $\sum_{\sigma \in S_n} \mathrm{sgn}(\sigma) \prod_{i=1}^n \lambda_{\sigma(i)}^{i-1}$.

Note that this determinant is non-zero on the interior of the Weyl chamber $\mathbf{R}^n_\geq$. The significance of this determinant for us lies in the identity

$$(3.9) \qquad \partial_{\lambda_i} \Delta_n = \sum_{1 \leq j \leq n : i \neq j} \frac{\Delta_n}{\lambda_i - \lambda_j}$$

which can be used to cancel off the second term in (3.7). Indeed, we have

Exercise 3.1.8. Let ρ be a smooth solution to (3.6) in the interior of $\mathbf{R}^n_\geq$, and write

$$(3.10) \qquad \rho = \Delta_n u$$

in this interior. Show that u obeys the linear heat equation

$$\partial_t u = \frac{1}{2} \sum_{i=1}^{n} \partial^2_{\lambda_i} u$$

in the interior of $\mathbf{R}^n_\geq$. (*Hint:* You may need to exploit the identity $\frac{1}{(a-b)(a-c)} + \frac{1}{(b-a)(b-c)} + \frac{1}{(c-a)(c-b)} = 0$ for distinct a, b, c. Equivalently, you may need to first establish that the Vandermonde determinant is a harmonic function.)

Let ρ be the density function of the $(\lambda_1, \ldots, \lambda_n)$, as in (3.1.6). Recall that the Wiener random matrix A_t has a smooth distribution in the space V of Hermitian matrices, while the space of matrices in V with non-simple spectrum has codimension 3 by Exercise 1.3.10. On the other hand, the non-simple spectrum only has codimension 1 in the Weyl chamber (being the boundary of this cone). Because of this, we see that ρ vanishes to at least second order on the boundary of this cone (with correspondingly higher vanishing on higher codimension facets of this boundary). Thus, the function u in Exercise 3.1.8 vanishes to first order on this boundary (again with correspondingly higher vanishing on higher codimension facets). Thus, if we extend ρ symmetrically across the cone to all of $\mathbf{R}^n$, and extend the function u antisymmetrically, then the equation (3.6) and the factorisation (3.10) extend (in the distributional sense) to all of $\mathbf{R}^n$. Extending (3.1.8) to this domain (and being somewhat careful with various issues involving distributions), we now see that u obeys the linear heat equation on all of $\mathbf{R}^n$.

Now suppose that the initial matrix A_0 had a deterministic spectrum $\nu = (\nu_1, \ldots, \nu_n)$, which to avoid technicalities we will assume to be in the interior of the Weyl chamber (the boundary case then being obtainable by a limiting argument). Then ρ is initially the Dirac delta function at ν, extended symmetrically. Hence, u is initially $\frac{1}{\Delta_n(\nu)}$ times the Dirac delta function at ν, extended antisymmetrically:

$$u(0, \lambda) = \frac{1}{\Delta_n(\nu)} \sum_{\sigma \in S_n} \text{sgn}(\sigma) \delta_{\lambda - \sigma(\nu)}.$$

Using the fundamental solution for the heat equation in n dimensions, we conclude that

$$u(t, \lambda) = \frac{1}{(2\pi t)^{n/2}} \sum_{\sigma \in S_n} \text{sgn}(\sigma) e^{-|\lambda - \sigma(\nu)|^2 / 2t}.$$

By the Leibniz formula for determinants

$$\det((a_{ij})_{1 \le i,j \le n}) = \sum_{\sigma \in S_n} \text{sgn}(\sigma) \prod_{i=1}^{n} a_{i\sigma(i)},$$

we can express the sum here as a determinant of the matrix

$$(e^{-(\lambda_i - \nu_j)^2 / 2t})_{1 \le i,j \le n}.$$

Applying (3.10), we conclude

Theorem 3.1.18 (Johansson formula). *Let A_0 be a Hermitian matrix with simple spectrum $\nu = (\nu_1, \ldots, \nu_n)$, let $t > 0$, and let $A_t = A_0 + t^{1/2}G$ where G is drawn from GUE. Then the spectrum $\lambda = (\lambda_1, \ldots, \lambda_n)$ of A_t has probability density function*

$$(3.11) \qquad \rho(t, \lambda) = \frac{1}{(2\pi t)^{n/2}} \frac{\Delta_n(\lambda)}{\Delta_n(\nu)} \det(e^{-(\lambda_i - \nu_j)^2 / 2t})_{1 \le i,j \le n}$$

on $\mathbf{R}^n_{\ge}$.

This formula is given explicitly in [**Jo2001**], who cites [**BrHi1996**] as inspiration. (One can also check by hand that (3.11) satisfies the Dyson equation (3.6).)

We will be particularly interested in the case when $A_0 = 0$ and $t = 1$, so that we are studying the probability density function of the eigenvalues $(\lambda_1(G), \ldots, \lambda_n(G))$ of a GUE matrix G. The Johansson formula does not directly apply here, because ν is vanishing. However, we can investigate the limit of (3.11) in the limit as $\nu \to 0$ inside the Weyl chamber; the Lipschitz nature of the eigenvalue operations $A \mapsto \lambda_i(A)$ (from the Weyl inequalities) tell us that if (3.11) converges locally uniformly as $\nu \to 0$ for λ in the interior of $\mathbf{R}^n_{\ge}$, then the limit will indeed[8] be the probability density function for $\nu = 0$.

---

[8]Note from continuity that the density function cannot assign any mass to the boundary of the Weyl chamber, and in fact must vanish to at least second order by the previous discussion.

Exercise 3.1.9. Show that as $\nu \to 0$, we have the identities

$$\det(e^{-(\lambda_i - \nu_j)^2/2})_{1 \le i,j \le n} = e^{-|\lambda|^2/2} e^{-|\nu|^2/2} \det(e^{\lambda_i \nu_j})_{1 \le i,j \le n}$$

and

$$\det(e^{\lambda_i \nu_j})_{1 \le i,j \le n} = \frac{1}{1! \ldots n!} \Delta_n(\lambda) \Delta_n(\nu) + o(\Delta_n(\nu))$$

locally uniformly in λ. (*Hint:* For the second identity, use Taylor expansion and the Leibniz formula for determinants, noting the left-hand side vanishes whenever $\Delta_n(\nu)$ vanishes and so can be treated by the (smooth) factor theorem.)

From the above exercise, we conclude the fundamental *Ginibre formula* [**Gi1965**]

(3.12) $$\rho(\lambda) = \frac{1}{(2\pi)^{n/2} 1! \ldots n!} e^{-|\lambda|^2/2} |\Delta_n(\lambda)|^2$$

for the density function for the spectrum $(\lambda_1(G), \ldots, \lambda_n(G))$ of a GUE matrix G.

This formula can be derived by a variety of other means; we sketch one such way below.

Exercise 3.1.10. For this exercise, assume that it is known that (3.12) is indeed a probability distribution on the Weyl chamber $\mathbf{R}^n_{\ge}$ (if not, one would have to replace the constant $(2\pi)^{n/2}$ by an unspecified normalisation factor depending only on n). Let $D = \operatorname{diag}(\lambda_1, \ldots, \lambda_n)$ be drawn at random using the distribution (3.12), and let U be drawn at random from Haar measure on $U(n)$. Show that the probability density function of UDU^* at a matrix A with simple spectrum is equal to $c_n e^{-\|A\|_F^2/2}$ for some constant $c_n > 0$. (*Hint*: Use unitary invariance to reduce to the case when A is diagonal. Now take a small ε and consider what U and D must be in order for UDU^* to lie within ε of A in the Frobenius norm, performing first order calculations only (i.e. linearising and ignoring all terms of order $o(\varepsilon)$).)

Conclude that (3.12) must be the probability density function of the spectrum of a GUE matrix.

Exercise 3.1.11. Verify by hand that the self-similar extension

$$\rho(t, x) := t^{-n^2/2} \rho(x/\sqrt{t})$$

of the function (3.12) obeys the Dyson PDE (3.6). Why is this consistent with (3.12) being the density function for the spectrum of GUE?

Remark 3.1.19. Similar explicit formulae exist for other invariant ensembles, such as the *Gaussian Orthogonal Ensemble* GOE and the *Gaussian Symplectic Ensemble* GSE. One can also replace the exponent in density functions such as $e^{-\|A\|_F^2/2}$ with more general expressions than quadratic expressions of A. We will, however, not detail these formulae in this text (with the exception of the spectral distribution law for random iid Gaussian matrices, which we discuss in Section 2.6).

3.2. The Golden-Thompson inequality

Let A, B be two Hermitian $n \times n$ matrices. When A and B commute, we have the identity

$$e^{A+B} = e^A e^B.$$

When A and B do not commute, the situation is more complicated; we have the *Baker-Campbell-Hausdorff formula* (or more precisely, the closely related *Zassenhaus formula*)

$$e^{A+B} = e^A e^B e^{-\frac{1}{2}[A,B]} \ldots$$

where the infinite product here is explicit but very messy. On the other hand, taking determinants we still have the identity

$$\det(e^{A+B}) = \det(e^A e^B).$$

An identity in a somewhat similar spirit (which Percy Deift has half-jokingly termed "the most important identity in mathematics") is the *Sylvester determinant theorem*

$$(3.13) \qquad \det(1 + AB) = \det(1 + BA)$$

whenever A, B are $n \times k$ and $k \times n$ matrices respectively (or more generally, A and B could be linear operators with sufficiently good spectral properties that make both sides equal). Note that the left-hand side is an $n \times n$ determinant, while the right-hand side is a $k \times k$ determinant; this formula is particularly useful when computing determinants of large matrices (or of operators), as one can often use it to transform such determinants into much smaller determinants. In particular, the asymptotic behaviour of $n \times n$ determinants as $n \to \infty$ can be converted via this formula to determinants of a fixed size (independent of n), which is often a more favourable situation to analyse. Unsurprisingly, this trick is particularly useful for understanding the asymptotic behaviour of determinantal processes.

There are many ways to prove (3.13). One is to observe first that when A, B are invertible square matrices of the same size, that $1 + BA$ and $1 + AB$ are conjugate to each other and thus clearly have the same determinant; a density argument then removes the invertibility hypothesis, and a padding-by-zeroes argument then extends the square case to the rectangular case. Another is to proceed via the spectral theorem, noting that AB and BA have the same non-zero eigenvalues.

By rescaling, one obtains the variant identity

$$\det(z + AB) = z^{n-k} \det(z + BA),$$

which essentially relates the characteristic polynomial of AB with that of BA. When $n = k$, a comparison of coefficients this already gives important basic identities such as $\operatorname{tr}(AB) = \operatorname{tr}(BA)$ and $\det(AB) = \det(BA)$; when n is larger than k, an inspection of the z^{n-k} coefficient similarly gives the *Cauchy-Binet formula*

$$(3.14) \qquad \det(BA) = \sum_{S \in \binom{[n]}{k}} \det(A_{S \times [k]}) \det(B_{[k] \times S})$$

where S ranges over all k-element subsets of $[n] := \{1, \ldots, n\}$, $A_{S \times [k]}$ is the $k \times k$ minor of A coming from the rows S, and $B_{[k] \times S}$ is similarly the $k \times k$ minor coming from the columns S. Unsurprisingly, the Cauchy-Binet formula is also quite useful when performing computations on determinantal processes.

There is another very nice relationship between e^{A+B} and $e^A e^B$, namely the *Golden-Thompson inequality* [**Go1965, Th1965**]

$$(3.15) \qquad \operatorname{tr}(e^{A+B}) \le \operatorname{tr}(e^A e^B).$$

The remarkable thing about this inequality is that no commutativity hypotheses whatsoever on the matrices A, B are required. Note that the right-hand side can be rearranged using the cyclic property of trace as $\operatorname{tr}(e^{B/2} e^A e^{B/2})$; the expression inside the trace is positive definite so the right-hand side is positive[9].

To get a sense of how delicate the Golden-Thompson inequality is, let us expand both sides to fourth order in A, B. The left-hand side expands as

$$\operatorname{tr} 1 + \operatorname{tr}(A + B) + \frac{1}{2} \operatorname{tr}(A^2 + AB + BA + B^2) + \frac{1}{6} \operatorname{tr}(A + B)^3$$
$$+ \frac{1}{24} \operatorname{tr}(A + B)^4 + \ldots$$

while the right-hand side expands as

$$\operatorname{tr} 1 + \operatorname{tr}(A + B) + \frac{1}{2} \operatorname{tr}(A^2 + 2AB + B^2)$$
$$+ \frac{1}{6} \operatorname{tr}(A^3 + 3A^2 B + 3AB^2 + B^3)$$
$$+ \frac{1}{24} \operatorname{tr}(A^4 + 4A^3 B + 6A^2 B^2 + 4AB^3 + B^4) + \ldots.$$

---

[9]In contrast, the obvious extension of the Golden-Thompson inequality to three or more Hermitian matrices fails dramatically; there is no reason why expressions such as $\operatorname{tr}(e^A e^B e^C)$ need to be positive or even real.

Using the cyclic property of trace $\mathrm{tr}(AB) = \mathrm{tr}(BA)$, one can verify that all terms up to third order agree. Turning to the fourth order terms, one sees after expanding out $(A + B)^4$ and using the cyclic property of trace as much as possible, we see that the fourth order terms *almost* agree, but the left-hand side contains a term $\frac{1}{12}\,\mathrm{tr}(ABAB)$ whose counterpart on the right-hand side is $\frac{1}{12}\,\mathrm{tr}(ABBA)$. The difference between the two can be factorised (again using the cyclic property of trace) as $-\frac{1}{24}\,\mathrm{tr}[A, B]^2$. Since $[A, B] := AB - BA$ is skew-Hermitian, $-[A, B]^2$ is positive definite, and so we have proven the Golden-Thompson inequality to fourth order[10].

Intuitively, the Golden-Thompson inequality is asserting that interactions between a pair A, B of non-commuting Hermitian matrices are strongest when cross-interactions are kept to a minimum, so that all the A factors lie on one side of a product and all the B factors lie on the other. Indeed, this theme will be running through the proof of this inequality, to which we now turn.

The proof of the Golden-Thompson inequality relies on the somewhat magical power of the *tensor power trick* (see [**Ta2008**, §1.9]). For any even integer $p = 2, 4, 6, \ldots$ and any $n \times n$ matrix A (not necessarily Hermitian), we define the *p-Schatten norm* $\|A\|_p$ of A by the formula[11]

$$\|A\|_p := (\mathrm{tr}(AA^*)^{p/2})^{1/p}.$$

This norm can be viewed as a non-commutative analogue of the ℓ^p norm; indeed, the p-Schatten norm of a *diagonal* matrix is just the ℓ^p norm of the coefficients.

Note that the 2-Schatten norm

$$\|A\|_2 := (\mathrm{tr}(AA^*))^{1/2}$$

is the Hilbert space norm associated to the *Frobenius inner product* (or *Hilbert-Schmidt inner product*)

$$\langle A, B \rangle := \mathrm{tr}(AB^*).$$

This is clearly a non-negative Hermitian inner product, so by the Cauchy-Schwarz inequality we conclude that

$$|\,\mathrm{tr}(A_1 A_2^*)| \le \|A_1\|_2 \|A_2\|_2$$

for any $n \times n$ matrices A_1, A_2. As $\|A_2\|_2 = \|A_2^*\|_2$, we conclude, in particular, that

$$|\,\mathrm{tr}(A_1 A_2)| \le \|A_1\|_2 \|A_2\|_2.$$

---

[10]One could also have used the Cauchy-Schwarz inequality for the Frobenius norm to establish this; see below.

[11]This formula in fact defines a norm for any $p \ge 1$; see Exercise 1.3.22(vi). However, we will only need the even integer case here.

We can iterate this and establish the *non-commutative Hölder inequality*

$$(3.16) \qquad |\operatorname{tr}(A_1 A_2 \ldots A_p)| \le \|A_1\|_p \|A_2\|_p \ldots \|A_p\|_p$$

whenever $p = 2, 4, 8, \ldots$ is an even power of 2 (compare with Exercise 1.3.9). Indeed, we induct on p, the case $p = 2$ already having been established. If $p \ge 4$ is a power of 2, then by the induction hypothesis (grouping $A_1 \ldots A_p$ into $p/2$ pairs) we can bound

$$(3.17) \qquad |\operatorname{tr}(A_1 A_2 \ldots A_p)| \le \|A_1 A_2\|_{p/2} \|A_3 A_4\|_{p/2} \ldots \|A_{p-1} A_p\|_{p/2}.$$

On the other hand, we may expand

$$\|A_1 A_2\|_{p/2}^{p/2} = \operatorname{tr} A_1 A_2 A_2^* A_1^* \ldots A_1 A_2 A_2^* A_1^*.$$

We use the cyclic property of trace to move the rightmost A_1^* factor to the left. Applying the induction hypothesis again, we conclude that

$$\|A_1 A_2\|_{p/2}^{p/2} \le \|A_1^* A_1\|_{p/2} \|A_2 A_2^*\|_{p/2} \ldots \|A_1^* A_1\|_{p/2} \|A_2 A_2^*\|_{p/2}.$$

But from the cyclic property of trace again, we have $\|A_1^* A_1\|_{p/2} = \|A_1\|_p^2$ and $\|A_2 A_2^*\|_{p/2} = \|A_2\|_p^2$. We conclude that

$$\|A_1 A_2\|_{p/2} \le \|A_1\|_p \|A_2\|_p$$

and similarly for $\|A_3 A_4\|_{p/2}$, etc. Inserting this into (3.17) we obtain (3.16).

Remark 3.2.1. Though we will not need to do so here, it is interesting to note that one can use the tensor power trick to amplify (3.16) for p equal to a power of two, to obtain (3.16) for all positive integers p, at least when the A_i are all Hermitian (again, compare with Exercise 1.3.9). Indeed, pick a large integer m and let N be the integer part of $2^m/p$. Then expand the left-hand side of (3.16) as $\operatorname{tr}(A_1^{1/N} \ldots A_1^{1/N} A_2^{1/N} \ldots A_p^{1/N} \ldots A_p^{1/N})$ and apply (3.16) with p replaced by 2^m to bound this by $\|A_1^{1/N}\|_{2^m}^N \ldots \|A_p^{1/N}\|_{2^m}^N \|1\|_{2^m}^{2^m - pN}$. Sending $m \to \infty$ (noting that $2^m = (1 + o(1))Np$) we obtain the claim.

Specialising (3.16) to the case where $A_1 = \cdots = A_p = AB$ for some Hermitian matrices A, B, we conclude that

$$\operatorname{tr}((AB)^p) \le \|AB\|_p^p$$

and hence by cyclic permutation

$$\operatorname{tr}((AB)^p) \le \operatorname{tr}((A^2 B^2)^{p/2})$$

for any $p = 2, 4, \ldots$. Iterating this we conclude that

$$(3.18) \qquad \operatorname{tr}((AB)^p) \le \operatorname{tr}(A^p B^p).$$

Applying this with A, B replaced by $e^{A/p}$ and $e^{B/p}$, respectively, we obtain

$$\operatorname{tr}((e^{A/p} e^{B/p})^p) \le \operatorname{tr}(e^A e^B).$$

Now we send $p \to \infty$. Since $e^{A/p} = 1 + A/p + O(1/p^2)$ and $e^{B/p} = 1 + B/p + O(1/p^2)$, we have $e^{A/p}e^{B/p} = e^{(A+B)/p+O(1/p^2)}$, and so the left-hand side is $\operatorname{tr}(e^{A+B+O(1/p)})$; taking the limit as $p \to \infty$ we obtain the Golden-Thompson inequality[12]

If we stop the iteration at an earlier point, then the same argument gives the inequality

$$\|e^{A+B}\|_p \leq \|e^A e^B\|_p$$

for $p = 2, 4, 8, \ldots$ a power of two; one can view the original Golden-Thompson inequality as the $p = 1$ endpoint of this case in some sense[13]. In the limit $p \to \infty$, we obtain in particular the operator norm inequality

(3.19) $$\|e^{A+B}\|_{\text{op}} \leq \|e^A e^B\|_{\text{op}}$$

This inequality has a nice consequence:

Corollary 3.2.2. *Let A, B be Hermitian matrices. If $e^A \leq e^B$ (i.e., $e^B - e^A$ is positive semi-definite), then $A \leq B$.*

Proof. Since $e^A \leq e^B$, we have $\langle e^A x, x \rangle \leq \langle e^B x, x \rangle$ for all vectors x, or in other words, $\|e^{A/2}x\| \leq \|e^{B/2}x\|$ for all x. This implies that $e^{A/2}e^{-B/2}$ is a contraction, i.e., $\|e^{A/2}e^{-B/2}\|_{\text{op}} \leq 1$. By (3.19), we conclude that $\|e^{(A-B)/2}\|_{\text{op}} \leq 1$, thus $(A-B)/2 \leq 0$, and the claim follows.   □

Exercise 3.2.1. Reverse the above argument and conclude that (3.2.2) is in fact equivalent to (3.19).

It is remarkably tricky to try to prove Corollary 3.2.2 directly. Here is a somewhat messy proof. By the fundamental theorem of calculus, it suffices to show that whenever $A(t)$ is a Hermitian matrix depending smoothly on a real parameter with $\frac{d}{dt}e^{A(t)} \geq 0$, then $\frac{d}{dt}A(t) \geq 0$. Indeed, Corollary 3.2.2 follows from this claim by setting $A(t) := \log(e^A + t(e^B - e^A))$ and concluding that $A(1) \geq A(0)$.

To obtain this claim, we use the *Duhamel formula*

$$\frac{d}{dt}e^{A(t)} = \int_0^1 e^{(1-s)A(t)}(\frac{d}{dt}A(t))e^{sA(t)} \, ds.$$

This formula can be proven by Taylor expansion, or by carefully approximating $e^{A(t)}$ by $(1 + A(t)/N)^N$; alternatively, one can integrate the identity

$$\frac{\partial}{\partial s}(e^{-sA(t)}\frac{\partial}{\partial t}e^{sA(t)}) = e^{-sA(t)}(\frac{\partial}{\partial t}A(t))e^{sA(t)},$$

_____
[12]See also [**Ve2008**] for a slight variant of this proof.
[13]In fact, the Golden-Thompson inequality is true in any operator norm; see [**Bh1997**, Theorem 9.3.7].

which follows from the product rule and by interchanging the s and t derivatives at a key juncture. We rearrange the Duhamel formula as

$$\frac{d}{dt}e^{A(t)} = e^{A(t)/2}\left(\int_{-1/2}^{1/2} e^{sA(t)}\left(\frac{d}{dt}A(t)\right)e^{-sA(t)} \, ds\right)e^{A(t)/2}.$$

Using the basic identity $e^A B e^{-A} = e^{\mathrm{ad}(A)}B$, we thus have

$$\frac{d}{dt}e^{A(t)} = e^{A(t)/2}\left[\left(\int_{-1/2}^{1/2} e^{s\,\mathrm{ad}(A(t))} \, ds\right)\left(\frac{d}{dt}A(t)\right)\right]e^{A(t)/2};$$

formally evaluating the integral, we obtain

$$\frac{d}{dt}e^{A(t)} = e^{A(t)/2}\left[\frac{\sinh(\mathrm{ad}(A(t))/2)}{\mathrm{ad}(A(t))/2}\left(\frac{d}{dt}A(t)\right)\right]e^{A(t)/2},$$

and thus

$$\frac{d}{dt}A(t) = \frac{\mathrm{ad}(A(t))/2}{\sinh(\mathrm{ad}(A(t))/2)}\left(e^{-A(t)/2}\left(\frac{d}{dt}e^{A(t)}\right)e^{-A(t)/2}\right).$$

As $\frac{d}{dt}e^{A(t)}$ was positive semi-definite by hypothesis, $e^{-A(t)/2}(\frac{d}{dt}e^{A(t)})e^{-A(t)/2}$ is also. It thus suffices to show that for any Hermitian A, the operator $\frac{\mathrm{ad}(A)}{\sinh(\mathrm{ad}(A))}$ preserves the property of being semi-definite.

Note that for any real ξ, the operator $e^{2\pi i \xi\,\mathrm{ad}(A)}$ maps a positive semi-definite matrix B to another positive semi-definite matrix, namely $e^{2\pi i \xi A}Be^{-2\pi i \xi A}$. By the Fourier inversion formula, it thus suffices to show that the kernel $F(x) := \frac{x}{\sinh(x)}$ is positive semi-definite in the sense that it has non-negative Fourier transform (this is a special case of *Bochner's theorem*). But a routine (but somewhat tedious) application of contour integration shows that the Fourier transform $\hat{F}(\xi) = \int_{\mathbf{R}} e^{-2\pi i x \xi}F(x) \, dx$ is given by the formula $\hat{F}(\xi) = \frac{1}{8\cosh^2(\pi^2\xi)}$, and the claim follows.

Because of the Golden-Thompson inequality, many applications of the *exponential moment method* in commutative probability theory can be extended without difficulty to the non-commutative case, as was observed in [**AhWi2002**]. For instance, consider (a special case of) the *Chernoff inequality*

$$\mathbf{P}(X_1 + \cdots + X_N \geq \lambda) \leq \max(e^{-\lambda^2/4}, e^{-\lambda\sigma/2})$$

for any $\lambda > 0$, where $X_1, \ldots, X_N \equiv X$ are iid scalar random variables taking values in $[-1, 1]$ of mean zero and with total variance σ^2 (i.e., each factor has variance σ^2/N). We briefly recall the standard proof of this inequality from Section 2.1. We first use Markov's inequality to obtain

$$\mathbf{P}(X_1 + \cdots + X_N \geq \lambda) \leq e^{-t\lambda}\mathbf{E}e^{t(X_1 + \cdots + X_N)}$$

for some parameter $t > 0$ to be optimised later. In the scalar case, we can factor $e^{t(X_1 + \cdots + X_N)}$ as $e^{tX_1} \ldots e^{tX_N}$ and then use the iid hypothesis to write the right-hand side as

$$e^{-t\lambda}(\mathbf{E}e^{tX})^N.$$

An elementary Taylor series computation then reveals the bound $\mathbf{E}e^{tX} \leq \exp(t^2\sigma^2/N)$ when $0 \leq t \leq 1$; inserting this bound and optimising in t we obtain the claim.

Now suppose that $X_1, \ldots, X_N \equiv X$ are iid $n \times n$ Hermitian matrices. One can try to adapt the above method to control the size of the sum $X_1 + \cdots + X_N$. The key point is then to bound expressions such as

$$\mathbf{E}\,\mathrm{tr}\,e^{t(X_1 + \cdots + X_N)}.$$

As $X_1, \ldots, X_N$ need not commute, we cannot separate the product completely. But by Golden-Thompson, we can bound this expression by

$$\mathbf{E}\,\mathrm{tr}\,e^{t(X_1 + \cdots + X_{N-1})}e^{tX_N},$$

which by independence we can then factorise as

$$\mathrm{tr}(\mathbf{E}e^{t(X_1 + \cdots + X_{N-1})})(\mathbf{E}e^{tX_N}).$$

As the matrices involved are positive definite, we can then take out the final factor in the operator norm:

$$\|\mathbf{E}e^{tX_n}\|_{\mathrm{op}}\,\mathrm{tr}\,\mathbf{E}e^{t(X_1 + \cdots + X_{N-1})}.$$

Iterating this procedure, we can eventually obtain the bound

$$\mathbf{E}\,\mathrm{tr}\,e^{t(X_1 + \cdots + X_N)} \leq \|\mathbf{E}e^{tX}\|_{\mathrm{op}}^N.$$

Combining this with the rest of the Chernoff inequality argument, we can establish a matrix generalisation

$$\mathbf{P}(\|X_1 + \cdots + X_N\|_{\mathrm{op}} \geq \lambda) \leq n\max(e^{-\lambda^2/4}, e^{-\lambda\sigma/2})$$

of the Chernoff inequality, under the assumption that the $X_1, \ldots, X_N$ are iid with mean zero, have operator norm bounded by 1, and have total variance $\sum_{i=1}^{N} \|\mathbf{E}X_i^2\|_{\mathrm{op}}$ equal to σ^2; see for instance [**Ve2008**] for details.

Further discussion of the use of the Golden-Thompson inequality and its variants to non-commutative Chernoff-type inequalities can be found in [**Gr2009**], [**Ve2008**], [**Tr2010**]. It seems that the use of this inequality may be quite useful in simplifying the proofs of several of the basic estimates in this subject.

3.3. The Dyson and Airy kernels of GUE via semiclassical analysis

Let n be a large integer, and let M_n be the *Gaussian Unitary Ensemble* (GUE), i.e., the random Hermitian matrix with probability distribution

$$C_n e^{-\operatorname{tr}(M_n^2)/2} dM_n$$

where dM_n is a Haar measure on Hermitian matrices and C_n is the normalisation constant required to make the distribution of unit mass. The eigenvalues $\lambda_1 < \cdots < \lambda_n$ of this matrix are then a coupled family of n real random variables. For any $1 \le k \le n$, we can define the *k-point correlation function* $\rho_k(x_1, \ldots, x_k)$ to be the unique symmetric measure on $\mathbf{R}^k$ such that

$$\int_{\mathbf{R}^k} F(x_1, \ldots, x_k) \rho_k(x_1, \ldots, x_k) = \mathbf{E} \sum_{1 \le i_1 < \cdots < i_k \le n} F(\lambda_{i_1}, \ldots, \lambda_{i_k}).$$

A standard computation (given for instance in Section 2.6 gives the *Ginibre formula* [**Gi1965**]

$$\rho_n(x_1, \ldots, x_n) = C_n' \big(\prod_{1 \le i < j \le n} |x_i - x_j|^2 \big) e^{-\sum_{j=1}^n |x_j|^2/2}$$

for the n-point correlation function, where C_n' is another normalisation constant. Using Vandermonde determinants, one can rewrite this expression in determinantal form as

$$\rho_n(x_1, \ldots, x_n) = C_n'' \det(K_n(x_i, x_j))_{1 \le i,j \le n}$$

where the kernel K_n is given by

$$K_n(x, y) := \sum_{k=0}^{n-1} \phi_k(x) \phi_k(y)$$

where $\phi_k(x) := P_k(x) e^{-x^2/4}$ and $P_0, P_1, \ldots$ are the (L^2-normalised) *Hermite polynomials* (thus the ϕ_k are an orthonormal family, with each P_k being a polynomial of degree k). Integrating out one or more of the variables, one is led to the *Gaudin-Mehta formula*[14]

(3.20) $$\rho_k(x_1, \ldots, x_k) = \det(K_n(x_i, x_j))_{1 \le i,j \le k}.$$

Again, see Section 2.6 for details.

The functions $\phi_k(x)$ can be viewed as an orthonormal basis of eigenfunctions for the *harmonic oscillator operator*

$$L\phi := \big(-\frac{d^2}{dx^2} + \frac{x^2}{4} \big)\phi;$$

---

[14]In particular, the normalisation constant C_n'' in the previous formula turns out to simply be equal to 1.

indeed it is a classical fact that

$$L\phi_k = (k + \frac{1}{2})\phi_k.$$

As such, the kernel K_n can be viewed as the integral kernel of the spectral projection operator $1_{(-\infty,n+\frac{1}{2}]}(L)$.

From (3.20) we see that the fine-scale structure of the eigenvalues of GUE are controlled by the asymptotics of K_n as $n \to \infty$. The two main asymptotics of interest are given by the following lemmas:

Lemma 3.3.1 (Asymptotics of K_n in the bulk). *Let $x_0 \in (-2,2)$, and let $\rho_{sc}(x_0) := \frac{1}{2\pi}(4 - x_0^2)_+^{1/2}$ be the semicircular law density at x_0. Then, we have*

(3.21)
$$K_n(x_0\sqrt{n} + \frac{y}{\sqrt{n}\rho_{sc}(x_0)}, x_0\sqrt{n} + \frac{z}{\sqrt{n}\rho_{sc}(x_0)})$$
$$\to \frac{\sin(\pi(y - z))}{\pi(y - z)}$$

as $n \to \infty$ for any fixed $y, z \in \mathbf{R}$ (removing the singularity at $y = z$ in the usual manner).

Lemma 3.3.2 (Asymptotics of K_n at the edge). *We have*

(3.22)
$$K_n(2\sqrt{n} + \frac{y}{n^{1/6}}, 2\sqrt{n} + \frac{z}{n^{1/6}})$$
$$\to \frac{\mathrm{Ai}(y)\,\mathrm{Ai}'(z) - \mathrm{Ai}'(y)\,\mathrm{Ai}(z)}{y - z}$$

as $n \to \infty$ for any fixed $y, z \in \mathbf{R}$, where Ai is the Airy function

$$\mathrm{Ai}(x) := \frac{1}{\pi}\int_0^\infty \cos(\frac{t^3}{3} + tx)\,dt$$

and again removing the singularity at $y = z$ in the usual manner.

The proof of these asymptotics usually proceeds via computing the asymptotics of Hermite polynomials, together with the Christoffel-Darboux formula; this is, for instance, the approach taken in Section 2.6. However, there is a slightly different approach that is closer in spirit to the methods of semi-classical analysis. For sake of completeness, we will discuss this approach here, although to focus on the main ideas, the derivation will not be completely rigorous[15].

---

[15]In particular, we will ignore issues such as convergence of integrals or of operators, or (removable) singularities in kernels caused by zeroes in the denominator. For a rigorous approach to these asymptotics in the discrete setting, see [**Ol2008**].

3.3.1. The bulk asymptotics. We begin with the bulk asymptotics; see Lemma 3.3.1. Fix x_0 in the bulk region $(-2, 2)$. Applying the change of variables

$$x = x_0\sqrt{n} + \frac{y}{\sqrt{n}\rho_{sc}(x_0)}$$

we see that the harmonic oscillator L becomes

$$-n\rho_{sc}(x_0)^2\frac{d^2}{dy^2} + \frac{1}{4}(x_0\sqrt{n} + \frac{y}{\sqrt{n}\rho_{sc}(x_0)})^2.$$

Since K_n is the integral kernel of the spectral projection to the region $L \leq n + \frac{1}{2}$, we conclude that the left-hand side of (3.21) (as a function of y, z) is the integral kernel of the spectral projection to the region

$$-n\rho_{sc}(x_0)^2\frac{d^2}{dy^2} + \frac{1}{4}(x_0\sqrt{n} + \frac{y}{\sqrt{n}\rho_{sc}(x_0)})^2 \leq n + \frac{1}{2}.$$

Isolating out the top order terms in n, we can rearrange this as

$$-\frac{d^2}{dy^2} \leq \pi^2 + o(1).$$

Thus, in the limit $n \to \infty$, we expect (heuristically, at least) that the left-hand side of (3.21) to converge as $n \to \infty$ to the integral kernel of the spectral projection to the region

$$-\frac{d^2}{dy^2} \leq \pi^2.$$

Introducing the Fourier dual variable ξ to y, as manifested by the Fourier transform

$$\hat{f}(\xi) = \int_{\mathbf{R}} e^{-2\pi i\xi y} f(y)\, dy$$

and its inverse

$$\check{F}(y) = \int_{\mathbf{R}} e^{2\pi i\xi y} F(\xi)\, d\xi,$$

then we (heuristically) have $\frac{d}{dy} = 2\pi i\xi$, and so we are now projecting to the region

(3.23) $$|\xi|^2 \leq 1/4,$$

i.e., we are restricting the Fourier variable to the interval $[-1/2, 1/2]$. Back in physical space, the associated projection P thus takes the form

$$\begin{aligned}
Pf(y) &= \int_{[-1/2,1/2]} e^{2\pi i\xi y}\hat{f}(\xi)\, d\xi \\
&= \int_{\mathbf{R}}\int_{[-1/2,1/2]} e^{2\pi i\xi y}e^{-2\pi i\xi z}\, d\xi f(z)\, dz \\
&= \int_{\mathbf{R}} \frac{\sin(\pi(y-z))}{y-z} f(z)\, dz
\end{aligned}$$

and the claim follows.

Remark 3.3.3. From a semiclassical perspective, the original spectral projection $L \leq n + \frac{1}{2}$ can be expressed in phase space (using the dual frequency variable η to x) as the ellipse

$$(3.24) \qquad 4\pi^2\eta^2 + \frac{x^2}{4} \leq n + \frac{1}{2},$$

which after the indicated change of variables becomes the elongated ellipse

$$\xi^2 + \frac{1}{2n\rho_{sc}(x_0)(4 - x_0^2)}y + \frac{1}{4n^2\rho_{sc}(x_0)^2(4 - x_0^2)}y^2$$
$$\leq \frac{1}{4} + \frac{1}{2n(4 - x_0^2)},$$

which converges (in some suitably weak sense) to the strip (3.23) as $n \to \infty$.

3.3.2. The edge asymptotics. A similar (heuristic) argument gives the edge asymptotics, Lemma 3.3.2. Starting with the change of variables

$$x = 2\sqrt{n} + \frac{y}{n^{1/6}}$$

the harmonic oscillator L now becomes

$$-n^{1/3}\frac{d^2}{dy^2} + \frac{1}{4}(2\sqrt{n} + \frac{y}{n^{1/6}})^2.$$

Thus, the left-hand side of (3.22) becomes the kernel of the spectral projection to the region

$$-n^{1/3}\frac{d^2}{dy^2} + \frac{1}{4}(2\sqrt{n} + \frac{y}{n^{1/6}})^2 \leq n + \frac{1}{2}.$$

Expanding out, computing all terms of size $n^{1/3}$ or larger, and rearranging, this (heuristically) becomes

$$-\frac{d^2}{dy^2} + y \leq o(1)$$

and so, heuristically at least, we expect (3.22) to converge to the kernel of the projection to the region

$$(3.25) \qquad -\frac{d^2}{dy^2} + y \leq 0.$$

To compute this, we again pass to the Fourier variable ξ, converting the above to

$$4\pi^2\xi^2 + \frac{1}{2\pi i}\frac{d}{d\xi} \leq 0$$

using the usual Fourier-analytic correspondences between multiplication and differentiation. If we then use the integrating factor transformation

$$F(\xi) \mapsto e^{8\pi^3 i \xi^3/3}F(\xi),$$

we can convert the above region to

$$\frac{1}{2\pi i}\frac{d}{d\xi} \le 0,$$

which on undoing the Fourier transformation becomes

$$y \le 0,$$

and the spectral projection operation for this is simply the spatial multiplier $1_{(-\infty,0]}$. Thus, informally at least, we see that the spectral projection P to the region (3.25) is given by the formula

$$P = M^{-1}1_{(-\infty,0]}M$$

where the Fourier multiplier M is given by the formula

$$\widehat{Mf}(\xi) := e^{8\pi^3 i\xi^3/3}\hat{f}(\xi).$$

In other words (ignoring issues about convergence of the integrals),

$$
\begin{aligned}
Mf(y) &= \int_{\mathbf{R}}(\int_{\mathbf{R}} e^{2\pi i y\xi}e^{8\pi^3 i\xi^3/3}e^{-2\pi iz\xi} \, d\xi)f(z) \, dz \\
&= 2\int_{\mathbf{R}}(\int_0^\infty \cos(2\pi(y-z)\xi + 8\pi^3\xi^3/3) \, d\xi)f(z) \, dz \\
&= \frac{1}{\pi}\int_{\mathbf{R}}(\int_0^\infty \cos(t(y-z)+t^3/3) \, dt)f(z) \, dz \\
&= \int_{\mathbf{R}} \mathrm{Ai}(y-z)f(z) \, dz,
\end{aligned}
$$

and similarly,

$$M^{-1}f(z) = \int_{\mathbf{R}} \mathrm{Ai}(y-z)f(y) \, dy$$

(this reflects the unitary nature of M). We thus see (formally, at least) that

$$Pf(y) = \int_{\mathbf{R}}(\int_{(-\infty,0]} \mathrm{Ai}(y-w)\,\mathrm{Ai}(z-w) \, dw)f(z) \, dz.$$

To simplify this expression we perform some computations closely related to the ones above. From the Fourier representation

$$
\begin{aligned}
\mathrm{Ai}(y) &= \frac{1}{\pi}\int_0^\infty \cos(ty+t^3/3) \, dt \\
&= \int_{\mathbf{R}} e^{2\pi i y\xi}e^{8\pi i\xi^3/3} \, d\xi
\end{aligned}
$$

we see that

$$\widehat{Ai}(\xi) = e^{8\pi^3 i\xi^3/3},$$

which means that

$$(4\pi^2\xi^2 + \frac{1}{2\pi i}\frac{d}{d\xi})\widehat{Ai}(\xi) = 0$$

and thus

$$(-\frac{d^2}{dy^2} + y)\,\mathrm{Ai}(y) = 0,$$

thus *Ai* obeys the *Airy equation*

$$\mathrm{Ai}''(y) = y\,\mathrm{Ai}(y).$$

Using this, one soon computes that

$$\frac{d}{dw}\,\frac{\mathrm{Ai}(y-w)\,\mathrm{Ai}'(z-w) - \mathrm{Ai}'(y-w)\,\mathrm{Ai}(z-w)}{y-z} = \mathrm{Ai}(y-w)\,\mathrm{Ai}(z-w).$$

Also, stationary phase asymptotics tell us that $\mathrm{Ai}(y)$ decays exponentially fast as $y \to +\infty$, and hence $\mathrm{Ai}(y-w)$ decays exponentially fast as $w \to -\infty$ for fixed y; similarly for $\mathrm{Ai}'(z-w), \mathrm{Ai}'(y-w), \mathrm{Ai}(z-w)$. From the fundamental theorem of calculus, we conclude that

$$\int_{(-\infty,0]} \mathrm{Ai}(y-w)\,\mathrm{Ai}(z-w)\,dw = \frac{\mathrm{Ai}(y)\,\mathrm{Ai}'(z) - \mathrm{Ai}'(y)\,\mathrm{Ai}(z)}{y-z},$$

(this is a continuous analogue of the Christoffel-Darboux formula), and the claim follows.

Remark 3.3.4. As in the bulk case, one can take a semi-classical analysis perspective and track what is going on in phase space. With the scaling we have selected, the ellipse (3.24) has become

$$4\pi^2 n^{1/3}\xi^2 + \frac{(2\sqrt{n} + y/n^{1/6})^2}{4} \le n + \frac{1}{2},$$

which we can rearrange as the eccentric ellipse

$$4\pi^2\xi^2 + y \le \frac{1}{2n^{1/3}} - \frac{y^2}{4n^{2/3}}$$

which is converging as $n \to \infty$ to the parabolic region

$$4\pi^2\xi^2 + y \le 0$$

which can then be shifted to the half-plane $y \le 0$ by the parabolic shear transformation $(y,\xi) \mapsto (y + 4\pi^2\xi^2, \xi)$, which is the canonical relation of the Fourier multiplier M. (The rapid decay of the kernel Ai of M at $+\infty$ is then reflected in the fact that this transformation only shears to the right and not the left.)

Remark 3.3.5. Presumably one should also be able to apply the same heuristics to other invariant ensembles, such as those given by probability distributions of the form

$$C_n e^{-\mathrm{tr}(P(M_n))}dM_n$$

for some potential function P. Certainly one can soon get to an orthogonal polynomial formulation of the determinantal kernel for such ensembles, but

I do not know if the projection operators for such kernels can be viewed as spectral projections to a phase space region as was the case for GUE. But if one could do this, this would provide a heuristic explanation as to the universality phenomenon for such ensembles, as Taylor expansion shows that all (reasonably smooth) regions of phase space converge to universal limits (such as a strip or paraboloid) after rescaling around either a non-critical point or a critical point of the region with the appropriate normalisation.

3.4. The mesoscopic structure of GUE eigenvalues

In this section we give a heuristic model of the *mesoscopic* structure of the eigenvalues $\lambda_1 \leq \cdots \leq \lambda_n$ of the $n \times n$ *Gaussian Unitary Ensemble* (GUE), where n is a large integer. From Section 2.6, the probability density of these eigenvalues is given by the *Ginibre distribution*

$$\frac{1}{Z_n} e^{-H(\lambda)} \, d\lambda$$

where $d\lambda = d\lambda_1 \ldots d\lambda_n$ is Lebesgue measure on the Weyl chamber

$$\{(\lambda_1, \ldots, \lambda_n) \in \mathbf{R}^n : \lambda_1 \leq \cdots \leq \lambda_n\},$$

Z_n is a constant, and the Hamiltonian H is given by the formula

$$H(\lambda_1, \ldots, \lambda_n) := \sum_{j=1}^{n} \frac{\lambda_j^2}{2} - 2 \sum_{1 \leq i < j \leq n} \log |\lambda_i - \lambda_j|.$$

As we saw in Section 2.4, at the macroscopic scale of $\sqrt{n}$, the eigenvalues λ_j are distributed according to the *Wigner semicircle law*

$$\rho_{sc}(x) := \frac{1}{2\pi} (4 - x^2)_+^{1/2}.$$

Indeed, if one defines the *classical location* γ_i^{cl} of the ith eigenvalue to be the unique solution in $[-2\sqrt{n}, 2\sqrt{n}]$ to the equation

$$\int_{-2\sqrt{n}}^{\gamma_i^{\mathrm{cl}}/\sqrt{n}} \rho_{sc}(x) \, dx = \frac{i}{n},$$

then it is known that the random variable λ_i is quite close to γ_i^{cl}. Indeed, a result of Gustavsson [**Gu2005**] shows that, in the bulk region when $\varepsilon n < i < (1 - \varepsilon)n$ for some fixed $\varepsilon > 0$, λ_i is distributed asymptotically as a Gaussian random variable with mean γ_i^{cl} and variance $\sqrt{\frac{\log n}{\pi}} \times \frac{1}{\sqrt{n}\rho_{sc}(\gamma_i^{\mathrm{cl}})}$. Note that from the semicircular law, the factor $\frac{1}{\sqrt{n}\rho_{sc}(\gamma_i^{\mathrm{cl}})}$ is the mean eigenvalue spacing.

At the other extreme, at the microscopic scale of the mean eigenvalue spacing (which is comparable to $1/\sqrt{n}$ in the bulk, but can be as large as $n^{-1/6}$ at the edge), the eigenvalues are asymptotically distributed with respect to a special determinantal point process, namely the Dyson sine process in the bulk (and the Airy process on the edge), as discussed in Section 3.3.

We now focus on the *mesoscopic* structure of the eigenvalues, in which one involves scales that are intermediate between the microscopic scale $1/\sqrt{n}$ and the macroscopic scale $\sqrt{n}$, for instance, in correlating the eigenvalues λ_i and λ_j in the regime $|i - j| \sim n^\theta$ for some $0 < \theta < 1$. Here, there is a surprising phenomenon; there is quite a long-range correlation between such eigenvalues. The results from [**Gu2005**] show that both λ_i and λ_j behave asymptotically like Gaussian random variables, but a further result from the same paper shows that the correlation between these two random variables is asymptotic to $1 - \theta$ (in the bulk, at least); thus, for instance, adjacent eigenvalues λ_{i+1} and λ_i are almost perfectly correlated (which makes sense, as their spacing is much less than either of their standard deviations), but that even very distant eigenvalues, such as $\lambda_{n/4}$ and $\lambda_{3n/4}$, have a correlation comparable to $1/\log n$. One way to get a sense of this is to look at the trace

$$\lambda_1 + \cdots + \lambda_n.$$

This is also the sum of the diagonal entries of a GUE matrix, and is thus normally distributed with a variance of n. In contrast, each of the λ_i (in the bulk, at least) has a variance comparable to $\log n/n$. In order for these two facts to be consistent, the average correlation between pairs of eigenvalues then has to be of the order of $1/\log n$.

In this section we will use a heuristic way to see this correlation, based on Taylor expansion of the convex Hamiltonian $H(\lambda)$ around the minimum γ, which gives a conceptual probabilistic model for the mesoscopic structure of the GUE eigenvalues. While this heuristic is in no way rigorous, it does seem to explain many of the features currently known or conjectured about GUE, and seems likely to extend also to other models.

3.4.1. Fekete points. It is easy to see that the Hamiltonian $H(\lambda)$ is convex in the Weyl chamber, and goes to infinity on the boundary of this chamber, so it must have a unique minimum, at a set of points $\gamma = (\gamma_1, \ldots, \gamma_n)$ known as the *Fekete points*. At the minimum, we have $\nabla H(\gamma) = 0$, which expands to become the set of conditions

$$(3.26) \qquad\qquad \gamma_j - 2 \sum_{i \neq j} \frac{1}{\gamma_j - \gamma_i} = 0$$

for all $1 \leq j \leq n$. To solve these conditions, we introduce the monic degree n polynomial

$$P(x) := \prod_{i=1}^{n} (x - \gamma_i).$$

Differentiating this polynomial, we observe that

$$(3.27) \qquad P'(x) = P(x) \sum_{i=1}^{n} \frac{1}{x - \gamma_i}$$

and

$$P''(x) = P(x) \sum_{1 \leq i,j \leq n: i \neq j} \frac{1}{x - \gamma_i} \frac{1}{x - \gamma_j}.$$

Using the identity

$$\frac{1}{x - \gamma_i} \frac{1}{x - \gamma_j} = \frac{1}{x - \gamma_i} \frac{1}{\gamma_i - \gamma_j} + \frac{1}{x - \gamma_j} \frac{1}{\gamma_j - \gamma_i}$$

followed by (3.26), we can rearrange this as

$$P''(x) = P(x) \sum_{1 \leq i \leq n: i \neq j} \frac{\gamma_i}{x - \gamma_i}.$$

Comparing this with (3.27), we conclude that

$$P''(x) = xP'(x) - nP(x),$$

or in other words, that P is the nth *Hermite polyomial*

$$P(x) = H_n(x) := (-1)^n e^{x^2/2} \frac{d}{dx^2} e^{-x^2/2}.$$

Thus, the Fekete points γ_i are nothing more than the zeroes of the nth Hermite polynomial.

Heuristically, one can study these zeroes by looking at the function

$$\phi(x) := P(x) e^{-x^2/4}$$

which solves the eigenfunction equation

$$\phi''(x) + (n - \frac{x^2}{4})\phi(x) = 0.$$

Comparing this equation with the harmonic oscillator equation $\phi''(x) + k^2\phi(x) = 0$, which has plane wave solutions $\phi(x) = A\cos(kx + \theta)$ for k^2 positive and exponentially decaying solutions for k^2 negative, we are led (heuristically, at least) to conclude that ϕ is concentrated in the region where $n - \frac{x^2}{4}$ is positive (i.e., inside the interval $[-2\sqrt{n}, 2\sqrt{n}]$) and will oscillate at frequency roughly $\sqrt{n - \frac{x^2}{4}}$ inside this region. As such, we expect the Fekete points γ_i to obey the same spacing law as the classical locations γ_i^{cl}; indeed, it is possible to show that $\gamma_i = \gamma_i^{\text{cl}} + O(1/\sqrt{n})$ in the bulk (with some standard modifications at the edge). In particular, we have the heuristic

$$(3.28) \qquad \gamma_i - \gamma_j \approx (i - j)/\sqrt{n}$$

for i, j in the bulk.

Remark 3.4.1. If one works with the *circular unitary ensemble* (CUE) instead of the GUE, in which M_n is drawn from the unitary $n \times n$ matrices using Haar measure, the Fekete points become equally spaced around the unit circle, so that this heuristic essentially becomes exact.

3.4.2. Taylor expansion. Now we expand around the Fekete points by making the ansatz

$$\lambda_i = \gamma_i + x_i,$$

thus the results of [**Gu2005**] predict that each x_i is normally distributed with standard deviation $O(\sqrt{\log n}/\sqrt{n})$ (in the bulk). We Taylor expand

$$H(\lambda) = H(\gamma) + \nabla H(\gamma)(x) + \frac{1}{2}\nabla^2 H(\gamma)(x,x) + \dots.$$

We heuristically drop the cubic and higher order terms. The constant term $H(\gamma)$ can be absorbed into the partition constant Z_n, while the linear term vanishes by the property $\nabla H(\gamma)$ of the Fekete points. We are thus lead to a quadratic (i.e., Gaussian) model

$$\frac{1}{Z'_n} e^{-\frac{1}{2}\nabla^2 H(\gamma)(x,x)} \, dx$$

for the probability distribution of the shifts x_i, where Z'_n is the appropriate normalisation constant.

Direct computation allows us to expand the quadratic form $\frac{1}{2}\nabla^2 H(\gamma)$ as

$$\frac{1}{2}\nabla^2 H(\gamma)(x,x) = \sum_{j=1}^{n} \frac{x_j^2}{2} + \sum_{1 \leq i < j \leq n} \frac{(x_i - x_j)^2}{(\gamma_i - \gamma_j)^2}.$$

The Taylor expansion is not particularly accurate when j and i are too close, say $j = i + O(\log^{O(1)} n)$, but we will ignore this issue as it should only affect the microscopic behaviour rather than the mesoscopic behaviour. This models the x_i as (coupled) Gaussian random variables whose covariance matrix can in principle be explicitly computed by inverting the matrix of the quadratic form. Instead of doing this precisely, we shall instead work heuristically (and somewhat inaccurately) by re-expressing the quadratic form in the Haar basis. For simplicity, let us assume that n is a power of 2. Then the Haar basis consists of the basis vector

$$\psi_0 := \frac{1}{\sqrt{n}}(1, \dots, 1)$$

together with the basis vectors

$$\psi_I := \frac{1}{\sqrt{|I|}}(1_{I_l} - 1_{I_r})$$

for every discrete dyadic interval $I \subset \{1, \dots, n\}$ of length between 2 and n, where I_l and I_r are the left and right halves of I, and $1_{I_l}, 1_{I_r} \in \mathbf{R}^n$ are the vectors that are one on I_l, I_r, respectively, and zero elsewhere. These form an orthonormal basis of $\mathbf{R}^n$, thus we can write

$$x = \xi_0 \psi_0 + \sum_I \xi_I \psi_I$$

for some coefficients ξ_0, ξ_I.

From orthonormality we have

$$\sum_{j=1}^n \frac{x_j^2}{2} = \xi_0^2 + \sum_I \xi_I^2$$

and we have

$$\sum_{1 \le i < j \le n} \frac{(x_i - x_j)^2}{(\gamma_i - \gamma_j)^2} = \sum_{I,J} \xi_I \xi_J c_{I,J}$$

where the matrix coefficients $c_{I,J}$ are given by

$$c_{I,J} := \sum_{1 \le i < j \le n} \frac{(\psi_I(i) - \psi_I(j))(\psi_J(i) - \psi_J(j))}{(\gamma_i - \gamma_j)^2}.$$

A standard heuristic wavelet computation using (3.28) suggests that $c_{I,J}$ is small unless I and J are actually equal, in which case one has

$$c_{I,I} \sim \frac{n}{|I|}$$

(in the bulk, at least). Actually, the decay of the $c_{I,J}$ away from the diagonal $I = J$ is not so large, because the Haar wavelets ψ_I have poor moment and regularity properties. But one could in principle use much smoother and much more balanced wavelets, in which case the decay should be much faster.

This suggests that the GUE distribution could be modeled by the distribution

(3.29) $$\frac{1}{Z_n'} e^{-\xi_0^2/2} e^{-C \sum_I \frac{n}{|I|} \xi_I^2} d\xi$$

for some absolute constant C; thus we may model $\xi_0 \equiv N(0,1)$ and $\xi_I \equiv C' \sqrt{|I|} \sqrt{n} g_I$ for some iid Gaussians $g_I \equiv N(0,1)$ independent of ξ_0. We then have as a model

$$x_i = \frac{\xi_0}{\sqrt{n}} + \frac{C'}{\sqrt{n}} \sum_I (1_{I_l}(i) - 1_{I_r}(i)) g_I$$

for the fluctuations of the eigenvalues (in the bulk, at least), leading of course to the model

(3.30) $$\lambda_i = \gamma_i + \frac{\xi_0}{\sqrt{n}} + \frac{C'}{\sqrt{n}} \sum_I (1_{I_l}(i) - 1_{I_r}(i)) g_I$$

for the fluctuations themselves. This model does not capture the microscopic behaviour of the eigenvalues such as the sine kernel (indeed, as noted before, the contribution of the very short I (which corresponds to very small values of $|j - i|$) is inaccurate), but appears to be a good model to describe the mesoscopic behaviour. For instance, observe that for each i there are $\sim \log n$ independent normalised Gaussians in the above sum, and so this model is consistent with the result of Gustavsson that each λ_i is Gaussian with standard deviation $\sim \frac{\sqrt{\log n}}{\sqrt{n}}$. Also, if $|i - j| \sim n^\theta$, then the expansions (3.30) of λ_i, λ_j share about $(1 - \theta) \log n$ of the $\log n$ terms in the sum in common, which is consistent with the further result of Gustavsson that the correlation between such eigenvalues is comparable to $1 - \theta$.

If one looks at the gap $\lambda_{i+1} - \lambda_i$ using (3.30) (and replacing the Haar cutoff $1_{I_l}(i) - 1_{I_r}(i)$ by something smoother for the purposes of computing the gap), one is led to a heuristic of the form

$$\lambda_{i+1} - \lambda_i = \frac{1}{\rho_{sc}(\gamma_i / \sqrt{n})} \frac{1}{\sqrt{n}} + \frac{C'}{\sqrt{n}} \sum_I (1_{I_l}(i) - 1_{I_r}(i)) \frac{g_I}{|I|}.$$

The dominant terms here are the first term and the contribution of the very short intervals I. At present, this model cannot be accurate, because it predicts that the gap can sometimes be negative; the contribution of the very short intervals must instead be replaced by some other model that gives sine process behaviour, but we do not know of an easy way to set up a plausible such model.

On the other hand, the model suggests that the gaps are largely decoupled from each other, and have Gaussian tails. Standard heuristics then suggest that of the $\sim n$ gaps in the bulk, the largest one should be comparable to $\sqrt{\frac{\log n}{n}}$, which was indeed established recently in [**BeBo2010**].

Given any probability measure $\mu = \rho \, dx$ on $\mathbf{R}^n$ (or on the Weyl chamber) with a smooth non-zero density, one can can create an associated heat flow on other smooth probability measures $f \, dx$ by performing gradient flow with respect to the Dirichlet form

$$D(f \, dx) := \frac{1}{2} \int_{\mathbf{R}^n} |\nabla \frac{f}{\rho}|^2 \, d\mu.$$

Using the ansatz (3.29), this flow decouples into a system of independent Ornstein-Uhlenbeck processes

$$d\xi_0 = -\xi_0 dt + dW_0$$

and

$$dg_I = C'' \frac{n}{|I|} (-g_I dt + dW_I)$$

where dW_0, dW_I are independent Wiener processes (i.e., Brownian motion). This is a toy model for the Dyson Brownian motion (see Section 3.1). In this model, we see that the mixing time for each g_I is $O(|I|/n)$; thus, the large-scale variables (g_I for large I) evolve very slowly by Dyson Brownian motion, taking as long as $O(1)$ to reach equilibrium, while the fine scale modes (g_I for small I) can achieve equilibrium in as brief a time as $O(1/n)$, with the intermediate modes taking an intermediate amount of time to reach equilibrium. It is precisely this picture that underlies the Erdos-Schlein-Yau approach [**ErScYa2009**] to universality for Wigner matrices via the local equilibrium flow, in which the measure (3.29) is given an additional (artificial) weight, roughly of the shape $e^{-n^{1-\varepsilon}(\xi_0^2 + \sum_I \xi_I^2)}$, in order to make equilibrium achieved globally in just time $O(n^{1-\varepsilon})$, leading to a local log-Sobolev type inequality that ensures convergence of the local statistics once one controls a Dirichlet form connected to the local equilibrium measure, and then one can use the localisation of eigenvalues provided by a local semicircle law to control that Dirichlet form in turn for measures that have undergone Dyson Brownian motion.

Bibliography

[AhWi2002] R. Ahlswede, A. Winter, *Strong converse for identification via quantum channels*, IEEE Trans. Information Theory 48 (2002), 568–579.

[AlKrVu2002] N. Alon, M. Krivelevich, V. Vu, *On the concentration of eigenvalues of random symmetric matrices*, Israel J. Math. **131** (2002), 259–267.

[AnGuZi2010] G. Anderson, A. Guionnet, O. Zeitouni, An introduction to Random Matrices, Cambridge Studies in Advanced Mathematics, 118. Cambridge University Press, Cambridge, 2010.

[At1982] M. F. Atiyah, *Convexity and commuting Hamiltonians*, Bull. London Math. Soc. **14** (1982), no. 1, 1–5.

[Ba1993] Z. D. Bai, *Convergence rate of expected spectral distributions of large random matrices. I. Wigner matrices*, Ann. Probab. **21** (1993), no. 2, 625–648.

[Ba1993b] Z. D. Bai, *Convergence rate of expected spectral distributions of large random matrices. II. Sample covariance matrices.* Ann. Probab. **21** (1993), no. 2, 649–672.

[Ba1997] Z. D. Bai, *Circular law*, Ann. Probab. **25** (1997), no. 1, 494–529.

[BaSi2010] Z. Bai, J. Silverstein, Spectral analysis of large dimensional random matrices. Second edition. Springer Series in Statistics. Springer, New York, 2010. xvi+551 pp.

[BaYi1988] Z.D. Bai, Y.Q. Yin, *Necessary and sufficient conditions for almost sure convergence of the largest eigenvalue of a Wigner matrix*, Ann. Probab. **16** (1988), no. 4, 1729–1741.

[BaHa1984] A. D. Barbour, P. Hall, *Stein's method and the Berry-Esseen theorem*, Austral. J. Statist. **26** (1984), no. 1, 8–15.

[BeBo2010] G. Ben Arous, P. Bourgade, *Extreme gaps between eigenvalues of random matrices*, `arXiv:1010.1294`.

[BeCoDyLiTi2010] H. Bercovici, B. Collins, K. Dykema, W. Li, D. Timotin, *Intersections of Schubert varieties and eigenvalue inequalities in an arbitrary finite factor*, J. Funct. Anal. **258** (2010), no. 5, 1579–1627.

[Bh1997] R. Bhatia, Matrix analysis, Graduate Texts in Mathematics, 169. Springer-Verlag, New York, 1997.

[Bi2003] P. Biane, *Free probability for probabilists*, Quantum Probability Communications, Volume XI, 55–72, World Scientific (2003).

[BiLe2001] P. Biane, F. Lehner, *Computation of some examples of Brown's spectral measure in free probability*, Colloq. Math. **90** (2001), no. 2, 181–211.

[BoCaCh2008] C. Bordenave, P. Caputo, D. Chafai, *Circular law theorem for random Markov matrices*, `arXiv:0808.1502`.

[BoVuWo2010] J. Bourgain, V. Vu, P. Wood, *On the singularity probability of discrete random matrices*, J. Funct. Anal. **258** (2010), no. 2, 559–603.

[BrHi1996] E. Brézin, S. Hikami, *Correlations of nearby levels induced by a random potential*, Nuclear Phys. B **479** (1996), no. 3, 697–706.

[Br1986] L. G. Brown, *Lidskii's theorem in the type II case*, Geometric methods in operator algebras (Kyoto, 1983), 1–35, Pitman Res. Notes Math. Ser., 123, Longman Sci. Tech., Harlow, 1986.

[De1999] P. Deift, Orthogonal polynomials and random matrices: a Riemann-Hilbert approach. Courant Lecture Notes in Mathematics, 3. New York University, Courant Institute of Mathematical Sciences, New York; American Mathematical Society, Providence, RI, 1999.

[DeGi2007] P. Deift, D. Gioev, *Universality at the edge of the spectrum for unitary, orthogonal, and symplectic ensembles of random matrices*, Comm. Pure Appl. Math. **60** (2007), no. 6, 867–910.

[Dy1962] F. Dyson, *A Brownian-motion model for the eigenvalues of a random matrix*, J. Mathematical Phys. **3** (1962) 1191–1198.

[Ed1988] A. Edelman, *Eigenvalues and condition numbers of random matrices*, SIAM J. Matrix Anal. Appl. **9** (1988), 543–560.

[Ed1996] A. Edelman, *The probability that a random real Gaussian matrix has k real eigenvalues, related distributions, and the circular law*, J. Multivariate Anal. **60** (1997), no. 2, 203–232.

[Er1945] P. Erdős, *On a lemma of Littlewood and Offord*, Bull. Amer. Math. Soc. **51**, (1945) 898–902.

[ErScYa2008] L. Erdős, B. Schlein, H.-T. Yau, *Local semicircle law and complete delocalization for Wigner random matrices*, Comm. Math. Phys. **287** (2009), 641–655.

[ErScYa2009] L. Erdős, B. Schlein, H.-T. Yau, *Universality of Random Matrices and Local Relaxation Flow*, Invent. Math. **185** (2011), 75-119.

[Fe1969] P. Federbush, *Partially alternative derivation of a result of Nelson*, J. Math. Phys. **10** (1969) 50–52.

[Fo2010] Forrester, P. J. Log-gases and random matrices. London Mathematical Society Monographs Series, 34. Princeton University Press, Princeton, NJ, 2010.

[Ga2007] J. Garnett, Bounded analytic functions. Revised first edition. Graduate Texts in Mathematics, 236. Springer, New York, 2007.

[Ge1980] S. Geman, *A limit theorem for the norm of random matrices*, Ann. Probab. **8** (1980), no. 2, 252–261.

[Gi1965] J. Ginibre, *Statistical ensembles of complex, quaternion, and real matrices*, J. Mathematical Phys. **6** (1965), 440–449.

[Gi1984] V. L. Girko, *The circular law*, Teor. Veroyatnost. i Primenen. **29** (1984), no. 4, 669–679.

[Gi2004] V. L. Girko, *The strong circular law. Twenty years later. II*, Random Oper. Stochastic Equations 12 (2004), no. 3, 255–312.

[Go1965] S. Golden, *Lower bounds for the Helmholtz function*, Phys. Rev. **137** (1965), B1127–B1128.

[GoTi2007] F. Götze, A. Tikhomirov, *The circular law for random matrices*, Ann. Probab. **38** (2010), 1444–1491.

[Go2008] W. T. Gowers, *Quasirandom groups*, Combin. Probab. Comput. **17** (2008), no. 3, 363–387.

[Gr1975] L. Gross, *Logarithmic Sobolev inequalities*, Amer. J. Math. **97** (1975), no. 4, 1061–1083.

[Gr2009] D. Gross, *Recovering low-rank matrices from few coefficients in any basis*, `arXiv:0910.1879`.

[Gr1999] D. Grabiner, *Brownian motion in a Weyl chamber, non-colliding particles, and random matrices*, Ann. Inst. H. Poincaré Probab. Statist. **35** (1999), no. 2, 177–204.

[GuSt1982] V. Guillemin, S. Sternberg, *Convexity properties of the moment mapping*, Invent. Math. **67** (1982), no. 3, 491–513.

[Gu2009] A. Guionnet, Large random matrices: lectures on macroscopic asymptotics. Lectures from the 36th Probability Summer School held in Saint-Flour, 2006. Lecture Notes in Mathematics, 1957. Springer-Verlag, Berlin, 2009.

[Gu2009b] A. Guionnet, Grandes matrices aléatoires et théorèmes d'universalité (d'aprés Erdős, Schlein, Tao, Vu, et Yau), Séminaire Bourbaki, 62éme anée, 2009–2010, no. 1019.

[GuKrZe2009] A. Guionnet, M. Krishnapur, O. Zeitouni, *The single ring theorem*, `arXiv:0909.2214`.

[GuZe2000] A. Guionnet, O. Zeitouni, *Concentration of the spectral measure for large matrices*, Electron. Comm. Probab. **5** (2000), 119–136.

[Gu2005] J. Gustavsson, *Gaussian fluctuations of eigenvalues in the GUE*, Ann. Inst. H. Poincaré Probab. Statist. **41** (2005), no. 2, 151–178.

[Ha1977] G. Halász,*Estimates for the concentration function of combinatorial number theory and probability*, Period. Math. Hungar. **8** (1977), no. 3–4, 197–211.

[HeRo1995] U. Helmke, J. Rosenthal, *Eigenvalue inequalities and Schubert calculus*, Math. Nachr. **171** (1995), 207–225.

[Ho1954] A. Horn, *Doubly stochastic matrices and the diagonal of a rotation matrix*, Amer. J. Math. **76** (1954), 620–630.

[Jo2001] K. Johansson, *Universality of the local spacing distribution in certain ensembles of Hermitian Wigner matrices*, Comm. Math. Phys. 215 (2001), no. 3, 683–705.

[Jo1982] D. Jonsson, *Some limit theorems for the eigenvalues of a sample covariance matrix*, J. Multivariate Anal. 12 (1982), no. 1, 1–38.

[Ka1985] J. P. Kahane. Some random series of functions. Second edition, Cambridge Studies in Advanced Math. 5 (1985), Cambridge University Press.

[KaKoSz1995] J. Kahn, J. Komlós, E. Szemerédi, *On the probability that a random ±1-matrix is singular*, J. Amer. Math. Soc. **8** (1995), no. 1, 223–240.

[Ka2002] O. Kallenberg, Foundations of modern probability. Second edition. Probability and its Applications (New York). Springer-Verlag, New York, 2002.

[KeVa2007] J. Keller, J. Vanden-Broeck, *Stirling's formula derived simply*, `arXiv:0711.4412`

[Kh2009] O. Khorunzhiy, *High moments of large Wigner random matrices and asymptotic properties of the spectral norm*, `arXiv:0907.3743`.

[Klyachko] A. Klyachko, *Stable bundles, representation theory and Hermitian operators*, Selecta Math. (N.S.) **4** (1998), no. 3, 419–445.

[KnTa2001] A. Knutson, T. Tao, *Honeycombs and sums of Hermitian matrices*, Notices Amer. Math. Soc. **48** (2001), no. 2, 175–186.

[KnTaWo2004] A. Knutson, T. Tao, C. Woodward, *The honeycomb model of $GL_n(\mathbf{C})$ tensor products. II. Puzzles determine facets of the Littlewood-Richardson cone*, J. Amer. Math. Soc. **17** (2004), no. 1, 19–48.

[Ko1967] J. Komlós, *On the determinant of (0,1) matrices*, Studia Sci. Math. Hungar **2** (1967), 7–21.

[KuNi2006] L. Kuipers; H. Niederreiter, Uniform Distribution of Sequences, Dover Publishing, 2006.

[La2005] R. Latała, *Some estimates of norms of random matrices*, Proc. Amer. Math. Soc. 133 (2005), no. 5, 1273–1282.

[Le2001] M. Ledoux, The concentration of measure phenomenon, Mathematical Surveys and Monographs, 89. American Mathematical Society, Providence, RI, 2001.

[Le1995] M. Ledoux, *On Talagrand's deviation inequalities for product measures*, ESAIM Probab. Statist. **1** (1995/97), 63–87.

[Li1922] J.W. Lindeberg, *Eine neue Herleitung des Exponentialgesetzes in der Wahrschein-lichkeitsrechnung*, Math. Zeit. **15** (1922), pp. 211–225.

[LiPaRuTo2005] A. E. Litvak, A. Pajor, M. Rudelson, N. Tomczak-Jaegermann, *Smallest singular value of random matrices and geometry of random polytopes*, Adv. Math. **195** (2005), no. 2, 491–523.

[Ly2009] A. Lytova, L. Pastur, *Central limit theorem for linear eigenvalue statistics of random matrices with independent entries*, Ann. Probab. **37** (2009), no. 5, 1778–1840.

[Me2004] M. Meckes, *Concentration of norms and eigenvalues of random matrices*, J. Funct. Anal. **211** (2004), no. 2, 508–524.

[Me2004] M. Mehta, Random matrices. Third edition. Pure and Applied Mathematics (Amsterdam), 142. Elsevier/Academic Press, Amsterdam, 2004.

[MeGa1960] M. L. Mehta, M. Gaudin, *On the density of eigenvalues of a random matrix*, Nuclear Phys. **18** (1960) 420–427.

[Ol2008] G. Olshanski, *Difference operators and determinantal point processes*, arXiv:0810.3751.

[PaZh2010] G. Pan, W. Zhou, *Circular law, extreme singular values and potential theory*, J. Multivariate Anal. **101** (2010), no. 3, 645–656.

[Pa1973] L. Pastur, *On the spectrum of random matrices*, Teoret. Mat.Fiz. **10**, 102–112 (1973).

[Pe2006] S. Péché, *The largest eigenvalue of small rank perturbations of Hermitian random matrices*, Probab. Theory Related Fields **134** (2006), no. 1, 127–173.

[Pe2009] S. Péché, *Universality results for the largest eigenvalues of some sample covariance matrix ensembles*, Probab. Theory Related Fields **143** (2009), no. 3–4, 481–516.

[PeSo2007] S. Péché, A. Soshnikov, *On the lower bound of the spectral norm of symmetric random matrices with independent entries*, Electron. Commun. Probab. **13** (2008), 280–290.

[Ru2008] M. Rudelson, *Invertibility of random matrices: norm of the inverse*, Ann. of Math. (2) **168** (2008), no. 2, 575–600.

[RuVe2008] M. Rudelson, R. Vershynin, *The Littlewood-Offord problem and invertibility of random matrices*, Adv. Math. **218** (2008), no. 2, 600–633.

[RuVe2009] M. Rudelson, R. Vershynin, *The least singular value of a random square matrix is $O(n^{-1/2})$*, C. R. Math. Acad. Sci. Paris **346** (2008), no. 15–16, 893–896.

[Ru2007] A. Ruzmaikina, *Universality of the edge distribution of eigenvalues of Wigner random matrices with polynomially decaying distributions of entries*, Comm. Math. Phys. **261** (2006), no. 2, 277–296.

[Sn2002] P. Sniady, *Random regularization of Brown spectral measure*, J. Funct. Anal. **193** (2002), no. 2, 291–313.

[So1999] A. Soshnikov, *Universality at the edge of the spectrum in Wigner random matrices*, Comm. Math. Phys. **207** (1999), no. 3, 697–733.

[So2004] A. Soshnikov, *Poisson statistics for the largest eigenvalues of Wigner random matrices with heavy tails*, Electron. Comm. Probab. **9** (2004), 82–91.

[SoFy2005] A. Soshnikov, Y. Fyodorov, *On the largest singular values of random matrices with independent Cauchy entries*, J. Math. Phys. **46** (2005), no. 3, 033302, 15 pp.

[Sp] R. Speicher, `www.mast.queensu.ca/~speicher/survey.html`

[St1970] C. Stein, *A bound for the error in the normal approximation to the distribution of a sum of dependent random variables*, Proc. Wh. Berk. Symp. Math. Star. Hob. 11 (1970), 583–602.

[Ta1995] M. Talagrand, *Concentration of measure and isoperimetric inequalities in product spaces*, Inst. Hautes Études Sci. Publ. Math. No. **81** (1995), 73–205.

[Ta2005] M. Talagrand, The generic chaining. Upper and lower bounds of stochastic processes. Springer Monographs in Mathematics. Springer-Verlag, Berlin, 2005.

[Ta2008] T. Tao, Structure and Randomness, American Mathematical Society, 2008.

[Ta2009] T. Tao, Poincaré's legacies, Vol. I., American Mathematical Society, 2009.

[Ta2009b] T. Tao, Poincaré's legacies, Vol. II., American Mathematical Society, 2009.

[Ta2010] T. Tao, An epsilon of room, Vol. I., American Mathematical Society, 2010.

[Ta2010b] T. Tao, An epsilon of room, Vol. II., American Mathematical Society, 2010.

[Ta2011] T. Tao, An introduction to measure theory, American Mathematical Society, 2011.

[TaVu2007] T. Tao, V. Vu, *On the singularity probability of random Bernoulli matrices*, J. Amer. Math. Soc. **20** (2007), no. 3, 603–628.

[TaVu2008] T. Tao, V. Vu, *Random matrices: the circular law*, Commun. Contemp. Math. **10** (2008), no. 2, 261–307.

[TaVu2009] T. Tao, V. Vu, *Inverse Littlewood-Offord theorems and the condition number of random discrete matrices*, Ann. of Math. (2) **169** (2009), no. 2, 595–632.

[TaVu2009b] T. Tao, V. Vu, *Random matrices: Universality of local eigenvalue statistics*, Acta Math. **206** (2011), 127–204.

[TaVu2009c] T. Tao, V. Vu, *Random matrices: Universality of local eigenvalue statistics up to the edge*, Comm. Math. Phys. **298** (2010), 549–572.

[TaVu2010] T. Tao, V. Vu, *Random matrices: the distribution of the smallest singular values*, Geom. Funct. Anal. **20** (2010), no. 1, 260–297.

[TaVuKr2010] T. Tao, V. Vu, M. Krishnapur, *Random matrices: Universality of ESDs and the circular law*, Annals of Probability **38** (2010), no. 5, 2023–206.

[Th1965] C. Thompson, *Inequality with applications in statistical mechanics*, Journal of Mathematical Physics **6** (1965), 1812–1813.

[To1994] B. Totaro, *Tensor products of semistables are semistable*, Geometry and analysis on complex manifolds, 242–250, World Sci. Publ., River Edge, NJ, 1994.

[TrWi2002] C. Tracy, H. Widom, *Distribution functions for largest eigenvalues and their applications*, Proceedings of the International Congress of Mathematicians, Vol. I (Beijing, 2002), 587–596, Higher Ed. Press, Beijing, 2002.

[Tr1991] L. N. Trefethen, *Pseudospectra of matrices*, Numerical analysis 1991 (Dundee, 1991), 234–266, Pitman Res. Notes Math. Ser., 260, Longman Sci. Tech., Harlow, 1992.

[Tr2010] J. Tropp, *User-friendly tail bounds for sums of random matrices*, arXiv:1004.4389.

[Tr1984] H. Trotter, *Eigenvalue distributions of large Hermitian matrices; Wigner's semicircle law and a theorem of Kac, Murdock, and Szegö*, Adv. in Math. **54**(1):67–82, 1984.

[Ve2008] R. Vershynin, *Golden-Thompson inequality*, www-personal.umich.edu/~romanv/teaching/reading-group/golden-thompson.pdf

[Vo1991] D. Voiculescu, *Limit laws for random matrices and free products*, Invent. Math. **104** (1991), no. 1, 201–220.

[Vu2007] V. Vu, *Spectral norm of random matrices*, Combinatorica **27** (2007), no. 6, 721–736.

[Wo2003] T. H. Wolff, Lectures on Harmonic Analysis, American Mathematical Society, University Lecture Series vol. 29, 2003.

[YiBaKr1988] Y. Q. Yin, Z. D. Bai, P. R. Krishnaiah, *On the limit of the largest eigenvalue of the large-dimensional sample covariance matrix*, Probab. Theory Related Fields **78** (1988), no. 4, 509–521.

Index

Titles in This Series